Biological Control of Plant Pathogens

Biological Control of Plant Pathogens

Syed Abuzar
Head of Regulatory Affairs & Registrations-Asia Pacific, Rovensa Group

Nassreen Fatima Kacho
Krishi Vigyan Kendra, Kargil-1, SKUAST-K, Ladakh, India

Pravin Pati Tripathi
Plant Protection, Krishi Vigyan Kendra, East Kameng, Arunachal Pradesh, India

Prajna Samal
Department of Plant Pathology, Uttar Banga Krishi Viswavidyalaya, Pundibari, CoochBehar, West Bengal, India

Mohd Ashaq
Department of Botany, Govt Degree College, Thannamandi, J&K, India

Zuby Gohar Ansari
Department of Plant Physiology, B.M. College of Agriculture, Jaswadi Road Khandwa, Madhya Pradesh, India

Mahendra Kumar Tiwari
Government Hussainabad Inter College, Lucknow, U.P. India

CRC Press
Taylor & Francis Group
Boca Raton London New York

CRC Press is an imprint of the Taylor & Francis Group, an **informa** business

-EPH-

Elite Publishing House

First published 2026
by CRC Press
4 Park Square, Milton Park, Abingdon, Oxon, OX14 4RN

and by CRC Press
2385 NW Executive Center Drive, Suite 320, Boca Raton FL 33431

British Library Cataloguing-in-Publication Data
A catalogue record for this book is available from the British Library

Print edition not for sale in India

ISBN: 9781041247258 (hbk)
ISBN: 9781041264644 (pbk)
ISBN: 9781003750420 (ebk)

DOI: 10.1201/9781003750420

Typeset in A Caslon Pro
by Elite Publishing House, Delhi

-EPH-

Contents

Preface

Biological control has emerged as a cornerstone in sustainable agriculture, offering a viable and environmentally friendly alternative to chemical-based strategies for managing plant pathogens. As the agricultural sector grapples with the challenges posed by pesticide resistance, ecological imbalances, and the pressing need for sustainability, biological control presents itself as an effective and eco-conscious approach to plant disease management. The integration of beneficial microorganisms and their mechanisms into crop production systems represents a shift toward a more harmonious coexistence with the natural ecosystem, where the intricate interplay of biological interactions is harnessed to protect and sustain plant health.

This textbook, "Biological Control of Plant Pathogens," is a comprehensive endeavour to encapsulate the scientific principles, practical applications, and the latest advancements in the field of biological control. Designed to cater to students, researchers, and practitioners, this book aims to provide a thorough understanding of the subject, equipping readers with the knowledge required to explore, implement, and innovate in this vital area of plant pathology. The text delves deeply into the foundational principles of biological control, mechanisms of action, and the complex relationships among plants, pathogens, and biocontrol agents. Moreover, it addresses the practical aspects of commercial production, application strategies, and integration into modern farming systems, ensuring a balanced perspective that bridges theoretical knowledge and real-world practice.

The first chapter lays the groundwork by introducing the concept of biological control. Definitions, historical perspectives, and the principles of plant disease management are discussed to establish a firm understanding of the topic. Biological control is examined through the lens of its significance in agricultural sustainability, highlighting its role in reducing chemical dependence and minimizing environmental impact. Additionally, the chapter delves into the merits and demerits of this approach, providing a balanced view of its potential and limitations.

The second chapter explores the diverse biological interactions that form the basis of biocontrol mechanisms. It delves into competition, mycoparasitism, and exploitation of hypovirulence, illustrating how these processes contribute to the suppression of plant pathogens. The chapter also examines rhizosphere colonization, competitive saprophytic ability, antibiosis, and induced resistance, shedding light on the complex and dynamic interactions within the soil microbiome and plant systems. Mycorrhizal associations and their contributions to plant health and pathogen control are also discussed, alongside operational mechanisms that define the relevance of these interactions in practical applications.

The third chapter addresses the factors that govern the success and efficacy of biological control. It emphasizes the critical role of the physical environment, agroecosystem dynamics, and cultural practices in shaping the effectiveness of biocontrol agents. The chapter also discusses the relationships between pathogens and antagonists, offering insights into the ecological and biochemical dynamics that underpin their interactions. This section provides a nuanced understanding of the operational mechanisms that influence biological control, highlighting the need for a holistic approach to maximize its potential.

The fourth chapter focuses on the relationships between pathogens, antagonists, and biocontrol agents. It discusses the ecological and biochemical dynamics of these relationships and compares approaches involving resident versus introduced antagonists. The chapter examines strategies for controlling soil-borne and foliar diseases, emphasizing the compatibility of bioagents with agrochemicals and other microbial antagonists. By addressing these critical aspects, this section aims to provide practical insights for implementing biocontrol strategies in diverse agricultural contexts.

The fifth chapter delves into the commercial production and delivery systems of biocontrol agents. It outlines the techniques and processes involved in the mass production of antagonists, highlighting challenges related to scalability and economic feasibility. The chapter also explores delivery systems and application strategies, emphasizing the importance of monitoring and efficacy assessment. This section bridges the gap between laboratory research and field application, ensuring that biocontrol agents can be effectively utilized in agricultural practices.

The sixth chapter integrates biological control into modern farming systems, exploring its role in Integrated Disease Management (IDM), Integrated Pest Management (IPM), and organic farming. This chapter underscores the versatility of biological control as a key component of sustainable agriculture. It highlights its compatibility with various farming practices and its potential to enhance crop productivity while reducing environmental impact. The integration of biocontrol agents into holistic management systems represents a significant step toward achieving sustainability in agriculture.

The seventh chapter examines the market applications of biocontrol agents, focusing on biopesticides. It provides an overview of commercially available products and their applications, offering case studies to illustrate their effectiveness. This chapter also discusses regulatory standards and quality control measures, emphasizing the need for robust frameworks to ensure the efficacy and safety of biocontrol products. By addressing these critical aspects, this section aims to provide practical insights for researchers, manufacturers, and policymakers involved in the development and deployment of biopesticides.

Writing this textbook has been a journey through the rich and evolving field of biological control. The content is meticulously structured to cater to a wide audience, from undergraduate students and academic researchers to extension workers and policymakers. Each chapter is crafted to provide a blend of theoretical foundations, practical insights, and cutting-edge research, ensuring that the book remains relevant and valuable across diverse contexts. The aim is to equip readers with the knowledge and tools necessary to understand, apply, and innovate in the field of biological control.

This textbook also acknowledges the challenges and limitations associated with biological control, offering critical perspectives on its efficacy, adaptability, and scalability. While the benefits of biological control are well-documented, the field continues to face significant hurdles, including variability in field performance, compatibility with existing agricultural practices, and the need for standardized regulatory frameworks. By addressing these issues, the book aims to foster a deeper understanding of the complexities involved in biological control and encourage the development of innovative solutions.

We are deeply grateful to the researchers, educators, and practitioners who have contributed to the advancement of biological control. Their groundbreaking work has provided the foundation for this textbook, and their dedication to sustainable agriculture continues to inspire and drive progress in the field. We also extend my appreciation to the institutions and organizations that support research and development in biological control, recognizing their vital role in shaping the future of agriculture.

We hope that this textbook will serve as a valuable resource for understanding the principles and applications of biological control. By fostering knowledge and innovation in this field, we can move closer to achieving sustainable agricultural systems that prioritize ecological balance, economic viability, and environmental health. Biological control is not just a method of managing plant pathogens—it is a commitment to preserving the integrity of our ecosystems and ensuring the long-term resilience of our agricultural systems. We invite readers to embark on this journey of discovery, exploration, and innovation, as we work together to unlock the full potential of biological control in plant pathology.

Authors

About the Authors

Syed Abuzar Ph.D; RAC, possesses a distinguished career spanning 18 years in agricultural sciences, with expertise in Plant Pathology, Nematology, Biocontrol Agents, and Integrated Pest Management (IPM) in Agriculture. Currently serving as the Head of Regulatory Affairs & Registrations-Asia Pacific at Rovensa Group, based in Malaysia, he has previously held roles at prestigious institutions such as ICAR, DBT, and UGC, contributing significantly to various projects. Dr. Syed has been honored with awards from esteemed scientific societies, recognizing his substantial contributions to the field and his advocacy for the farming community. He actively engages as a reviewer for numerous reputable national and international journals, leveraging his expertise to advance scholarly discourse. Additionally, he maintains active membership in several global scientific communities, including the American Chemical Society, (ACS) USA, and the Regulatory Affairs Professional Society (RAPS) UK, further demonstrating his commitment to advancing scientific knowledge and regulatory practices.

Nassreen Fatima Kacho is presently working as a Senior Scientist and Head at KVK Kargil-1, SKUAST-K, Ladakh. She completed her B.Sc. in Agriculture from SKUAST Jammu and her Postgraduate and Ph.D. degrees in Plant Pathology from SKUAST-K, Shalimar, Srinagar.

She joined as an Assistant Professor cum Junior Scientist in 2009 at the Regional Agriculture Research Station, Kargil, SKUAST-K. With 15 years of experience in research and extension activities, her areas of interest include organic plant protection, mushroom cultivation, and biofertilizers. Dr. Kacho has published more than 40 research articles in reputed journals, 7 book chapters, and 11 extension bulletins. Additionally, she has contributed various popular articles on agricultural issues to newspapers and magazines. She has been recognized with the Women Scientist Award and holds one patent.

Pravin Pati Tripathi has graduated with B.Sc. (Agriculture) from Purvanchal University Jaunpur, (UP) and completed a Master's Degree in Plant Pathology from T. D. College Jaunpur. He worked in sugarcane industries as a Cane Development Officer in MP from 2006 to 2009. During that period, he have encouraged many unemployed youth toward profitable agriculture. Later he joined Krishi Vigyan Kendra as a Subject Matter Specialist (SMS) in the discipline of Plant Protection in 2013 in Arunachal Pradesh. He is a member of the Society for Upliftment of Rural Economy and Agricultural Technology Development Society. He is the recipient of an award viz. Best Extension Scientist Award 2017 and 2018. Received a Commendation certificate from District administration on the occasion of Independence Day 2023 due to his visible performance in Apiculture.

Prajna Samal is a research scholar currently pursuing her Ph.D. at Uttar Banga Krishi Viswavidyalaya. She completed her B.Sc. (Hons.) in Agriculture from Odisha University of Agriculture and Technology in 2020 and earned her M.Sc. in Agriculture with a specialization in Plant Pathology from the College of Agriculture, Central Agricultural University, Imphal, in 2022. She has successfully cleared the ICAR (AICE-JRF/SRF (Ph.D.)-2023) exam and the ICAR-NET. She has published two research papers in an international journal and one in a national journal. Additionally, she has contributed a few book chapters.

Mohd Ashaq is an Associate Professor of Botany in the Higher Education Department of the Jammu and Kashmir Government presently posted at Govt Degree College Thannamandi, J&K, India. He completed his M.Sc., M.Phil., and Ph. D degrees in Botany with a specialization in Plant Virology from Aligarh Muslim University, Aligarh, India in 1992, 1993, and 1997 respectively. Dr. Ashaq served in Eritrea for 8 years between 2000-2008, finally as Director of Research at Eritrea Institute of Technology, Asmara, State of Eritrea (2006-2008). Dr. Ashaq has 19 books, 60 book chapters, 40 Research papers, 26 patents, and over 50 popular articles to his credit. He has edited over 100 issues of magazines and newsletters for various organizations. His publications on Refugees and volunteerism have been translated into French, Portuguese, and Spanish. He is the commissioned member of 5 of 6 IUCN (International Union of Conservation of Nature) commissions. He has presented papers and served as a resource person, invited speaker, and keynote speaker in over 150 conferences/seminars/workshops, etc. He has organized dozens of international and national conferences/seminars, workshops, and FDPs. He is a Reviewer/Editorial board member/fellow of about three dozen national and international peer-reviewed Scopus/web of Science journals and societies besides 15 development organizations.

Zuby Gohar Ansari holds a Doctorate in Plant Physiology and currently serves as an Assistant Professor at B.M. College of Agriculture, Khandwa under RVSKVV University, Gwalior. Her research focuses on abiotic stress in legume crops during research work. Dr. Zuby has received numerous fellowships and awards and is a life member of the Andhra Pradesh Journal of Agricultural Sciences (Society of Agricultural Scientists, SVAC, Tirupati), Astha Foundation, Meerut (U.P.), Environment and Social Development Association (ESDA), New Delhi, All India Agricultural Students Association. She is currently mentoring one M.Sc. student at the B.M. College of Agriculture under the Plant Pathology Department and has authored over 14 research articles in peer-reviewed national and international journals, along with several book chapters and books.

Mahendra Kumar Tiwari is currently serving as a Senior Agriculture Teacher at Government Hussainabad Inter College, Lucknow. He holds academic qualifications including M.Sc. (Agricultural Economics), Ph.D. (Agricultural Economics), and B.Ed. in Agriculture. His special interests lie in promoting horticulture, agriculture, animal husbandry, dairy, cooperatives, organic farming, and milk production within his district, and he has always been dedicated to these pursuits.

Dr. Tiwari has three years of experience in providing critical training in crop science and horticulture to trainee teachers at the Government Creative Training College, Lucknow, now upgraded to the College of Teacher Education, Lucknow (CTE), which operates under the State Council of Educational Research and Training, Uttar Pradesh, Lucknow (SCERT). He has consistently stood in strong support of farmers in his district, Lucknow. Additionally, he has been actively involved with the farmers of Malihabad and Mal, regions nationally renowned as the Green Belt, particularly for the world-famous Dussehri mangoes and other fruits. Alongside this, he has dedicated the last 30 years to promoting scientific agriculture and horticulture by imparting knowledge about innovative research, studies, and practical insights, focusing on agriculture, animal husbandry, dairy, cooperatives, organic farming, and milk production. Dr. Tiwari conducted in-depth research on the topic *"Impact of Dairy Cooperatives on Income and Employment of Weaker Sections in the Area of Operation Flood Programme in District Lucknow (U.P.)"* and earned his Doctorate (Vidya-Vachaspati) in Agricultural Economics from C.S.J.M. University, Kanpur, Uttar Pradesh, India.

Chapter - 1

Concept of Biological Control and its Foundations

Biological control, commonly referred to as biocontrol, stands as one of the most sustainable and ecologically sensitive approaches in the realm of pest management, providing a viable alternative to synthetic chemical pesticides that have historically dominated agricultural and ecological practices. The concept is rooted in the natural ecological interactions between organisms, where predators, parasitoids, pathogens, and competitors naturally regulate the population of other species. Over time, humans have increasingly harnessed these natural dynamics to manage pest populations, particularly in agricultural settings, forestry, and even urban ecosystems. This practice involves utilizing natural enemies—organisms such as insects, nematodes, fungi, bacteria, and viruses—to suppress pest populations, thereby reducing their impact on crops, native plants, and ecosystems at large. The significance of biological control is underpinned by the ever-growing challenges posed by traditional chemical pesticides, including the development of resistance in pest populations, non-target effects, environmental contamination, and long-term ecological imbalances. These challenges have spurred the search for more sustainable and environmentally friendly pest management methods, with biological control emerging as a critical component of integrated pest management (IPM) strategies. The foundations of biological control are deeply embedded in ecological principles, particularly the understanding of population dynamics, predator-prey relationships, and the complex interdependencies that characterize ecosystems.

Historically, the roots of biological control can be traced back thousands of years, with records from ancient China documenting the use of ants to control pests in citrus orchards. These early practices were based on empirical observations of natural processes, long before the underlying scientific principles were understood. The formalization of biological control as a scientific discipline began in the late 19th and early 20th centuries, marked by groundbreaking successes such as the introduction of the vedalia beetle (*Rodolia cardinalis*) to control cottony cushion scale (*Icerya purchasi*) in California citrus orchards. This landmark achievement demonstrated the potential of classical biological control, which involves the deliberate introduction of natural enemies from a pest's native range into an area where the pest has become problematic. The vedalia beetle's remarkable success not only highlighted the efficacy of biological control but also paved the way for its widespread adoption as a sustainable pest management strategy. Since then, numerous classical biological control programs have been implemented worldwide, targeting a wide array of pests and invasive species that threaten agriculture, forestry, and biodiversity.

In addition to classical biological control, other approaches have emerged, including augmentative and conservation biological control. Augmentative biological control involves the mass production and periodic release of natural enemies to enhance their populations in a target area. This method is particularly effective in environments where natural enemy populations are insufficient to achieve effective pest suppression. Two primary forms of augmentative biological control exist: inundative releases, where large numbers of natural enemies are released to achieve immediate pest control, and inoculative releases, where smaller numbers are introduced to establish a long-term population that provides sustained control. On the other hand, conservation biological control focuses on preserving and enhancing existing populations of natural enemies through habitat management practices. This approach recognizes the importance of providing suitable habitats, food resources, and refuge for natural enemies, ensuring their survival and effectiveness in regulating pest populations.

The scientific foundations of biological control are deeply rooted in ecology, particularly the study of interactions between species and their environments. Ecological theories, such as the concepts of carrying capacity, predator-prey dynamics, and ecological niches, provide critical insights into the mechanisms underlying biological control. For instance, understanding the factors that influence the population growth rates of pests and their natural enemies is essential for designing effective biocontrol programs. Similarly, knowledge of the specific ecological requirements and behaviors of natural enemies, such as their host preferences, reproductive strategies, and dispersal capabilities, informs their selection and deployment in biological control efforts. Advances in molecular biology, genomics, and biotechnology have further revolutionized the field, enabling researchers to identify and develop more effective biocontrol agents. For example, molecular tools allow for the precise identification of natural enemies and their target pests, ensuring compatibility and minimizing non-target effects.

Despite its numerous advantages, biological control is not without challenges. The success of biocontrol programs depends on meticulous planning, rigorous testing, and careful monitoring to ensure that the introduction of natural enemies does not lead to unintended ecological consequences. Non-target effects, where biocontrol agents impact species other than the target pest, represent a significant concern. Such effects can disrupt native ecosystems, alter species interactions, and compromise biodiversity. To address these risks, strict regulatory frameworks govern the introduction and use of biological control agents, emphasizing comprehensive risk assessment, quarantine procedures, and post-release monitoring. These measures aim to balance the benefits of biological control with the need to protect ecological integrity and prevent unintended harm.

The role of biological control extends beyond agriculture to address challenges in forestry, urban landscapes, and natural ecosystems. In forestry, biological control is employed to manage pests such as bark beetles and gypsy moths, which pose significant threats to timber production and forest health. In urban settings, biocontrol agents are used to combat pests like mosquitoes, which transmit diseases such as malaria and dengue. Furthermore, biological control plays a crucial role in the management of invasive species, which are among the leading drivers of biodiversity loss worldwide. By targeting invasive pests and plants, biocontrol programs contribute to the restoration of native ecosystems and the preservation of ecological balance.

In the context of global challenges such as climate change, population growth, and the intensification of agricultural practices, the importance of biological control is more pronounced than ever. Climate change, in particular, has profound implications for pest dynamics and the effectiveness of biological control. Changes in temperature, precipitation, and other climatic factors can influence the distribution, abundance, and interactions of pests and their natural enemies, necessitating adaptive approaches to biocontrol. Additionally, the increasing demand for food and fiber driven by population growth underscores the need for sustainable agricultural practices that minimize environmental impacts while ensuring productivity. Biological control aligns with these objectives, offering a path toward more resilient and sustainable food systems.

Integrated Pest Management (IPM) serves as a holistic framework that incorporates biological control alongside other pest management strategies, such as cultural, physical, and chemical methods. By integrating multiple tactics, IPM seeks to achieve effective pest suppression while minimizing reliance on chemical pesticides. Biological control plays a central role in IPM, providing a natural and sustainable means of reducing pest populations and enhancing the resilience of agroecosystems. The integration of biocontrol into IPM also facilitates the transition to more environmentally friendly farming practices, contributing to the goals of sustainable agriculture and conservation.

The future of biological control lies in its continued integration with emerging technologies and interdisciplinary research. Innovations in biotechnology, such as genetic engineering and RNA interference, hold promise for the development of novel biocontrol agents with enhanced specificity and efficacy. Similarly, advancements in data science, remote sensing, and machine learning offer new tools for monitoring pest populations, predicting outbreaks, and optimizing biocontrol interventions. Global collaboration and knowledge-sharing are also critical for the advancement of biological control, as pests and invasive species often transcend national borders. International organizations, research institutions, and policymakers must work together to develop and implement biocontrol strategies that address shared challenges and promote sustainable pest management practices.

1. Definitions and Importance of Biological Control in Plant Pathology

1.1 Definitions of Biological Control

Biological control, commonly referred to as biocontrol, is a method of managing plant pathogens, pests, weeds, and other harmful organisms by employing natural organisms or their derivatives, forming the foundation of sustainable and environmentally friendly pest management practices. Over the years, this concept has evolved significantly, reflecting the advancements in agricultural sciences, ecology, and molecular biology, as well as the increasing complexity of modern agricultural ecosystems. Originally defined in a narrower sense, biological control was primarily associated with the introduction of exotic natural enemies, such as predators, parasitoids, and pathogens, to suppress invasive pest populations. However, the term has expanded to encompass a broader array of strategies, including the use of indigenous natural enemies, the application of microbial biocontrol agents, the enhancement of natural enemy habitats, and even genetic modifications of both the pests and their controlling agents to achieve desired outcomes. This evolution of the definition underscores the dynamic nature of biological control as a discipline that bridges traditional agricultural practices and cutting-edge technological innovations.

The classical definition of biological control can be traced back to the late 19th century, during which the term was first coined to describe the deliberate introduction of natural enemies from one region to another to suppress invasive species. A pivotal example is the successful use of the vedalia beetle (Rodolia cardinalis) to control the cottony cushion scale (Icerya purchasi) in California's citrus orchards. This event, often celebrated as the inception of modern biological control, set a precedent for the effective use of exotic natural enemies in pest management and highlighted the importance of understanding ecological interactions. The classical approach remains a cornerstone of biological control, particularly in managing invasive species that lack natural enemies in their introduced environments. By restoring the ecological balance, classical biological control contributes to the long-term suppression of pests without the need for continuous human intervention.

In an ecological context, biological control is defined as the manipulation of natural populations and interactions to maintain pest populations at levels below those that cause economic or ecological damage. This perspective emphasizes the role of natural enemies within the broader framework of ecosystem dynamics, where predators, parasitoids, pathogens, and competitors regulate the populations of their prey or hosts. By aligning with the natural processes that govern population dynamics, biological control not only addresses immediate pest problems but also fosters the resilience and stability of ecosystems. This ecological understanding forms the basis for integrating biological control into sustainable agricultural systems, where the goal is to balance productivity with environmental conservation.

Modern definitions of biological control have expanded to include a wide range of practices and approaches that leverage advancements in science and technology. For instance, microbial biological control involves the use of beneficial microorganisms, such as bacteria, fungi, and viruses, to suppress plant pathogens and pests. These microorganisms can act through various mechanisms, including competition for resources, production of antimicrobial compounds, induction of host resistance, and direct parasitism or predation. The development of microbial biocontrol agents has been facilitated by advances in molecular biology, genomics, and biotechnology, which have enabled the identification, characterization, and mass production of effective strains. Similarly, the use of RNA interference (RNAi) and other genetic engineering techniques to enhance the specificity and efficacy of biocontrol agents represents a significant leap in the field, broadening the scope of biological control to include targeted genetic interventions.

Another important dimension of biological control is its role within the framework of Integrated Pest Management (IPM). IPM emphasizes the integration of multiple pest management strategies, including biological, cultural, physical, and chemical methods, to achieve sustainable and effective pest control. Within this framework, biological control serves as a key component, providing a natural and environmentally friendly alternative to chemical pesticides. By reducing reliance on synthetic inputs, IPM enhances the sustainability of agricultural systems while minimizing the risks of pesticide resistance, environmental contamination, and non-target effects. The integration of biological control into IPM also underscores its compatibility with other pest management strategies, highlighting its versatility and adaptability to diverse agricultural and ecological contexts.

Biological control also encompasses conservation biological control, which focuses on preserving and enhancing the effectiveness of existing natural enemy populations through habitat management and other practices. This approach recognizes the importance of providing suitable habitats, food resources, and refuges for natural enemies to thrive and exert their regulatory effects on pest populations. For example, planting flowering plants that provide nectar and pollen can support populations of predatory insects and parasitoids, while reducing the use of broad-spectrum pesticides can help maintain the diversity and abundance of beneficial organisms. Conservation biological control aligns with the principles of agroecology, promoting biodiversity and ecosystem services as integral components of sustainable agriculture.

In the context of sustainable agriculture, biological control is often defined as a strategy that reduces the reliance on synthetic inputs while maintaining or enhancing crop productivity and environmental health. This definition highlights the dual objectives of biological control: to provide effective pest management and to contribute to the broader goals of sustainability and environmental conservation. By addressing the root causes of pest problems, such as habitat loss, monoculture practices, and excessive use of chemical inputs, biological control offers a holistic approach to pest management that aligns with the principles of sustainability. Furthermore, biological control supports the transition to organic and regenerative agricultural systems, where the use of synthetic pesticides and fertilizers is minimized or eliminated.

In addition to its applications in agriculture, biological control plays a crucial role in managing invasive species that threaten natural ecosystems. Invasive species are among the leading drivers of biodiversity loss, and their management often requires innovative and ecologically sensitive approaches. Biological control provides a solution by restoring the natural enemies of invasive species, thereby reducing their populations and mitigating their impact on native ecosystems. This application of biological control underscores its significance beyond agriculture, highlighting its potential to address global challenges such as biodiversity conservation, habitat restoration, and ecosystem resilience. The definition of biological control has also been shaped by advancements in monitoring, evaluation, and risk assessment. Modern biocontrol programs are guided by rigorous scientific methodologies that assess the safety, efficacy, and ecological impact of biocontrol agents before their deployment. These assessments ensure that biological control interventions are both effective and environmentally responsible, minimizing the risks of non-target effects and unintended ecological consequences. The development of standardized protocols and regulatory frameworks has further enhanced the credibility and reliability of biological control, paving the way for its widespread adoption in diverse contexts.

1.1.1 Traditional Definition of Biological Control

The traditional definition of biological control centers on the concept of using living organisms to suppress plant pathogens and pests. It focuses on ecological mechanisms such as predation, parasitism, competition, and antibiosis, wherein the natural antagonist directly or indirectly affects the growth or activity of the pathogen.

Core Concept and Mechanisms

1. **Predation**: Predatory fungi or protozoa consume nematodes or other microbial pests, reducing their populations. For example, fungi like *Arthrobotrys spp.* are known to trap nematodes using specialized hyphal structures.
2. **Parasitism**: In parasitism, one organism benefits at the expense of the host, such as the fungus *Trichoderma harzianum*, which parasitizes and degrades the cell walls of pathogenic fungi like *Rhizoctonia solani.*

3. **Competition**: Antagonistic microorganisms outcompete pathogens for nutrients and space. For instance, *Pseudomonas fluorescens* competes with soil-borne pathogens by colonizing the rhizosphere more efficiently.
4. **Antibiosis**: Antagonists produce chemical substances that inhibit or kill pathogens. Bacteria such as *Bacillus subtilis* secrete lipopeptides that exhibit strong antifungal properties.

Historical Context and the Evolution of Biological Control

The concept and practice of biological control have their roots in agricultural and ecological traditions dating back centuries, but the formalization of biological control as a scientific discipline emerged prominently in the late 19th century. During this time, researchers began documenting the suppression of plant pathogens, pests, and weeds through the use of naturally occurring organisms, marking the dawn of an era where ecological principles were consciously applied to pest management. One of the earliest documented successes in this field involved the application of fungal antagonists to combat diseases in fruit crops, providing a practical demonstration of the potential of biological control. These initial experiments laid a crucial foundation, igniting systematic efforts to explore the role of natural enemies in maintaining ecological balance and reducing the burden of pest infestations in agricultural systems. Notably, the introduction of the vedalia beetle (*Rodolia cardinalis*) to control the invasive cottony cushion scale (*Icerya purchasi*) in California during the late 1800s is often heralded as the first landmark success in classical biological control. This intervention not only salvaged the citrus industry in California but also demonstrated the potential of biological control to address invasive pest problems on a large scale, paving the way for its adoption worldwide.

The historical context of biological control reflects a growing awareness of the limitations of traditional pest management methods, particularly the excessive reliance on chemical pesticides. Before the advent of biological control, synthetic pesticides were the primary means of combating pests and diseases in agricultural settings. While effective in the short term, these chemicals often pose significant environmental and health risks, including the contamination of soil and water, the decline of non-target organisms, and the development of resistance in pest populations. The ecological crises resulting from these practices underscored the urgent need for alternative approaches that aligned more closely with the principles of sustainability and ecosystem health. Biological control emerged as a natural solution, offering a way to harness the regulatory functions of ecosystems without introducing harmful synthetic compounds. Early pioneers in the field, driven by the vision of reducing humanity's ecological footprint, began systematically studying the interactions between pests and their natural enemies, documenting the conditions under which these relationships could be manipulated for practical benefits.

Over time, the field of biological control expanded beyond classical approaches to include augmentative and conservation strategies. Augmentative biological control involves the mass production and periodic release of natural enemies to enhance their population densities in areas where they are otherwise insufficient. This strategy gained traction in the mid-20th century as advancements in mass-rearing techniques and insectary operations enabled the large-scale deployment of beneficial organisms. For example, the release of predatory mites to control spider mite infestations in greenhouse crops became a standard practice in many parts of the world, demonstrating the scalability and adaptability of augmentative biological control. Conservation biological control, on the other hand, focuses on creating conditions that support the survival and effectiveness of natural enemy populations already present in the environment. This approach emphasizes the importance of preserving habitats, providing food resources, and minimizing disturbances such as pesticide use that could harm beneficial organisms. By integrating these strategies, biological control evolved into a versatile and multifaceted discipline capable of addressing diverse pest management challenges across agricultural, forestry, and urban systems.

Despite its successes, biological control faced challenges and limitations that shaped its historical development. Early programs often lacked the scientific rigor and ecological understanding required to predict the outcomes of introducing natural enemies into new environments. In some cases, the release of biocontrol agents led to unintended ecological consequences, such as the displacement of native species or the disruption of non-target populations. These experiences highlighted the need for comprehensive risk assessments and post-release monitoring to ensure the safety and efficacy of biological control interventions. As a result, the field adopted more stringent scientific and regulatory frameworks, incorporating advancements in ecology, population biology, and genetics to improve the precision and predictability of biocontrol programs. Today, biological control is recognized not only as a practical tool for pest management but also as a model of applied ecology that demonstrates the potential of working with natural processes to achieve sustainable outcomes.

Strengths and Limitations of the Traditional Definition of Biological Control

The traditional definition of biological control, which focuses on the use of living organisms to suppress pest populations, remains a cornerstone of the field, providing the conceptual framework upon which modern practices are built. One of its greatest strengths lies in its ecological foundation, emphasizing the natural interactions that occur within ecosystems. This perspective underscores the importance of maintaining ecological balance, where predators, parasitoids, and pathogens regulate the abundance of their prey or hosts, preventing any single species from dominating the system. By aligning with these natural processes, biological control offers a sustainable alternative to chemical pesticides, reducing the need for synthetic inputs and their associated environmental risks. The chemical-free approach inherent in the traditional definition minimizes the contamination of soil, water, and air, protecting non-target organisms and preserving biodiversity. Furthermore, biological control supports the health and resilience of agroecosystems by promoting the diversity and abundance of beneficial organisms, creating a more robust and self-regulating system.

Another significant strength of the traditional definition is its compatibility with the principles of sustainability and integrated pest management (IPM). By prioritizing natural solutions, biological control contributes to the broader goals of reducing reliance on non-renewable resources, enhancing soil health, and fostering ecosystem services. The use of biological control aligns with global efforts to address pressing environmental challenges, such as climate change, habitat loss, and the decline of pollinator populations. Moreover, biological control is particularly well-suited to organic farming systems, where the use of synthetic pesticides and fertilizers is restricted. In these contexts, biological control serves as a vital component of pest management strategies, enabling farmers to maintain productivity while adhering to organic standards.

However, the traditional definition of biological control also has its limitations, particularly in the context of modern agriculture and scientific advancements. One of the key criticisms is its lack of specificity in capturing the intricate relationships and mechanisms involved in biocontrol. The traditional approach often focuses on the observable outcomes of natural enemy introductions, such as the suppression of pest populations, without fully addressing the underlying ecological and molecular processes. For example, the roles of secondary metabolites, microbial interactions, and genetic factors in mediating biocontrol effects are not explicitly accounted for in the traditional definition. This gap limits the ability to design and optimize biocontrol strategies based on a mechanistic understanding of these complex dynamics.

Another limitation of the traditional definition is its exclusion of modern innovations and technologies that have expanded the scope of biological control. Advances in molecular biology, genomics, and biotechnology have enabled the development of novel biocontrol agents and techniques, such as genetically modified organisms (GMOs), RNA interference (RNAi), and formulated microbial products. These innovations offer new possibilities for enhancing the specificity, efficacy, and stability of biological control interventions, addressing challenges such as resistance development and environmental variability. However, the traditional definition does not encompass these advancements, potentially limiting its relevance in the context of contemporary scientific and practical applications.

1.1.2 Ecological Definition

The ecological definition of biological control expands the traditional perspective by situating biocontrol within the context of entire ecosystems. It emphasizes the role of interactions between plants, pathogens, and their natural enemies in restoring balance and promoting agricultural sustainability.

Key Features

1. **Microbial Diversity**: The ecological approach underscores the importance of maintaining a diverse microbial community in the rhizosphere and phyllosphere. Such diversity enhances the resilience of the ecosystem to pathogen invasions.
2. **Ecosystem Services**: Biocontrol agents contribute to broader ecological processes, such as nutrient cycling, organic matter decomposition, and the maintenance of soil structure.

Examples of Ecological Biocontrol

1. **Nutrient Cycling**: Mycorrhizal fungi improve nutrient uptake in plants while simultaneously suppressing root pathogens by altering the microbial community structure.
2. **Disease Suppression in Polyculture Systems**: Diverse cropping systems often harbor a range of biocontrol agents, reducing the dominance of specific pathogens.

Benefits of Ecological Biocontrol

1. **Sustainability**: It aligns with the principles of agroecology, promoting long-term agricultural health without depleting resources.
2. **Adaptability**: Ecological biocontrol can be tailored to different environments, making it applicable to a wide range of agricultural systems.

Challenges

1. **Complexity of Interactions**: Managing biocontrol agents within ecosystems requires an understanding of complex interactions, which can be challenging to predict or control.
2. **Time-Intensive**: Ecological biocontrol often requires a longer time frame to establish effective pathogen suppression compared to chemical methods.

1.1.3 Modern Definitions

Modern definitions of biological control incorporate advancements in biotechnology, molecular biology, and precision agriculture. These definitions reflect the increasing sophistication of biocontrol practices, which now include engineered organisms, microbial metabolites, and advanced formulations designed for targeted and enhanced pathogen suppression.

Biotechnology and Genetic Engineering

1. **Engineered Organisms**: Advances in genetic engineering have enabled the development of microorganisms with enhanced biocontrol properties. For example, strains of *Pseudomonas fluorescens* have been engineered to produce additional antifungal compounds, improving their efficacy against root pathogens.
2. **CRISPR Technology**: Tools like CRISPR are being explored to edit genes in biocontrol agents, enhancing their adaptability and performance under diverse environmental conditions.

Metabolomics and Formulations

1. **Microbial Metabolites**: Modern biocontrol agents often rely on specific metabolites, such as antibiotics, siderophores, and enzymes, that directly inhibit pathogen activity.
2. **Advanced Delivery Systems**: Encapsulation technologies and bioformulations extend the shelf life of biocontrol agents, making them more practical for large-scale agricultural applications.

Examples of Modern Biocontrol Innovations

1. **RNA Interference (RNAi)**: Emerging biocontrol strategies include the use of RNAi to target specific genes in pathogens, disrupting their life cycle without harming non-target organisms.
2. **Integrated Biocontrol Platforms**: Digital tools and sensors are being used to monitor pathogen levels and deploy biocontrol agents precisely where needed, enhancing efficiency and reducing waste.

Advantages of Modern Biocontrol

1. **Specificity**: Modern techniques allow for highly targeted interventions, reducing the risk of collateral damage to beneficial organisms.
2. **Scalability**: Innovations in production and formulation make biocontrol agents more accessible and practical for widespread use.

Limitations

1. **Cost**: Developing and deploying advanced biocontrol technologies can be expensive, posing a barrier for small-scale farmers.

2. **Regulatory Hurdles**: Modern biocontrol agents, especially genetically modified organisms, face strict regulatory scrutiny, which can delay their adoption.

Modern definitions reflect the evolution of biological control into a science-driven, technologically sophisticated field, capable of addressing the complex challenges of modern agriculture.

1.1.4 Institutional Perspectives

Institutional perspectives on biological control emphasize the regulatory, ethical, and practical aspects of its implementation. Leading organizations such as the Food and Agriculture Organization (FAO) and International Organization for Biological Control (IOBC) have played pivotal roles in shaping the global understanding and application of biocontrol.

FAO Definition

1. **Key Elements**: According to the FAO, biological control involves the use of living organisms to suppress pests, pathogens, or weeds, reducing their abundance or impact.
2. **Focus on Practicality**: The FAO emphasizes the role of biocontrol in achieving tangible outcomes in pest and disease management.

IOBC Definition

1. **Ethical Considerations**: The IOBC stresses the need to minimize unintended ecological impacts, such as the displacement of native species or harm to non-target organisms.
2. **Policy Integration**: The IOBC promotes the inclusion of biocontrol in national and international agricultural policies, supporting its adoption as a mainstream practice.

Role of Institutions in Advancing Biocontrol

1. **Research and Development**: Institutions provide funding and resources for the study of biocontrol agents and their mechanisms.
2. **Regulatory Frameworks**: They establish guidelines for the safe and effective use of biocontrol agents, ensuring their compatibility with environmental and public health standards.

Challenges in Institutional Adoption

1. **Knowledge Gaps**: Limited awareness among policymakers and practitioners can hinder the widespread adoption of biocontrol strategies.
2. **Market Barriers**: The commercialization of biocontrol agents faces challenges such as high production costs and competition from chemical pesticides.

1.2 Importance of Biological Control in Plant Pathology

Biological control has emerged as a transformative approach in the management of plant diseases, addressing critical agricultural challenges while aligning with the principles of sustainability. By leveraging natural antagonists and their mechanisms, biocontrol not only mitigates the impact of pathogens but also fosters long-term ecological and economic benefits.

1.2.1 Environmental Sustainability

Environmental sustainability is one of the most compelling advantages of biological control. By prioritizing natural solutions over synthetic interventions, biocontrol minimizes the adverse environmental impacts associated with chemical pesticides. This approach preserves ecosystems while maintaining agricultural productivity.

Reduction in Chemical Usage

1. **Minimizing Environmental Contamination**: Chemical pesticides often leach into the soil and water, contaminating ecosystems and threatening biodiversity. Biological control reduces this dependency, protecting water sources and aquatic organisms.
2. **Reduced Soil Degradation**: Excessive pesticide use leads to the accumulation of harmful residues in soil, affecting its structure and fertility. Biocontrol agents, such as *Trichoderma* species, not only manage pathogens but also contribute to soil health by enhancing microbial activity.
3. **Protecting Air Quality**: The application of volatile chemical pesticides contributes to air pollution. Biocontrol eliminates this issue, offering a cleaner alternative.

Preservation of Non-Target Species

1. **Protecting Beneficial Insects**: Pollinators, such as bees and butterflies, are critical for crop production. Unlike broad-spectrum chemical pesticides, biocontrol agents target specific pathogens, sparing beneficial insects.
2. **Safeguarding Soil Microbiota**: Biocontrol enhances microbial diversity, fostering a balanced ecosystem where beneficial microorganisms thrive alongside plants.
3. **Minimizing Collateral Damage**: Chemical sprays often impact non-target organisms, disrupting food chains and ecosystems. Biological control, being organism-specific, avoids such unintended consequences.

Promotion of Biodiversity

1. **Fostering Microbial Diversity**: Biocontrol agents, such as mycorrhizal fungi and rhizobacteria, contribute to a rich microbial ecosystem that supports plant health and suppresses pathogens.
2. **Enhancing Ecosystem Resilience**: Biodiverse systems are more resilient to invasive species and environmental stressors. Biological control strengthens this resilience by promoting natural checks and balances.
3. **Supporting Wildlife**: Reduced chemical usage ensures that habitats for terrestrial and aquatic wildlife remain intact, contributing to broader conservation goals.

Environmental sustainability through biological control highlights the harmony between agricultural practices and ecological conservation, offering a pathway to long-term productivity without compromising the health of ecosystems.

1.2.2 Addressing Pesticide Resistance

The overuse of chemical pesticides has led to the widespread development of resistance among plant pathogens. This escalating problem threatens the effectiveness of traditional disease management strategies. Biological control offers a sustainable alternative by deploying natural mechanisms that reduce the likelihood of resistance development.

Mechanisms of Resistance Mitigation

1. **Mycoparasitism**
   - Beneficial fungi, such as *Trichoderma harzianum*, parasitize pathogenic fungi by degrading their cell walls. This direct interaction is highly specific, reducing the pathogen's ability to develop resistance.
   - Mycoparasitism also suppresses pathogen reproduction, further curbing their populations.
2. **Antibiosis**
   - Biocontrol agents like *Bacillus subtilis* produce antimicrobial compounds that inhibit the growth of pathogens. These substances target multiple pathways in pathogens, making resistance development less likely.
   - For example, the lipopeptides produced by *Bacillus* species disrupt fungal membranes, effectively killing the pathogens.
3. **Induced Resistance**
   - Certain biocontrol agents trigger systemic acquired resistance (SAR) or induced systemic resistance (ISR) in plants. This activation of the plant's immune system reduces its vulnerability to future pathogen attacks.
   - *Pseudomonas fluorescens* is a well-known agent that enhances the plant's natural defense mechanisms.

Long-Term Benefits of Biological Control in Resistance Management

1. **Durability of Solutions**: Natural mechanisms are more resilient over time, offering a stable approach to disease management.
2. **Reduced Selection Pressure**: Unlike chemical pesticides, which apply intense selection pressure on pathogens, biological control agents allow for natural coexistence without creating "superpathogens."
3. **Sustainable Crop Protection**: By integrating biocontrol into agricultural practices, farmers can reduce reliance on fungicides and bactericides, preserving their effectiveness for critical cases.

1.2.3 Economic Benefits

In addition to its well-documented environmental and biological advantages, biological control offers substantial economic benefits that make it an attractive option for farmers, agricultural enterprises, and even policymakers focused on promoting sustainable agricultural practices. By reducing input costs, enhancing crop yield and quality, and opening up new market opportunities, biological control not only supports the profitability of farming systems but also fosters their long-term sustainability. One of the most significant economic advantages lies in its ability to lower input costs through the self-sustaining nature of many biocontrol agents. Unlike chemical pesticides, which often require frequent and costly reapplication, many biological control agents, such as predatory insects, parasitic nematodes, or beneficial microbes, establish themselves in the soil or plant environment and provide ongoing pest suppression. For example, species like *Bacillus subtilis* and *Trichoderma harzianum*, which are widely used as microbial biocontrol agents, colonize plant roots and create a lasting protective barrier against pathogens, reducing the need for repetitive interventions. This inherent sustainability of biocontrol agents translates into significant cost savings for farmers, especially when compared to the recurring expenses associated with chemical pesticide use.

Moreover, biological control reduces dependency on synthetic chemicals, which represent a significant expenditure in conventional farming systems. By minimizing or eliminating the need for chemical pesticides, farmers can substantially lower the costs associated with their purchase, storage, and application. This is particularly advantageous in the face of rising pesticide prices and increasing regulatory restrictions on chemical usage due to concerns about environmental contamination and human health. Additionally, biocontrol offers affordable solutions tailored to the needs of small-scale farmers, who often operate with limited financial resources. For smallholders, who constitute a significant portion of the global agricultural workforce, biological control provides a cost-effective alternative that enhances their capacity to manage pests and diseases without the financial burden of expensive chemical inputs. For instance, the use of naturally occurring predators such as *Cryptolaemus montrouzieri* (the mealybug destroyer) and parasitoids like *Encarsia formosa* for the control of pests in horticultural crops represents a low-cost yet highly effective approach that aligns with the economic realities of smallholder farmers.

Beyond cost savings, biological control also enhances agricultural productivity by reducing disease losses and improving crop quality. Effective biocontrol agents have been shown to significantly lower the incidence and severity of plant diseases, which are a major cause of yield losses in many crops. For example, the use of the entomopathogenic fungus *Beauveria bassiana* to manage insect pests in maize or the application of *Pseudomonas fluorescens* to suppress soil-borne pathogens in vegetable crops can result in notable yield increases, ensuring that farmers derive maximum value from their investments. Furthermore, crops grown using biological control practices often exhibit superior quality compared to those treated with chemical pesticides. The absence or reduction of chemical residues not only improves the health and safety of the produce but also enhances its appeal to health-conscious consumers. This improved quality frequently translates into higher market prices, further contributing to the profitability of farming systems. For instance, fruits and vegetables grown with biocontrol practices often command premium prices in organic markets, where consumers are willing to pay more for produce that meets stringent safety and sustainability standards.

Another economic advantage of biological control is its ability to provide consistency in productivity. By addressing disease outbreaks and pest infestations in a sustainable manner, biocontrol ensures that crop losses are minimized and agricultural systems remain resilient to fluctuations in pest pressure. This stability in production is particularly important for farmers who depend on reliable yields to meet market demands and secure their livelihoods. In the long term, the use of biological control contributes to the economic viability of farming enterprises by reducing the risks associated with pest and disease outbreaks and fostering agricultural systems that are better equipped to adapt to changing environmental conditions, including those driven by climate change.

Biological control also opens up new market opportunities for farmers, particularly in the context of the growing demand for organic and sustainably produced food. With increasing consumer awareness of the environmental and health impacts of conventional farming practices, there is a rising preference for food grown using methods that prioritize ecological balance and minimize chemical inputs. Biological control aligns perfectly with these consumer expectations, enabling farmers to tap into lucrative organic markets that offer higher returns for certified produce. For example, crops grown under biological control programs can be marketed as "pesticide-free" or "eco-friendly," appealing to a wide range of consumers seeking sustainable food options. In addition to capturing higher market prices, these products often benefit from branding and marketing opportunities that further enhance their value. Farmers who adopt biocontrol practices can also explore value-added products, such as organically certified grains, fruits, and vegetables, which fetch premium prices in domestic and international markets. The growing popularity of eco-labels and certifications provides an additional layer of economic benefit, allowing farmers to differentiate their products in competitive markets and build trust with consumers.

Export potential is another key economic benefit associated with biological control. Many countries, particularly in the European Union, have stringent regulations on pesticide residues in imported agricultural products. Crops grown using biocontrol methods are more likely to comply with these standards, creating opportunities for farmers to access high-value export markets. For instance, the use of *Metarhizium anisopliae*, a fungal biocontrol agent, to control locusts and grasshoppers in cereal crops has enabled farmers in regions such as Africa and Asia to produce residue-free grains that meet international export requirements. Similarly, the adoption of biological control in the production of tropical fruits such as bananas, pineapples, and mangoes has facilitated access to markets in Europe and North America, where consumers and regulators prioritize produce with minimal chemical residues. By aligning with global trade standards, biological control enhances the competitiveness of agricultural products on the international stage, boosting farm incomes and contributing to the economic development of farming communities.

1.2.4 Contribution to Integrated Disease Management

Integrated Disease Management (IDM) is a holistic approach to plant health that combines multiple strategies to achieve effective and sustainable control of plant pathogens while minimizing the environmental and economic impacts of disease outbreaks. As a key pillar of IDM, biological control offers a sustainable and complementary solution that integrates seamlessly with chemical, cultural, and physical methods to provide comprehensive disease management. The role of biological control in IDM is multifaceted, leveraging the synergistic interactions between biocontrol agents, resistant crop varieties, and cultural practices to enhance the overall efficacy of disease management programs. For example, biocontrol agents such as *Trichoderma spp.*, which are well-known for their antagonistic properties against soil-borne pathogens like *Rhizoctonia solani* and *Pythium spp.*, can significantly enhance the effectiveness of disease-resistant crop varieties. These agents suppress residual pathogen populations that might otherwise overcome the plant's genetic resistance, creating a layered defense mechanism that minimizes the risk of disease outbreaks. This integration of biological control with resistant varieties exemplifies the synergy achieved in IDM, where multiple strategies work together to provide superior protection compared to any single approach.

Another critical aspect of biological control's contribution to IDM lies in its compatibility with cultural practices. Agricultural practices such as crop rotation, intercropping, and the use of cover crops create environmental conditions that favor the proliferation and activity of beneficial organisms, including biocontrol agents. For instance, intercropping legumes with cereals has been shown to enhance the abundance of rhizosphere-associated biocontrol microbes like *Bacillus subtilis* and *Pseudomonas fluorescens*, which suppress pathogenic fungi and promote plant growth. Similarly, crop rotation disrupts the life cycles of soil-borne pathogens, providing an opportunity for biocontrol agents to establish themselves and exert their protective effects. These cultural practices not only reduce pathogen pressure but also create a conducive environment for the natural regulation of disease-causing organisms, aligning with the principles of ecological balance and sustainability central to IDM.

One of the most significant contributions of biological control to IDM is its potential to reduce dependency on chemical pesticides. The selective use of chemical pesticides, when combined with biocontrol agents, can result in effective disease management with minimal environmental impact. For instance, targeted applications of fungicides can be used to manage acute disease outbreaks, while biocontrol agents provide long-term suppression of pathogen populations. This complementary approach reduces the overall quantity of chemicals needed, thereby lowering the risks of pesticide resistance, non-target effects, and environmental contamination. Biological control agents like *Gliocladium virens* and *Streptomyces spp.* have demonstrated their ability to control diseases such as damping-off and wilt, allowing farmers to rely less on chemical treatments. By integrating biocontrol into IDM programs, farmers can achieve a balanced approach that optimizes disease control while preserving ecological integrity and reducing the footprint of agriculture on the environment.

In addition to its synergistic role, biological control offers a unique advantage in addressing the root causes of disease outbreaks, rather than merely treating their symptoms. Chemical pesticides, while effective at suppressing visible symptoms, often fail to address underlying factors such as imbalances in the microbial community or environmental stressors that predispose plants to disease. Biocontrol agents, on the other hand, work at the microbial and ecological levels to restore balance in the rhizosphere and phyllosphere. For example, biocontrol microbes like *Trichoderma harzianum* not only suppress pathogens through competition and antibiosis but also promote the proliferation of beneficial microorganisms that contribute to plant health. This holistic approach addresses the root causes of disease outbreaks, enhancing the overall health and resilience of the cropping system.

Furthermore, biological control contributes to the long-term resilience of agricultural systems, a critical goal of IDM. By fostering the establishment of diverse and functional microbial communities, biocontrol agents enhance the ability of cropping systems to withstand environmental stressors and pathogen pressures over time. For instance, the application of arbuscular mycorrhizal fungi (AMF) has been shown to improve plant nutrient uptake and tolerance to abiotic stresses, while simultaneously suppressing root pathogens. Such resilience-building effects ensure that crops are better equipped to cope with challenges such as climate variability, soil degradation, and emerging pathogens. Unlike chemical pesticides, which often require repeated applications and lose efficacy over time due to resistance development, biological control agents provide a sustainable and adaptive solution that strengthens the natural defense mechanisms of plants and ecosystems.

1.2.5 Enhancing Food Safety

Food safety is a critical concern in agriculture, and biological control directly contributes to the production of safer, healthier crops. By minimizing the use of chemical pesticides, biocontrol reduces the risks associated with pesticide residues and aligns with consumer preferences for clean, organic produce.

Consumer Health

1. **Reduced Residue Levels:** Crops grown using biological control methods have significantly lower levels of chemical residues, reducing health risks for consumers.
2. **Alignment with Public Health Goals:** Biocontrol supports initiatives aimed at reducing exposure to harmful chemicals in food.

Market Trends

1. **Meeting Consumer Demand:** The increasing popularity of organic and sustainably grown food underscores the value of biocontrol in modern agriculture.
2. **Enhancing Market Access:** By producing residue-free crops, farmers can access high-value markets and comply with export regulations.

1.3 Case Studies Illustrating the Importance of Biological Control

Biological control is not just a theoretical framework; its significance is best understood through practical applications demonstrating its efficacy in managing plant pathogens. Real-world case studies provide compelling evidence of how biological control agents (BCAs) have transformed disease management in various agricultural systems.

1.3.1 Control of Soil-Borne Pathogens

Soil-borne pathogens are among the most destructive agents in agriculture, causing substantial economic losses by affecting plant roots and lower stems. The fungus *Trichoderma harzianum* has emerged as a cornerstone in the biological control of these pathogens, offering an effective, eco-friendly solution to diseases caused by organisms like *Rhizoctonia solani* and *Fusarium oxysporum*.

Mechanisms of *Trichoderma harzianum* in Soil-Borne Disease Control

1. **Mycoparasitism**
    - *Trichoderma harzianum* attacks pathogenic fungi by producing enzymes such as chitinases and glucanases, which degrade the cell walls of the target fungi.
    - This process not only inhibits the growth of pathogens but also prevents their reproduction, significantly reducing the pathogen load in the soil.
    - For example, in tomato plants, *T. harzianum* has been shown to parasitize *Fusarium oxysporum*, thereby controlling Fusarium wilt.
2. **Antibiosis**
    - The fungus produces secondary metabolites like gliotoxin and viridin, which have strong antifungal properties.

- These metabolites inhibit the growth of pathogens such as *Rhizoctonia solani* by disrupting their cellular processes.
- Research has demonstrated that the application of *T. harzianum* to cucumber fields reduces the incidence of Rhizoctonia damping-off by more than 60%.

3. **Induced Systemic Resistance (ISR)**
   - *T. harzianum* interacts with plant roots to trigger systemic resistance mechanisms.
   - This ISR involves the activation of plant defense pathways, including the production of pathogenesis-related proteins and phytoalexins, which enhance the plant's ability to resist subsequent pathogen attacks.
   - Studies have shown that tomatoes treated with T. harzianum exhibit increased resistance to multiple pathogens, including *Pythium* and *Sclerotium rolfsii.*

Applications and Success Stories

1. **Tomato Production**: In tomato fields affected by Fusarium wilt, the application of *T. harzianum* has reduced disease incidence by up to 75%. This biocontrol agent also promotes root growth, further enhancing plant vigor.
2. **Cucumber Farming**: *T. harzianum* has been effectively used to manage damping-off in cucumbers, leading to healthier seedlings and higher yields.
3. **Peppers**: In pepper fields, the use of *T. harzianum* formulations has significantly reduced losses caused by root rot, while improving overall soil health.

Broader Implications

The use of *Trichoderma harzianum* illustrates how biological control can address soil-borne diseases sustainably. By replacing or supplementing chemical fungicides, it reduces environmental impact, enhances soil fertility, and supports long-term agricultural productivity.

1.3.2 Management of Foliar Diseases

Foliar diseases, which affect the leaves and aerial parts of plants, pose a significant challenge in agriculture, particularly in cereal and vegetable production. The bacterium *Pseudomonas fluorescens* has proven to be an effective biocontrol agent for managing foliar diseases, including bacterial blight caused by *Pseudomonas syringae.*

Mechanisms of *Pseudomonas fluorescens* in Foliar Disease Control

1. **Antibiosis**
   - *Pseudomonas fluorescens* produces a range of antimicrobial compounds, including phenazines, pyoluteorin, and 2,4-diacetylphloroglucinol (DAPG).
   - These compounds inhibit the growth of foliar pathogens by disrupting their cellular membranes and metabolic pathways.
   - For instance, the application of *P. fluorescens* on rice plants has significantly reduced bacterial blight caused by *P. syringae.*

2. **Rhizosphere Colonization and Phyllosphere Protection**
   - Although primarily a rhizosphere bacterium, *P. fluorescens* can colonize leaf surfaces (phyllosphere), creating a protective barrier against pathogens.
   - This colonization reduces the ability of foliar pathogens to establish themselves, effectively preventing infection.
3. **Induced Systemic Resistance (ISR)**
   - *Pseudomonas fluorescens* triggers ISR in plants by producing signaling molecules such as salicylic acid and jasmonates.
   - This systemic activation of the plant's defense mechanisms enhances its ability to resist a wide range of foliar pathogens.
   - In wheat plants, ISR induced by *P. fluorescens* has reduced the severity of leaf rust, improving crop health and yield.

Applications and Success Stories

1. **Rice Farming**: In rice cultivation, *Pseudomonas fluorescens* has been extensively used to manage bacterial blight, resulting in yield improvements of up to 30%.
2. **Wheat Production**: Wheat treated with *P. fluorescens* has shown increased resistance to fungal pathogens such as *Puccinia spp.*, responsible for rust diseases.
3. **Vegetable Crops**: The application of *P. fluorescens* on leafy vegetables, such as lettuce and spinach, has reduced the incidence of foliar diseases like bacterial leaf spot.

Broader Implications

The success of *Pseudomonas fluorescens* in managing foliar diseases highlights its versatility as a biocontrol agent. Its dual role in suppressing pathogens and enhancing plant health makes it a valuable tool in sustainable agriculture. Moreover, its compatibility with other disease management strategies, such as resistant crop varieties, enhances its utility in integrated pest and disease management programs.

2. Principles of Plant Disease Management with Bioagents

Plant disease management with bioagents has emerged as a sustainable and eco-friendly alternative to conventional methods that rely heavily on chemical pesticides. Biological agents, such as bacteria, fungi, viruses, and nematodes, are deployed to suppress the activity of pathogens, enhance plant defenses, and promote overall plant health. The principles of plant disease management with bioagents revolve around understanding the complex interactions between plants, pathogens, and antagonists while leveraging these natural processes to achieve effective disease control.

2.2 Principle of Enhancing Plant Defense Mechanisms

A core principle of disease management with bioagents lies in their ability to enhance the plant's innate defense mechanisms. Unlike direct pathogen suppression, this approach focuses on activating systemic plant defenses, making crops more resilient to both biotic (pathogens and pests) and abiotic (drought, salinity, etc.) stresses. The use of biological control agents (BCAs) to induce these defense mechanisms reflects a sustainable and proactive strategy for managing plant health. This principle is primarily realized through two pathways: **Induced Systemic Resistance (ISR)** and **Systemic Acquired Resistance (SAR).** Both mechanisms equip plants with the tools to better withstand pathogen attacks, reducing reliance on chemical inputs and enhancing overall agricultural sustainability.

2.2.1 Induced Systemic Resistance (ISR)

ISR is a plant defense mechanism activated by beneficial microbes, particularly those associated with the root environment. By priming plants for a faster and stronger defensive response, ISR ensures that plants are well-prepared to combat potential pathogen threats.

2.2.1.1 Mechanism of Action

ISR is predominantly triggered by root-colonizing microorganisms, such as *Pseudomonas fluorescens* and *Bacillus subtilis*. These microbes form a beneficial association with plant roots, initiating changes in plant physiology and gene expression that prime the plant's immune system.

1. **Root Colonization**
   - Beneficial microbes colonize the root surface, forming biofilms that act as protective layers against soil-borne pathogens.
   - This colonization triggers molecular and biochemical changes in the plant's root cells, enhancing nutrient uptake and initiating defense signaling.
2. **Priming Defense Responses**
   - The concept of "priming" refers to a heightened state of readiness in the plant's immune system. Although no defense is activated at the time of priming, the plant's defenses are prepared for rapid deployment upon pathogen attack.
   - ISR stimulates the production of defense-related compounds, including phytoalexins (antimicrobial chemicals) and pathogenesis-related (PR) proteins. These compounds inhibit pathogen growth and spread within the plant.
3. **Production of Reactive Oxygen Species (ROS)**
   - ROS, such as hydrogen peroxide and superoxide, are produced as part of ISR. These molecules serve dual roles: acting as antimicrobial agents and signaling molecules that activate other defense pathways.

2.2.1.2 Signaling Pathways

The effectiveness of ISR is mediated by specific signaling pathways, particularly those involving jasmonic acid (JA) and ethylene (ET).

1. **Jasmonic Acid Pathway**
   - JA is a key hormone involved in plant defense against necrotrophic pathogens (pathogens that kill plant cells and feed on dead tissue).
   - Beneficial microbes like *Bacillus subtilis* stimulate JA production, triggering the expression of genes responsible for producing PR proteins and antimicrobial compounds.

2. **Ethylene Pathway**
   - ET works synergistically with JA to coordinate the plant's response to stress.
   - Bioagents such as *Pseudomonas fluorescens* enhance ET levels in the plant, contributing to the activation of ISR.

Significance of ISR Pathways:

ISR is highly effective against a range of necrotrophic pathogens, including *Rhizoctonia solani* and *Sclerotium rolfsii*, as well as certain insect pests.

2.2.2 Systemic Acquired Resistance (SAR)

Systemic Acquired Resistance (SAR) is a defense mechanism that provides long-lasting and broad-spectrum resistance against a wide range of pathogens. Unlike ISR, which is primarily associated with beneficial microbes, SAR is triggered by localized pathogen infections, leading to systemic immunity throughout the plant.

2.2.2.1 Salicylic Acid Pathway

The salicylic acid (SA) pathway is central to SAR. Upon infection by a pathogen, SA levels increase at the infection site and throughout the plant, initiating systemic defense responses.

1. **Localized Infection and Signal Transduction**
   - When a pathogen infects a plant, localized SA production occurs at the site of infection.
   - This SA acts as a signal, moving through the plant's vascular system to uninfected tissues, where it primes them for potential pathogen attack.
2. **Activation of Pathogenesis-Related Proteins**
   - PR proteins, such as chitinases and β-1,3-glucanases, are upregulated during SAR.
   - These proteins have direct antimicrobial effects, breaking down the cell walls of fungi and bacteria, and creating an inhospitable environment for pathogens.
3. **Bioagent-Mediated SAR**
   - Certain bioagents, like *Trichoderma harzianum*, mimic pathogen attacks to activate SAR without causing actual damage. This mechanism primes the plant for future challenges, creating a state of "immune readiness."

2.2.2.2 Cross-Protection

SAR provides cross-protection, a phenomenon where resistance induced against one pathogen extends to other unrelated pathogens.

1. **Broad-Spectrum Immunity**
   - The activation of SAR protects against diverse pathogens, including fungi, bacteria, and viruses.
   - For example, SAR activated by *Bacillus subtilis* in wheat enhances resistance not only to fungal rusts but also to bacterial blights.

2. **Sustainability in Crop Management**
   - Cross-protection reduces the need for multiple chemical treatments, contributing to sustainable crop production.
   - This approach is particularly beneficial in regions where access to fungicides and other chemical treatments is limited.

2.2.3 Applications in Crop Protection

The practical application of ISR and SAR in agriculture demonstrates their potential to transform plant disease management. These mechanisms have been successfully utilized in various crops, providing effective and sustainable protection against a range of diseases.

2.2.3.1 Rice Cultivation

Rice is a staple crop worldwide and is particularly vulnerable to diseases such as sheath blight and blast. Biological control agents that induce ISR have proven effective in managing these diseases.

1. **ISR Against Sheath Blight**
   - *Pseudomonas fluorescens* is widely used in rice cultivation to induce ISR.
   - The bioagent triggers the production of antimicrobial compounds in the plant, reducing sheath blight incidence by 40-60%.
2. **Enhanced Plant Growth**
   - In addition to disease suppression, ISR improves root architecture and nutrient uptake in rice plants, leading to increased yields.

2.2.3.2 Wheat Production

Wheat is a major cereal crop that faces threats from fungal diseases such as rusts and blights. SAR-inducing bioagents have shown great promise in controlling these diseases.

1. **SAR Activation by *Bacillus subtilis***
   - *Bacillus subtilis* triggers SAR in wheat, resulting in enhanced resistance to rust diseases caused by *Puccinia spp.*
   - This reduces reliance on fungicides, contributing to environmentally friendly farming practices.
2. **Protection Against Abiotic Stress**
   - SAR also enhances the plant's ability to withstand abiotic stresses, such as drought and salinity, further contributing to crop resilience.

2.2.4 Advantages of Enhancing Plant Defense Mechanisms

The principle of enhancing plant defenses offers several advantages that make it an essential component of sustainable agriculture.

1. **Reduction in Chemical Dependency**
   - By activating the plant's natural defenses, ISR and SAR minimize the need for chemical pesticides, reducing environmental pollution and input costs.
2. **Long-Lasting Protection**
   - Both ISR and SAR provide systemic and durable immunity, reducing the frequency of interventions required to manage diseases.
3. **Broad-Spectrum Efficacy**
   - These mechanisms protect against multiple pathogens, making them versatile tools for integrated disease management.
4. **Compatibility with Other Practices**
   - Bioagents that enhance plant defenses can be seamlessly integrated with cultural, chemical, and physical control methods, enhancing the effectiveness of disease management programs.

2.2.5 Challenges and Future Perspectives

Despite their benefits, the practical application of ISR and SAR faces certain challenges.

1. **Environmental Variability**
   - The efficacy of ISR and SAR can be influenced by environmental factors such as soil conditions, temperature, and humidity.
2. **Pathogen Adaptation**
   - Over time, some pathogens may develop strategies to evade these defense mechanisms, necessitating ongoing research and innovation.
3. **Scalability and Cost**
   - Large-scale production and application of ISR- and SAR-inducing bioagents require significant investments in technology and infrastructure.

Future Directions:

- Advancements in genomics and metabolomics could lead to the development of bioagents with enhanced ISR and SAR-inducing capabilities.
- Precision agriculture tools, such as drones and IoT sensors, could optimize the application of these bioagents, ensuring maximum efficacy with minimal waste.

2.3 Principle of Ecological Balance and Sustainability

Biological control represents a harmonious approach to agriculture, firmly rooted in the principle of maintaining and enhancing ecological balance. Unlike conventional methods that often disrupt natural ecosystems, the use of biological control agents (BCAs) works with the environment to suppress pathogens while preserving biodiversity. By leveraging natural processes and organisms, biological control fosters a sustainable agricultural system that minimizes harm to non-target organisms, maintains soil and water quality, and ensures long-term productivity. This principle aligns with the global goals of sustainable farming, addressing both the ecological and economic challenges faced by modern agriculture.

2.3.1 Biodiversity and Disease Suppression

Biodiversity is a cornerstone of ecological stability and resilience. In the context of agriculture, a diverse microbial and biological community contributes to disease suppression by limiting the dominance of pathogenic species. Biological control not only helps maintain this diversity but also enhances it by introducing or encouraging beneficial organisms that complement existing ecosystems.

2.3.1.1 Diverse Microbial Communities

A healthy and diverse microbial community in the soil or plant environment plays a pivotal role in suppressing diseases. Pathogens are less likely to dominate in ecosystems with high microbial diversity due to competitive interactions, predation, and the occupation of ecological niches.

1. **Microbial Antagonism**
   - Diverse microbial populations compete with pathogens for nutrients, space, and resources, creating a natural barrier against pathogen proliferation.
   - For example, the presence of beneficial bacteria such as *Pseudomonas fluorescens* and fungi like *Trichoderma harzianum* in the rhizosphere suppresses soil-borne pathogens such as *Fusarium oxysporum* and *Rhizoctonia solani.*
2. **Role of Arbuscular Mycorrhizal Fungi (AMF)**
   - AMF forms mutualistic relationships with plant roots, enhancing nutrient uptake and improving soil structure.
   - By promoting microbial diversity in the rhizosphere, AMF reduce the survival and growth of pathogenic organisms. For instance, AMF have been shown to suppress root pathogens like *Phytophthora* and *Pythium.*
3. **Natural Checks and Balances**
   - In biodiverse ecosystems, predators, parasites, and competitors naturally regulate pathogen populations. This reduces the need for chemical interventions, aligning with sustainable agricultural practices.

2.3.1.2 Functional Redundancy

Functional redundancy refers to the presence of multiple organisms within an ecosystem that perform similar roles or functions. This redundancy ensures that even if one population is compromised, others can maintain the ecosystem's functionality.

1. **Stability in Disease Management**
   - The presence of multiple bioagents with similar disease-suppressing capabilities ensures consistent pathogen control, even under fluctuating environmental conditions.
   - For instance, if a specific strain of *Trichoderma* is affected by soil pH changes, other microbial antagonists like *Bacillus subtilis* can continue suppressing pathogens.

2. **Enhanced Resilience**
   - Functional redundancy creates a buffer against environmental disruptions, such as drought or flooding, which can impact microbial populations.
   - This resilience is particularly important in large-scale agricultural systems, where environmental variability can affect the efficacy of disease management strategies.
3. **Case Studies**
   - In vegetable crops, a combination of *Pseudomonas fluorescens*, *Bacillus subtilis*, and AMF has shown to suppress soil-borne diseases more effectively than single-agent applications.
   - In rice fields, the use of multiple bioagents enhances the suppression of sheath blight caused by *Rhizoctonia solani*.

2.3.2 Reduction of Chemical Inputs

The use of chemical pesticides and fungicides in agriculture has long been associated with adverse ecological effects, including soil degradation, water pollution, and harm to non-target organisms. Biological control provides a sustainable alternative by reducing the reliance on these chemicals and their associated negative impacts.

2.3.2.1 Minimizing Pesticide Use

Biological control agents offer targeted and efficient suppression of pathogens, significantly reducing the need for synthetic chemicals.

1. **Replacement of Fungicides**
   - BCAs like *Trichoderma spp.* are widely used as an alternative to chemical fungicides in managing soil-borne diseases.
   - These bioagents not only suppress pathogens but also enhance plant growth, offering dual benefits. For example, *Trichoderma* applications in tomato fields have reduced the use of fungicides by 50%, while improving crop yields.
2. **Economic Benefits**
   - Reducing chemical pesticide use lowers input costs for farmers, particularly in resource-limited settings.
   - Additionally, the absence of chemical residues enhances the marketability of produce, especially in regions with strict food safety standards.
3. **Compatibility with Organic Farming**
   - Biological control aligns seamlessly with organic farming systems, which prohibit or severely restrict the use of synthetic pesticides.

2.3.2.2 Preserving Non-Target Organisms

Chemical pesticides often have non-specific effects, harming beneficial organisms such as pollinators, predatory insects, and soil microbiota. Biological control agents, in contrast, are highly specific, targeting only the intended pathogens.

1. **Protection of Pollinators**
    - Pollinators like bees and butterflies are critical for crop production, particularly in fruit and vegetable farming. By using BCAs instead of broad-spectrum pesticides, farmers can protect these vital organisms.
2. **Conservation of Natural Enemies**
    - Predatory insects and parasitoids, which help control pests, are often affected by pesticide applications. Biological control ensures their survival, maintaining ecological balance.
3. **Microbial Community Health**
    - Chemical inputs can disrupt soil microbiomes, reducing their diversity and functionality. BCAs, on the other hand, enhance microbial diversity, contributing to long-term soil health.

2.3.3 Examples of Ecological Benefits

The ecological benefits of biological control are far-reaching, extending beyond disease suppression to include improvements in soil health, water quality, and overall ecosystem stability.

2.3.3.1 Soil Health

The use of biological control agents contributes significantly to the health and fertility of agricultural soils.

1. **Improved Soil Structure**
    - BCAs like *Bacillus subtilis* and *Trichoderma spp.* promote the formation of soil aggregates, enhancing soil aeration and water retention.
    - These improvements support root growth and microbial activity, creating a favorable environment for plant development.
2. **Nutrient Cycling**
    - Many BCAs facilitate the breakdown of organic matter, releasing nutrients that are readily available to plants.
    - For example, *Bacillus subtilis* produces enzymes that decompose plant residues, contributing to nutrient cycling and reducing the need for chemical fertilizers.
3. **Suppression of Soil-Borne Pathogens**
    - By colonizing the rhizosphere, BCAs outcompete pathogens for resources, reducing disease incidence while maintaining soil health.

2.3.3.2 Water Quality

The reduction of chemical inputs through biological control has significant implications for water quality.

1. **Reduced Chemical Runoff**
    - Synthetic pesticides and fertilizers often leach into nearby water bodies, contaminating aquatic ecosystems. By replacing these chemicals with BCAs, farmers can mitigate water pollution.

- Cleaner water sources benefit not only agricultural communities but also downstream ecosystems.

2. **Protection of Aquatic Life**
   - Reduced pesticide runoff decreases the toxicity levels in aquatic habitats, protecting fish, amphibians, and other organisms.
   - Biological control aligns with conservation efforts aimed at preserving aquatic biodiversity.
3. **Improved Groundwater Safety**
   - In regions reliant on groundwater for irrigation and drinking, the adoption of biological control reduces the risk of contamination by harmful agrochemicals.

2.3.4 Broader Implications for Sustainability

The principle of ecological balance and sustainability underscores the transformative potential of biological control in agriculture. By reducing chemical dependency, enhancing biodiversity, and protecting critical ecosystems, biological control serves as a model for sustainable farming practices that prioritize environmental health and long-term productivity.

1. **Resilience to Climate Change**
   - Biodiverse agricultural systems supported by biological control are more resilient to climate-induced stressors, such as extreme weather events and pest outbreaks.
2. **Alignment with Global Goals**
   - The use of biological control supports international efforts, such as the United Nations' Sustainable Development Goals (SDGs), particularly those related to sustainable agriculture, clean water, and biodiversity conservation.
3. **Economic and Social Benefits**
   - Beyond environmental advantages, biological control contributes to the economic sustainability of farming by reducing costs and opening markets for organic and sustainably produced goods.

2.4 Principle of Integration into Farming Systems

The integration of biological control into broader disease and pest management strategies represents a holistic and sustainable approach to agriculture. By combining biocontrol agents (BCAs) with cultural, chemical, and physical methods, farmers can create synergistic systems that enhance the efficacy of disease management while minimizing environmental impact. This principle underscores the need to consider biological control not as a standalone solution but as a critical component of Integrated Disease Management (IDM) and Integrated Pest Management (IPM).

2.4.1 Synergies with Other Management Practices

Biological control agents achieve optimal effectiveness when combined with complementary strategies that create conducive environments for their action. The integration of cultural practices, selective chemical applications, and physical controls enhances the efficacy of BCAs, while reducing the reliance on synthetic inputs.

2.4.1.1 Cultural Practices

Cultural practices form the foundation of sustainable farming systems, creating conditions that favor beneficial organisms and suppress pathogens. When integrated with biological control, these practices amplify the effectiveness of BCAs.

1. **Crop Rotation**
   - Rotating crops with non-host plants interrupts the life cycle of pathogens, reducing their population density in the soil.
   - For example, alternating cereals with legumes in rotations diminishes the incidence of soil-borne diseases like Fusarium wilt, enhancing the action of bioagents such as *Trichoderma harzianum*.
2. **Intercropping**
   - Growing diverse crops together creates a complex environment that inhibits pathogen spread.
   - Intercropping systems, such as maize and beans, provide a supportive habitat for bioagents while disrupting pathogen colonization.
3. **Mulching**
   - Organic mulches enrich the soil with nutrients, fostering microbial diversity and supporting bioagent populations.
   - For instance, mulch-treated fields demonstrate enhanced colonization by *Pseudomonas fluorescens*, improving disease suppression.

Examples:

- In rice fields, a combination of crop rotation and the application of *Pseudomonas fluorescens* reduces sheath blight incidence by 50%.
- Mulching with organic residues enhances the effectiveness of *Bacillus subtilis* in managing tomato root diseases.

2.4.1.2 Chemical Compatibility

The compatibility of biological control agents with chemical treatments is critical for integrated systems. Selective use of low-toxicity chemicals can complement BCAs, reducing the need for broad-spectrum pesticides while preserving ecological balance.

1. **Reduced Chemical Dosage**
   - By integrating BCAs, farmers can lower the application rates of chemical fungicides and bactericides.
   - For example, *Pseudomonas fluorescens* combined with half the recommended dose of fungicides effectively controls blast disease in rice.
2. **Selective Chemicals**
   - Low-toxicity chemicals, such as potassium bicarbonate and neem oil, are compatible with most BCAs, preserving their efficacy.

- These selective treatments ensure that beneficial organisms, including predators and parasitoids, remain unharmed.

3. **Preventing Resistance**
   - Alternating between BCAs and chemical applications reduces the selection pressure on pathogens, minimizing the risk of resistance development.

Examples:

» Studies have shown that combining *Trichoderma spp.* with reduced doses of captan fungicide achieves superior control of damping-off in cucumber seedlings.

» In wheat, *Bacillus subtilis* is compatible with systemic fungicides, jointly reducing rust disease severity.

2.4.1.3 Physical Controls

Physical methods, such as soil solarization and protective barriers, complement biological control by reducing the initial pathogen load and creating favorable conditions for BCAs.

1. **Soil Solarization**
   - Covering soil with transparent polyethylene sheets traps solar radiation, heating the soil and killing pathogens.
   - Post-solarization, bioagents like *Trichoderma harzianum* can quickly colonize the treated soil, preventing pathogen resurgence.
2. **Protective Nets**
   - Nets protect crops from pest vectors, indirectly reducing the transmission of vector-borne diseases.
   - Combined with BCAs, protective barriers enhance the overall health of crops by reducing stress from both pests and pathogens.
3. **Temperature and Humidity Management**
   - Controlling greenhouse environments ensures optimal conditions for BCA activity, enhancing their effectiveness.

Examples:

» Soil solarization followed by the application of *Bacillus subtilis* significantly reduces wilt diseases in vegetable crops.

» In greenhouses, protective nets combined with *Pseudomonas fluorescens* applications reduce the incidence of bacterial blight in lettuce.

2.4.2 Case Studies in Integrated Approaches

The integration of biological control with other management practices has been successfully implemented in various agricultural systems. These case studies highlight the real-world effectiveness of integrated approaches.

2.4.2.1 Rice Fields in Asia

Rice cultivation faces significant challenges from diseases such as blast, sheath blight, and bacterial leaf blight. Integrated approaches combining biological control agents with resistant varieties and selective chemical treatments have proven highly effective in managing these diseases.

1. **Combination of Strategies**
   - Resistant varieties reduce the susceptibility of rice plants to blast disease.
   - *Pseudomonas fluorescens* applications induce systemic resistance, further enhancing disease suppression.
   - Selective fungicides, applied at reduced doses, target residual pathogens without disrupting the efficacy of bioagents.
2. **Results**
   - Field studies have shown a 60% reduction in blast disease incidence when *Pseudomonas fluorescens* is integrated with IDM practices.
   - Yield improvements of 20-30% have been observed in fields employing integrated strategies.
3. **Environmental and Economic Benefits**
   - The reduced use of fungicides lowers production costs and minimizes environmental contamination.
   - Enhanced soil health and microbial diversity ensure long-term productivity.

2.4.2.2 Greenhouse Vegetables in Europe

Greenhouse environments, while protective, often create favourable conditions for the rapid spread of root and foliar diseases. Integrated systems combining biological control agents with cultural and physical practices have been effective in managing these challenges.

1. **Integration of *Trichoderma spp.* and AMF**
   - *Trichoderma spp.* suppress root pathogens like *Pythium* and *Rhizoctonia solani*, while arbuscular mycorrhizal fungi (AMF) enhance nutrient uptake and plant vigour.
   - Cultural practices, such as crop rotation and organic mulching, further support these bioagents.
2. **IPM Programs**
   - Physical controls, such as humidity regulation and the use of protective screens, reduce the incidence of foliar diseases.
   - Pesticide applications are limited to specific outbreaks, preserving beneficial organisms.
3. **Results**
   - Greenhouses employing integrated approaches report a 50-70% reduction in disease incidence.

- The combined application of *Trichoderma* and AMF improves plant health, leading to higher yields and better quality produce.

2.4.3 Broader Implications for Sustainable Agriculture

The integration of biological control into farming systems aligns with the broader goals of sustainable agriculture. By combining diverse strategies, farmers can achieve robust disease management while reducing their environmental footprint.

1. **Resilience to Climate Change**
   - Integrated systems enhance the resilience of agricultural ecosystems to climate-induced stressors, such as pest outbreaks and extreme weather events.
2. **Economic Viability**
   - The combination of cost-effective bioagents and reduced chemical inputs lowers production costs, increasing profitability for farmers.
3. **Adoption in Smallholder Systems**
   - Integrated approaches are particularly beneficial for smallholder farmers, offering low-cost, sustainable solutions that improve productivity and soil health.

2.5 Principle of Monitoring and Adaptation

The principle of monitoring and adaptation is essential for maximizing the effectiveness of plant disease management with bioagents. Unlike chemical treatments that often provide immediate but temporary solutions, biocontrol agents (BCAs) operate within dynamic ecosystems where environmental factors, pathogen behaviour, and plant responses continually evolve. Continuous monitoring ensures the efficacy of BCAs by assessing their performance and interactions within the system, while adaptive strategies allow adjustments to optimize their effectiveness under varying conditions. This approach ensures that biocontrol remains a sustainable and reliable method for disease management.

2.5.1 Monitoring Biocontrol Efficacy

Effective monitoring of biocontrol efficacy involves both qualitative and quantitative assessments. By systematically tracking the performance of bioagents and their impact on pathogens, farmers and researchers can identify areas for improvement and make informed decisions to enhance disease management strategies.

2.5.1.1 Quantitative Assessments

Quantitative assessments involve the measurement of bioagent populations, pathogen levels, and plant health parameters. These assessments provide objective data to evaluate the performance of biocontrol agents.

1. **Soil and Plant Tissue Testing**
   - Regular testing of soil and plant tissues helps measure the population dynamics of both bioagents and pathogens.

- These tests can reveal whether bioagent populations are thriving in the target environment and whether pathogen levels are being suppressed effectively.

2. **Molecular Tools for Detection**
   - **Polymerase Chain Reaction (PCR)**: PCR techniques can detect and quantify the DNA of bioagents like *Trichoderma spp.* and *Bacillus subtilis*. This precise method ensures that bioagents are present at effective levels.
   - **Enzyme-Linked Immunosorbent Assay (ELISA)**: ELISA assays track the presence of specific proteins or metabolites associated with bioagents, providing insights into their activity and functionality.
3. **Longitudinal Studies**
   - Periodic assessments over time help monitor trends in bioagent populations and pathogen suppression, offering a clearer picture of long-term efficacy.

Example:

In greenhouse cucumber cultivation, PCR testing has been used to monitor the colonization of *Pseudomonas fluorescens* in the rhizosphere, correlating its presence with reduced incidence of root rot.

2.5.1.2 Visual Indicators

While molecular tools provide precise data, visual observations remain a vital component of monitoring biocontrol efficacy. These indicators offer immediate feedback on the health of the crop and the effectiveness of the biocontrol strategy.

1. **Plant Health Observations**
   - Visual assessment of leaf color, stem vigor, and root development provides insights into overall plant health.
   - Healthy plants with minimal signs of stress indicate effective pathogen suppression and bioagent activity.
2. **Disease Incidence and Severity**
   - Field observations track the prevalence and severity of diseases in treated versus untreated areas.
   - A significant reduction in disease symptoms, such as wilting or leaf spots, highlights the success of biocontrol measures.
3. **Yield Analysis**
   - Crop yields provide a direct measure of the effectiveness of biocontrol strategies. Higher yields with reduced chemical inputs indicate successful integration of BCAs.

Example:

In tomato fields, visual monitoring of disease incidence, combined with yield analysis, has shown that *Trichoderma harzianum* applications reduce Fusarium wilt and improve fruit quality.

2.5.2 Adaptive Strategies

Adaptation is a critical component of sustainable biocontrol practices. By modifying strategies in response to environmental conditions, pathogen behaviour, and bioagent performance, farmers can optimize the efficacy of biocontrol systems.

2.5.2.1 Environmental Adjustments

Environmental factors such as temperature, humidity, and soil pH significantly influence the activity and survival of bioagents. Adjusting these factors can create favourable conditions for BCAs, enhancing their performance.

1. **Temperature Optimization**
    - Many bioagents, such as *Bacillus subtilis*, exhibit peak activity within specific temperature ranges. Maintaining these conditions ensures consistent pathogen suppression.
    - In cooler climates, greenhouse heating can be used to maintain optimal temperatures for bioagent activity.
2. **Humidity Control**
    - High humidity levels often promote fungal pathogen proliferation but can also enhance the colonization of BCAs like *Trichoderma spp.*
    - Adjusting humidity through irrigation or ventilation can balance these effects, maximizing the benefits of biocontrol.
3. **Soil pH Management**
    - Soil pH affects the survival and activity of bioagents. For example, *Pseudomonas fluorescens* thrives in slightly acidic to neutral soils.
    - Liming acidic soils or adding organic amendments can create favourable conditions for bioagent colonization.

Example:

In lettuce cultivation, adjusting irrigation schedules and soil pH has improved the efficacy of *Pseudomonas fluorescens* in suppressing bacterial leaf spot.

2.5.2.2 Rotating Bioagents

Pathogen resistance to biocontrol agents can emerge over time, particularly when a single bioagent is used repeatedly. Rotating or combining different bioagents helps maintain long-term efficacy and prevents resistance development.

1. **Multiple Bioagent Strategies**
    - Combining bioagents with complementary modes of action, such as *Trichoderma harzianum* (mycoparasitism) and *Bacillus subtilis* (antibiosis), provides comprehensive pathogen suppression.
    - This approach reduces the likelihood of pathogens adapting to a single mechanism of action.

2. **Seasonal Adjustments**
    - Rotating bioagents based on seasonal changes ensures that effective species are used under varying environmental conditions.
    - For instance, in temperate regions, bioagents suited to cooler conditions can be used in early spring, while those thriving in warmer temperatures can be applied in summer.
3. **Case Studies**
    - In strawberry farming, alternating *Bacillus subtilis* with *Pseudomonas fluorescens* applications has reduced the incidence of gray mold caused by *Botrytis cinerea.*

2.5.3 Examples of Adaptive Management

Adaptive management strategies demonstrate the practical benefits of monitoring and adjusting biocontrol practices. These examples highlight the flexibility and effectiveness of such approaches in diverse agricultural systems.

2.5.3.1 Greenhouse Tomatoes

1. **Problem**: High humidity in greenhouses often promotes fungal diseases such as powdery mildew, complicating disease management.
2. **Solution:**
    - Monitoring humidity levels revealed that *Trichoderma spp.* colonization was optimal at moderate humidity.
    - Adjusting the irrigation schedule and improving ventilation reduced humidity, enhancing bioagent efficacy.
3. **Results:**
    - Disease outbreaks decreased by 40%, and tomato yields improved by 25%, demonstrating the success of adaptive management.

2.5.3.2 Field Crops

1. **Problem**: Sandy soils in arid regions often present challenges for bioagent survival, reducing their effectiveness against nematodes and fungal pathogens.
2. **Solution:**
    - Soil solarization was used to reduce pathogen loads before bioagent application.
    - The integration of organic amendments improved soil moisture retention, creating a supportive environment for BCAs.
3. **Results:**
    - The combination of solarization and *Bacillus subtilis* applications reduced nematode populations by 60% and fungal infections by 50%.

2.5.4 Broader Implications for Sustainable Agriculture

The broader implications of biological control for sustainable agriculture lie in its ability to align pest and disease management strategies with the principles of ecological balance, resource efficiency, and long-term resilience. A key principle underpinning the success of biological control systems is the need for continuous monitoring and adaptation. Unlike static approaches, biocontrol systems are inherently dynamic, requiring ongoing assessment and refinement to optimize their performance in diverse and changing agricultural contexts. This adaptability not only ensures sustainable and effective disease management but also positions biological control as a central component of future-ready agricultural systems. One of the most significant implications of this adaptability is its potential to build resilience to climate variability. As global temperatures rise, precipitation patterns shift, and extreme weather events become more frequent, agricultural systems face unprecedented challenges. Pests and pathogens are expected to respond to these changes with altered distributions, life cycles, and population dynamics. Adaptive management of biocontrol systems, which involves monitoring environmental conditions and pest pressures and making informed adjustments to strategies, enables farmers to maintain effective pathogen suppression even under changing climatic conditions. For example, certain biocontrol agents, such as fungal pathogens like *Beauveria bassiana* and *Metarhizium anisopliae*, may need specific environmental conditions to thrive. Monitoring temperature, humidity, and other factors helps ensure that these agents remain viable and effective despite climatic fluctuations.

The integration of digital tools further amplifies the potential of biocontrol systems to contribute to sustainable agriculture. Emerging technologies such as remote sensing, Internet of Things (IoT)-based sensors, and data analytics provide farmers with real-time information on environmental variables, pest populations, and crop health. These tools allow for precise adjustments to biocontrol strategies, ensuring that interventions are timely, targeted, and resource-efficient. For instance, remote sensing can detect early signs of pest outbreaks or disease hotspots, enabling farmers to deploy biocontrol agents exactly where they are needed, minimizing waste and maximizing impact. IoT sensors installed in fields can continuously monitor soil moisture, temperature, and pathogen activity, providing actionable data that helps optimize the conditions for biocontrol agents to function effectively. The integration of these digital tools not only enhances the efficiency of biological control but also reduces the reliance on chemical inputs, aligning agricultural practices with the goals of sustainability and environmental conservation.

Another critical implication of biological control for sustainable agriculture is its role in empowering farmers through education and capacity building. Training programs that teach farmers to monitor and adapt biocontrol practices empower them to implement sustainable disease management strategies independently. This empowerment is particularly important for smallholder farmers, who often face resource constraints and limited access to external support. By equipping farmers with the knowledge and skills to assess pest and disease pressures, identify effective biocontrol agents, and adjust their strategies based on field conditions, these programs foster self-reliance and resilience. For example, farmer field schools that focus on integrated pest management (IPM) often include modules on biological control, emphasizing the importance of observation, experimentation, and adaptation. Such programs not only improve the effectiveness of biocontrol systems but also contribute to broader goals of rural development and poverty alleviation by enhancing farmers' productivity, incomes, and environmental stewardship.

3. History of Biological Control in Agriculture

The history of biological control in agriculture spans centuries, evolving from ancient practices to modern, science-driven approaches. It represents humanity's ongoing effort to harness natural processes for managing pests and pathogens, aiming to sustain agricultural productivity while preserving ecological balance.

3.1 Early Practices and Ancient Beginnings

The concept of biological control, while formally recognized only in recent centuries, has deep historical roots. For millennia, human societies relied on natural methods to manage agricultural pests and diseases. These methods were often based on keen observations of ecological interactions, particularly predator-prey relationships. Ancient farmers intuitively recognized the benefits of harnessing nature to maintain crop health and productivity. These early practices laid the foundation for modern biological control, demonstrating that the principles of sustainability and ecological balance have always been integral to agriculture.

3.1.1 Use of Predators

Predatory organisms have long been employed as natural allies in agriculture, often with remarkable efficacy. Early civilizations such as those in China, Egypt, and Rome made use of predators to control pests, showcasing an understanding of the natural balance within ecosystems.

3.1.1.1 China (300 BC)

One of the most well-documented examples of early biological control comes from ancient China, where farmers used predatory ants (*Oecophylla smaragdina*) to protect citrus orchards from insect pests.

1. **Ant Deployment in Citrus Orchards**
    - Farmers observed that predatory ants could efficiently eliminate harmful insects, such as caterpillars and borers, from their orchards.
    - To maximize the ants' effectiveness, they placed bamboo sticks between trees, enabling the ants to move freely throughout the orchard. This simple yet ingenious method created a natural pest control system.
2. **Cultural Significance**
    - The practice was so effective that it became a standard part of citrus farming in ancient China.
    - The integration of ants into farming systems reflects the early understanding of ecological harmony and the reliance on natural systems for agricultural productivity.
3. **Legacy and Modern Relevance**
    - The use of *Oecophylla smaragdina* is still practiced in some parts of Asia today, underscoring the lasting value of this ancient technique.
    - Modern entomologists consider this one of the earliest recorded examples of intentional biological control.

3.1.1.2 Egypt and Rome

The use of natural predators extended beyond Asia, with evidence of similar practices in ancient Egypt and Rome.

1. **Rodent Control in Egypt**
    - Ancient Egyptians faced the challenge of protecting stored grain from rodents. To address this, they domesticated cats and kept them near granaries.
    - Cats effectively reduced rodent populations, safeguarding food supplies and preventing economic losses.
    - This practice also had cultural significance, as cats became revered in Egyptian society for their role in maintaining agricultural prosperity.
2. **Insect Suppression in Rome**
    - Roman agricultural texts, such as those by Columella and Pliny the Elder, describe the deliberate use of insect predators to control agricultural pests.
    - For instance, farmers encouraged the presence of ladybird beetles (predators of aphids) in their fields, recognizing their role in reducing pest populations.
3. **Evolution of Agricultural Knowledge**
    - These practices reflect a growing awareness among ancient societies of the interconnectedness between crops, pests, and their natural enemies.
    - By observing and utilizing predator-prey relationships, these early farmers contributed to the foundational principles of biological control.

3.1.2 Cultural Practices Supporting Biological Control

Beyond the direct use of predators, ancient farming practices often incorporated cultural techniques that indirectly supported biological control. These methods created conditions favourable to natural enemies of pests, reducing the need for external interventions.

3.1.2.1 Crop Diversity

Crop diversity has been a cornerstone of traditional agriculture, providing multiple benefits, including pest suppression.

1. **Intercropping Systems**
    - Ancient farmers grew diverse crops together, disrupting pest life cycles and reducing the likelihood of severe infestations.
    - For example, intercropping cereals with legumes not only improved soil fertility but also supported predatory insects that fed on pest species.
    - The spatial complexity of intercropped fields made it harder for pests to locate their preferred host plants, effectively reducing their populations.

2. **Crop Rotation**
   - Rotating crops with different life cycles and pest susceptibilities was another common practice.
   - For instance, alternating grains with root vegetables disrupted the breeding cycles of soil-borne pests, indirectly supporting natural pest predators by reducing pest abundance.
3. **Modern Implications**
   - These traditional practices are still relevant today, forming the basis of integrated pest management (IPM) strategies.
   - By promoting crop diversity, farmers create resilient systems that naturally suppress pest populations.

3.1.2.2 Habitat Management

Habitat management involves creating and maintaining environments that support the natural enemies of agricultural pests. This approach was intuitively practiced by ancient farmers to enhance ecological stability in their fields.

1. **Providing Shelter**
   - Farmers recognized the importance of providing habitats for beneficial organisms such as birds, predatory insects, and amphibians.
   - Simple measures, such as leaving hedgerows or planting trees, created nesting sites and shelter for these natural predators.
2. **Attracting Beneficial Insects**
   - Ancient farmers often planted flowering plants near their crops to attract pollinators and predatory insects.
   - Nectar-producing plants served as a food source for beneficial insects, encouraging their presence in the field.
3. **Supporting Soil Biodiversity**
   - Practices like composting and applying organic manures enriched the soil, promoting the growth of beneficial microbes and nematodes that suppressed soil-borne pathogens.
   - This balance within the soil ecosystem reduced the need for external inputs to control pests and diseases.
4. **Sustainability and Legacy**
   - Habitat management remains a critical aspect of sustainable farming, bridging ancient practices with modern ecological farming principles.
   - These practices demonstrate the value of working with nature rather than against it, a philosophy central to biological control.

3.2 Scientific Foundations (17th to 19th Century)

The scientific study of biological control began during the 17th century and gained momentum in the 19th century with foundational discoveries in entomology and microbiology. This era marked the transition from observational practices to systematic investigations, providing the theoretical and practical framework for modern biological control. The contributions of early scientists during this period were pivotal in elucidating the mechanisms of natural pest control, shaping the development of biological control into a recognized scientific discipline.

3.2.1 Early Entomological Studies

Entomology, the study of insects, played a critical role in understanding the ecological dynamics of pests and their natural enemies. Observations and experiments conducted by pioneering entomologists laid the groundwork for biological control by identifying predation, parasitism, and other ecological interactions as viable methods for pest management.

3.2.1.1 Observations by Francesco Redi (1668)

Italian scientist Francesco Redi is often regarded as one of the earliest contributors to the scientific study of biological control. His groundbreaking observations in the 17th century provided insights into parasitism and its potential role in pest suppression.

1. **Study of Parasitism**
   - Redi observed wasps laying their eggs inside caterpillars. Upon hatching, the wasp larvae fed on the caterpillar, eventually killing it.
   - This interaction demonstrated the natural regulatory role of parasitoids in controlling pest populations.
2. **Significance of Redi's Work**
   - His studies highlighted the concept of biological control through parasitism, introducing a natural mechanism for managing pest outbreaks.
   - Redi's observations paved the way for further exploration of insect parasitism as a practical tool in agriculture.
3. **Broader Implications**
   - While Redi's work was primarily descriptive, it underscored the potential of harnessing natural predator-prey relationships for pest management.
   - His findings are considered foundational to the scientific exploration of biological control.

3.2.1.2 Riley's Work on Cottony Cushion Scale (1888)

The introduction of the vedalia beetle (*Rodolia cardinalis*) to control the cottony cushion scale (*Icerya purchasi*) in California is one of the most celebrated success stories in classical biological control.

1. **Problem in California Citrus Orchards**
   - In the late 19th century, California's citrus industry faced devastation due to an invasive pest, the cottony cushion scale.

- This pest, native to Australia, spread rapidly in the absence of its natural enemies, causing widespread crop losses.

2. **Charles Valentine Riley's Solution**
   - Riley, a prominent entomologist, introduced the vedalia beetle from Australia, where it was a natural predator of the cottony cushion scale.
   - The beetle successfully established itself in California, rapidly reducing the pest population and saving the citrus industry from collapse.
3. **Impact and Legacy**
   - This effort marked the first major success of classical biological control and set a precedent for the systematic introduction of natural enemies to manage invasive pests.
   - Riley's work established classical biological control as a viable and effective strategy for pest management, inspiring future programs worldwide.

3.2.2 Advancements in Microbiology

The field of microbiology emerged as a significant contributor to biological control during the 19th century. The discovery of pathogenic microorganisms and their role in regulating insect and plant pathogen populations provided new avenues for managing agricultural pests and diseases.

3.2.2.1 Discovery of Pathogenic Microorganisms

The identification of entomopathogenic fungi, bacteria, and viruses during the 19th century revolutionized biological control by introducing microbial agents as potential tools for pest management.

1. ***Beauveria bassiana* and Other Pathogenic Fungi**
   - In the early 1800s, *Beauveria bassiana* was identified as a fungus capable of infecting and killing insect pests, including silkworms.
   - This discovery demonstrated the potential of microbial agents to suppress pest populations effectively.
   - Subsequent studies revealed that *B. bassiana* could infect a wide range of insects, establishing its versatility as a biocontrol agent.
2. **Bacterial Pathogens**
   - The identification of bacterial pathogens, such as *Bacillus thuringiensis* (Bt), in the late 19th century further expanded the scope of microbial biological control.
   - Bt was found to produce toxic proteins that targeted specific insect pests, offering a selective and environmentally friendly alternative to chemical pesticides.
3. **Applications in Agriculture**
   - The discovery of entomopathogenic microorganisms enabled the development of microbial biopesticides, a cornerstone of modern biological control.
   - These agents provided effective solutions for managing insect pests in diverse cropping systems.

3.2.2.2 Louis Pasteur's Contributions

Louis Pasteur, renowned for his work on microbial diseases, played an indirect but critical role in advancing biological control. His research on the relationship between microbes and disease laid the foundation for understanding pathogen-based pest control.

1. **Pioneering Microbial Research**
   - Pasteur's studies on silkworm diseases, particularly *Pebrine* caused by *Nosema bombycis*, revealed the role of microorganisms in causing disease outbreaks.
   - His work demonstrated that microbial pathogens could be identified, studied, and potentially harnessed for pest management.
2. **Impact on Agricultural Practices**
   - While Pasteur's primary focus was on animal and human health, his findings inspired the application of microbial pathogens in agricultural systems.
   - The concept of utilizing microbial diseases to control insect pests and pathogens gained credibility through his research.
3. **Legacy in Biological Control**
   - Pasteur's contributions to microbiology provided a scientific framework for studying entomopathogenic fungi and bacteria, key agents in biological control.
   - His emphasis on systematic observation and experimentation continues to influence the development of biocontrol strategies.

3.3 Establishment of Biological Control as a Science (20th Century)

The 20th century marked a significant turning point in the history of biological control, with its establishment as a formal scientific discipline. During this period, advances in entomology, ecology, and microbiology converged, enabling the development of systematic biocontrol strategies. Governments, research institutions, and international organizations began to recognize the potential of biological control for sustainable agriculture, leading to the implementation of large-scale programs and the integration of biocontrol agents into comprehensive pest and disease management systems.

3.3.1 Classical Biological Control Programs

Classical biological control involves the introduction of natural enemies from a pest's native habitat to suppress its population in regions where it has become invasive. This approach was widely adopted in the 20th century, leading to several landmark successes that demonstrated its efficacy and established it as a foundational strategy in pest management.

3.3.1.1 Success Stories

The successful application of classical biological control during the 20th century highlights its potential to mitigate invasive pest outbreaks while reducing reliance on chemical pesticides.

1. **Control of Sugarcane Borers in Southeast Asia**
   - The introduction of the parasitoid wasp *Cotesia flavipes* was a groundbreaking effort to

manage sugarcane borers (*Chilo spp.*), which caused significant yield losses in Southeast Asia.

- *Cotesia flavipes* parasitizes the larvae of sugarcane borers, effectively reducing their population and preventing further crop damage.
- This program not only restored sugarcane productivity but also demonstrated the economic and ecological benefits of classical biological control.

2. **Release of *Cryptolaemus montrouzieri* for Mealybug Control**
   - Known as the "mealybug destroyer," *Cryptolaemus montrouzieri* is a predatory ladybird beetle introduced worldwide to combat mealybug infestations in fruit orchards.
   - Its effectiveness in controlling mealybugs on citrus, grapes, and ornamentals made it one of the most widely used biocontrol agents.
   - This success underscored the versatility of natural predators in managing diverse pest species across varying agroecosystems.
3. **Impact and Legacy**
   - These success stories inspired the adoption of classical biological control in other regions and crops, demonstrating that natural enemies could provide long-term, cost-effective pest management.
   - The ecological and economic benefits of these programs laid the foundation for modern biocontrol strategies.

3.3.1.2 Principles of Importation

The principles underlying classical biological control emphasize the importance of ecological compatibility and the careful selection of biocontrol agents.

1. **Native Range Exploration**
   - Biocontrol agents are sourced from a pest's native range, where they have co-evolved with the target pest. This ensures that the agent is effective in suppressing the pest population.
   - Rigorous screening processes are employed to identify natural enemies that are specific to the target pest, minimizing risks to non-target organisms.
2. **Introduction and Establishment**
   - Once selected, biocontrol agents are introduced into the target region and monitored to ensure their establishment and effectiveness.
   - The establishment of agents such as *Cotesia flavipes* and *Cryptolaemus montrouzieri* exemplifies the success of this approach.
3. **Ecological Balance**
   - Classical biological control seeks to restore ecological balance by reintroducing natural checks and balances into the invaded ecosystem.
   - This approach aligns with sustainable agricultural practices by reducing dependency on chemical inputs and supporting biodiversity.

3.3.2 Development of Microbial Biopesticides

The discovery and commercialization of microbial biopesticides during the 20th century represented a major advancement in biological control. Microbial agents such as bacteria and fungi offered targeted, environmentally friendly solutions to pest and disease management, complementing the classical approach.

3.3.2.1 *Bacillus thuringiensis* (Bt)

1. **Discovery and Applications**
    - Discovered in 1911, *Bacillus thuringiensis* (Bt) became one of the most widely used microbial insecticides.
    - Bt produces crystalline proteins (Cry toxins) that are toxic to specific insect pests, particularly lepidopterans (caterpillars). These proteins disrupt the digestive system of pests, causing death within hours.
2. **Specificity and Safety**
    - Bt's high specificity ensures that it targets only the intended pest species, leaving beneficial insects and other non-target organisms unharmed.
    - This safety profile made Bt a cornerstone of biological control, particularly in organic farming and IPM systems.
3. **Commercial Success**
    - Bt formulations were commercialized as biopesticides in the mid-20th century, becoming a key tool for managing pests in crops such as cotton, maize, and vegetables.
    - Genetically engineered Bt crops, introduced in the 1990s, further expanded its applications, providing built-in pest resistance.

3.3.2.2 Fungal Biocontrol Agents

1. **Key Entomopathogenic Fungi**
    - *Metarhizium anisopliae* and *Beauveria bassiana* are two of the most widely used fungal biocontrol agents.
    - These fungi infect and kill insect pests by penetrating their exoskeleton and disrupting their physiological functions.
2. **Applications in Agriculture**
    - *Beauveria bassiana* is used to control pests such as whiteflies, aphids, and thrips in horticultural crops.
    - *Metarhizium anisopliae* has been successfully deployed against soil-dwelling pests like termites and root grubs.
3. **Advantages of Fungal Agents**
    - Fungal biocontrol agents offer a sustainable alternative to chemical insecticides, reducing environmental contamination and promoting ecological balance.

- Their ability to persist in the soil provides long-term pest suppression.

3.3.3 Ecological Insights

The integration of ecological principles into biological control strategies during the 20th century enhanced their effectiveness and sustainability. These insights emphasized the importance of understanding ecosystem dynamics and leveraging them for comprehensive pest and disease management.

3.3.3.1 Integrated Pest Management (IPM)

1. **Concept and Evolution**
   - The concept of IPM emerged in the mid-20th century as a holistic approach to pest management. It integrates biological control with cultural, chemical, and physical methods to achieve sustainable pest suppression.
   - IPM prioritizes the use of biocontrol agents as the first line of defense, with chemical interventions employed only as a last resort.
2. **Role of Biocontrol Agents**
   - BCAs, including predators, parasitoids, and microbial agents, are central to IPM systems.
   - For example, *Cotesia flavipes* is a key component of IPM programs for sugarcane pests, while *Pseudomonas fluorescens* is used in rice-based IPM systems to suppress bacterial blight.
3. **Global Adoption**
   - IPM strategies have been adopted worldwide, reducing pesticide use, lowering production costs, and mitigating environmental impacts.

3.3.3.2 Rhizosphere Research

1. **Understanding Rhizosphere Dynamics**
   - The rhizosphere, the soil region surrounding plant roots, is a hotspot of microbial activity and interactions.
   - Research during the 20th century revealed the role of beneficial microorganisms such as *Pseudomonas fluorescens* and *Trichoderma spp.* in suppressing soil-borne pathogens.
2. **Applications in Disease Management**
   - *Pseudomonas fluorescens* produces antibiotics and siderophores that inhibit pathogens like *Fusarium* and *Rhizoctonia.*
   - *Trichoderma spp.* are effective against root rot and damping-off diseases, promoting plant growth while suppressing pathogens.
3. **Integration into Farming Systems**
   - Rhizosphere-based biocontrol agents have become integral to sustainable agriculture, contributing to soil health, nutrient cycling, and disease suppression.

3.4 Modern Advances and Global Adoption (21st Century)

The 21st century has witnessed transformative advancements in biological control, propelled by innovations in biotechnology, precision agriculture, and a global emphasis on sustainability. Increasing environmental concerns, coupled with the harmful effects of chemical pesticides, have fueled the adoption of biocontrol as a cornerstone of modern farming. Global initiatives and policy support have further accelerated its integration into agricultural systems worldwide.

3.4.1 Biotechnology and Genetic Engineering

Biotechnology and genetic engineering have significantly expanded the scope and effectiveness of biological control. By enhancing the traits of bioagents and exploring new microbial combinations, these advancements have addressed some of the limitations of traditional biocontrol methods, such as efficacy under variable environmental conditions and pest resistance.

3.4.1.1 Genetically Modified Bioagents

The application of genetic engineering has revolutionized biological control by creating enhanced bioagents that are more efficient, specific, and resilient.

1. **Enhancements in *Bacillus thuringiensis* (Bt)**
    - Genetically modified strains of *Bt* have been engineered to produce multiple Cry toxins, targeting a broader range of pest species.
    - These modifications also increase the efficacy of *Bt* against pests that have developed resistance to traditional strains, ensuring continued effectiveness.
    - For instance, stacked traits in genetically modified Bt cotton provide protection against multiple insect pests, including bollworms and armyworms.
2. **Development of Virus-Based Bioagents**
    - Advances in genetic engineering have enabled the modification of baculoviruses, naturally occurring viruses that infect insect pests.
    - Modified baculoviruses are designed to replicate faster and kill pests more efficiently, making them a viable alternative to chemical insecticides.
3. **Enhanced Traits for Environmental Resilience**
    - Genetic engineering has produced bioagents with improved resilience to extreme environmental conditions, such as high temperatures or salinity.
    - For example, modified strains of *Pseudomonas fluorescens* demonstrate enhanced activity in alkaline soils, broadening their applicability in diverse agroecosystems.

Implications:

Genetically modified bioagents address the challenges of pest resistance and environmental variability, ensuring consistent performance and expanding the reach of biological control.

3.4.1.2 Microbial Consortia

Microbial consortia represent a breakthrough in biological control, combining multiple bioagents to achieve synergistic effects in pest and disease suppression.

1. **Concept and Development**
   - Advances in microbiome research have revealed the complex interactions between microorganisms in the soil and plant rhizosphere.
   - By assembling compatible microbial strains with complementary functions, microbial consortia enhance pathogen suppression, plant growth, and soil health.
2. **Examples of Microbial Consortia**
   - A consortium combining *Trichoderma spp.* (antifungal), *Pseudomonas fluorescens* (antibiosis), and nitrogen-fixing bacteria has shown superior performance in managing root rot diseases while improving plant nutrition.
   - Another example is the integration of *Bacillus subtilis* with arbuscular mycorrhizal fungi (AMF), which provides dual benefits of disease suppression and enhanced nutrient uptake.
3. **Applications in Sustainable Farming**
   - Microbial consortia are particularly valuable in organic farming, where synthetic inputs are limited.
   - They offer comprehensive solutions for pest and disease management while improving overall crop resilience.

3.4.2 Precision Agriculture

Precision agriculture integrates advanced technologies to optimize farming practices, including the application of biological control agents. By leveraging data and automation, precision tools enhance the efficiency, accuracy, and scalability of biocontrol in modern farming systems.

3.4.2.1 Remote Sensing and IoT

1. **Role of Remote Sensing**
   - Satellite and drone-based remote sensing systems monitor pest populations and disease outbreaks in real time.
   - These systems provide high-resolution data on crop health, pest hotspots, and environmental conditions, enabling timely and targeted application of bioagents.
2. **Integration with IoT**
   - Internet of Things (IoT) devices, such as soil sensors and weather stations, collect continuous data on factors like soil moisture, temperature, and humidity.
   - This data helps optimize the timing and placement of biocontrol applications, maximizing their efficacy.

Example:

In vineyards, IoT-enabled systems detect the early stages of fungal infections and guide the application of *Trichoderma spp.* to prevent disease spread, reducing crop losses by up to 30%.

3.4.2.2 Automated Delivery Systems

Automation has transformed the deployment of biological control agents, making it more precise and efficient.

1. **Drone-Based Spraying**
   - Drones equipped with advanced spraying systems can distribute biopesticides uniformly over large areas.
   - This method minimizes waste and ensures that bioagents are delivered to target sites, such as the canopy or rhizosphere, with precision.
2. **Robotic Dispensers**
   - Robots are increasingly used in greenhouses and orchards to release predatory insects like ladybird beetles or parasitic wasps.
   - These systems ensure consistent distribution and reduce labour costs.
3. **Large-Scale Applications**
   - Automated systems enable the rapid scaling of biological control in commercial farming, addressing challenges of labour shortages and large-area pest outbreaks.

Impact:

Automation and precision tools enhance the adoption of biological control in intensive farming systems, ensuring cost-effectiveness and scalability.

3.4.3 Global Initiatives and Policy Support

The global push for sustainable agriculture has driven widespread adoption of biological control, supported by international organizations and national policies.

3.4.3.1 FAO's Role in Promoting Biocontrol

1. **Advocacy and Education**
   - The Food and Agriculture Organization (FAO) has been at the forefront of promoting biological control as a sustainable alternative to chemical pesticides.
   - The FAO's initiatives include training programs, knowledge sharing, and technical support for implementing biocontrol strategies in developing countries.
2. **Global Action Plans**
   - As part of its Integrated Pest Management (IPM) programs, the FAO advocates for the integration of biocontrol agents into national agricultural policies.
   - The organization has highlighted success stories, such as the use of *Cryptolaemus montrouzieri* in Asia, to encourage adoption worldwide.

3. **Collaboration with Governments**
   - The FAO collaborates with governments to establish regulatory frameworks for the safe and effective use of biocontrol agents.

3.4.3.2 Organic Farming Movements

1. **Rising Popularity of Organic Farming**
   - The global demand for organic produce has surged, driven by consumer awareness of health and environmental issues.
   - Biological control aligns with the principles of organic farming, providing natural solutions for pest and disease management.
2. **Certification and Market Opportunities**
   - Organic certification programs often require the exclusive use of biocontrol agents and other non-chemical inputs.
   - Farmers adopting biocontrol gain access to premium markets, boosting their profitability.
3. **Policy Incentives**
   - Many governments provide subsidies and technical support for organic farming, indirectly promoting the adoption of biocontrol.
   - For instance, the European Union's Common Agricultural Policy (CAP) includes provisions for biocontrol in its environmental sustainability goals.

3.5 Challenges and Future Prospects

Biological control has emerged as a critical component of sustainable agriculture, offering eco-friendly alternatives to chemical pest and disease management. Despite its successes, significant challenges hinder its widespread adoption and efficacy in diverse agricultural systems. Variability in field performance, production scalability, and regulatory hurdles are among the key obstacles that must be addressed. At the same time, rapid advancements in biotechnology, artificial intelligence, and integrated pest management hold immense promise for overcoming these barriers.

3.5.1 Challenges

Biological control operates within dynamic ecosystems, making it inherently complex and context-dependent. The following challenges illustrate the multifaceted nature of these difficulties.

3.5.1.1 Environmental Variability

Environmental factors such as temperature, humidity, and soil pH significantly influence the efficacy of biological control agents (BCAs).

1. **Sensitivity to Climatic Conditions**
   - Many bioagents, particularly microbial ones, have specific temperature and humidity ranges within which they are most effective.

- For instance, *Trichoderma spp.* thrive in moist soils with moderate temperatures but may become less active in arid or cold conditions.

2. **Soil and Water Quality**
   - Soil pH and nutrient levels can impact the survival and activity of bioagents. For example, *Pseudomonas fluorescens* performs poorly in highly alkaline soils.
   - Water quality, including salinity and contamination, also affects the viability of bioagents in irrigation systems.
3. **Unpredictable Field Conditions**
   - Unlike controlled laboratory environments, field conditions are highly variable, with fluctuations in weather, soil composition, and pest dynamics.
   - This unpredictability often results in inconsistent performance of biocontrol agents.

Mitigation Strategies:

» Research to identify and develop bioagents with broader environmental tolerances.

» Use of soil amendments and controlled irrigation to create favorable conditions for bioagent activity.

3.5.1.2 Pathogen Resistance

The evolution of resistance among target pathogens poses a significant challenge to the long-term efficacy of biological control strategies.

1. **Mechanisms of Resistance**
   - Pathogens may develop mechanisms to overcome bioagent activity, such as producing enzymes that degrade antimicrobial compounds produced by bioagents.
   - Over-reliance on a single bioagent or mode of action increases the risk of resistance development.
2. **Examples of Resistance**
   - Resistance to *Bacillus thuringiensis* (Bt) toxins has been reported in several insect pest populations, particularly in regions with extensive use of Bt crops.
   - Fungal pathogens like *Fusarium* have shown reduced susceptibility to antagonistic fungi in some cases.
3. **Impact on Biocontrol Programs**
   - Resistance undermines the reliability of biological control, necessitating the development of alternative strategies and new bioagents.

Mitigation Strategies:

» Rotation or combination of multiple bioagents to diversify modes of action and delay resistance development.

» Monitoring and early detection of resistance to adapt management practices proactively.

3.5.2 Future Directions

The challenges facing biological control present opportunities for innovation and improvement. The following advancements hold the potential to transform the field, making it more robust, efficient, and scalable.

3.5.2.1 Next-Generation Bioagents

Advances in synthetic biology, gene editing, and microbiome research are paving the way for the development of next-generation bioagents with enhanced capabilities.

1. **Synthetic Biology Applications**
   - Synthetic biology allows the creation of bioagents with tailored traits, such as enhanced resistance to environmental stressors or increased pathogen-targeting specificity.
   - For example, genetically modified *Pseudomonas fluorescens* strains can produce multiple antifungal compounds simultaneously, improving their efficacy against diverse pathogens.
2. **CRISPR and Gene Editing**
   - CRISPR-Cas9 technology enables precise modifications of microbial genomes to enhance their performance as biocontrol agents.
   - Edited strains of *Bacillus subtilis* have shown improved production of lipopeptides, increasing their effectiveness against soil-borne pathogens.
3. **Harnessing Microbiomes**
   - Research on plant and soil microbiomes is uncovering complex microbial interactions that can be leveraged for biocontrol.
   - Microbial consortia, designed to mimic natural microbiome communities, offer synergistic effects in pest and disease suppression.

Potential Impact:

- Next-generation bioagents will be more adaptable, effective, and sustainable, addressing many limitations of current biocontrol systems.

3.5.2.2 Integrated Approaches

The integration of biological control with advanced technologies and other management practices will enhance its efficiency and scalability.

1. **AI-Driven Pest Prediction Models**
   - Artificial intelligence (AI) algorithms can analyze weather, soil, and pest data to predict outbreaks and recommend the optimal timing for bioagent application.
   - For example, AI systems have been used to guide the release of parasitoids in greenhouse tomato cultivation, reducing pest populations by over 50%.

2. **Precision Agriculture Tools**
    - Remote sensing, drones, and IoT sensors enable precise application of bioagents, minimizing waste and ensuring coverage of targeted areas.
    - Precision tools are particularly valuable in large-scale farming, where manual application is impractical.
3. **Integration with IPM Systems**
    - Combining biological control with cultural, chemical, and physical methods enhances its effectiveness and resilience.
    - Integrated Pest Management (IPM) frameworks provide a holistic approach, ensuring that biocontrol agents are supported by complementary practices.

Examples:

» In rice fields, the combination of *Trichoderma spp.*, resistant varieties, and reduced chemical inputs has successfully suppressed blast disease.

» Drone-based applications of microbial consortia have improved pest management in maize and soybean crops.

Potential Impact:

» Integrated approaches will make biological control more accessible and effective across diverse agricultural systems, from smallholder farms to large commercial operations.

Broader Implications

The challenges and prospects of biological control highlight its dynamic and evolving nature. While environmental variability, resistance, and scalability issues remain significant hurdles, ongoing research and technological advancements offer solutions that promise to transform biological control into a reliable and universal tool for sustainable agriculture.

1. **Addressing Climate Change**
    - Biological control agents with enhanced resilience will play a critical role in mitigating the impacts of climate variability on pest and disease management.
2. **Global Food Security**
    - By reducing dependence on chemical pesticides and promoting sustainable practices, biological control contributes to long-term food security and environmental health.
3. **Collaboration and Innovation**
    - Cross-disciplinary collaboration among researchers, policymakers, and farmers will be essential for advancing biological control and ensuring its integration into global agricultural systems.

4. Merits and Demerits of Biological Control Approaches

Biological control has emerged as a vital strategy in sustainable agriculture, offering natural solutions to pest and disease management. While its adoption has brought significant environmental and economic benefits, it is not without challenges. The merits and demerits of biological control are essential for evaluating its role in modern agricultural systems and identifying areas for improvement.

4.1 Merits of Biological Control

Biological control is a cornerstone of sustainable agriculture, offering a range of benefits that promote environmental protection, economic stability, and public health. By leveraging natural ecological processes, biological control minimizes the negative impacts of synthetic inputs, enhances long-term productivity, and contributes to the preservation of biodiversity.

4.1.1 Environmental Sustainability

Biological control aligns closely with the principles of environmental sustainability, making it a preferred alternative to chemical-based pest and disease management systems.

4.1.1.1 Reduction in Chemical Pesticides

One of the most significant advantages of biological control is its ability to reduce the reliance on synthetic pesticides, which often have harmful effects on ecosystems.

1. **Minimizing Soil Contamination**
    - Synthetic pesticides tend to accumulate in soils, altering their microbial composition and fertility. Biocontrol agents, such as *Trichoderma spp.*, mitigate this by offering an eco-friendly solution for soil-borne disease management.
    - In tomato and cucumber farming, the application of *Trichoderma spp.* has reduced fungicide use by up to 40%, preserving soil health and productivity.
2. **Preventing Water Pollution**
    - The runoff of chemical pesticides into nearby water bodies poses significant risks to aquatic ecosystems. Biological control reduces this risk by eliminating or minimizing pesticide usage, safeguarding water quality.
3. **Air Quality Preservation**
    - Unlike chemical fumigants, biocontrol agents do not release toxic gases into the atmosphere, contributing to cleaner air and a healthier environment.

4.1.1.2 Preservation of Non-Target Species

Biological control agents are highly specific in their action, targeting only the intended pests and pathogens while leaving beneficial organisms unharmed.

1. **Protection of Pollinators**
    - Pollinators like bees, butterflies, and other beneficial insects play a vital role in agriculture. Broad-spectrum pesticides often harm these species, whereas biocontrol agents ensure their protection.

2. **Supporting Soil Biodiversity**
   - Soil microbes and arthropods essential for nutrient cycling and plant health are preserved when biocontrol agents are used instead of synthetic chemicals.
3. **Case Study:**
   - The introduction of *Cotesia flavipes* in sugarcane fields in Southeast Asia not only reduced pesticide use by 30% but also preserved natural predators, enhancing overall ecosystem health.

4.1.1.3 Mitigation of Pesticide Resistance

Continuous use of synthetic pesticides leads to the development of resistant pest populations, undermining their efficacy. Biological control offers a diverse range of mechanisms that reduce the likelihood of resistance development.

1. **Multiple Modes of Action**
   - Biocontrol agents employ strategies such as parasitism, competition, and antibiosis, making it harder for pests to develop resistance.
   - For example, *Trichoderma spp.* and *Pseudomonas fluorescens* attack pathogens through different pathways, ensuring robust disease suppression.
2. **Long-Term Efficacy**
   - The use of bioagents delays or prevents the onset of resistance, ensuring consistent pest and disease management over time.

4.1.2 Economic Benefits

Biological control not only protects the environment but also provides significant economic advantages, making it an attractive option for farmers and the agricultural industry.

4.1.2.1 Cost-Effectiveness in the Long Term

Although the initial cost of implementing biological control systems may be higher than chemical alternatives, the long-term economic benefits are substantial.

1. **Reduced Input Costs**
   - Once established, many biocontrol agents, such as parasitoids and predators, sustain themselves in the ecosystem, eliminating the need for repeated applications.
   - For instance, *Cryptolaemus montrouzieri* effectively manages mealybug infestations in citrus orchards with minimal recurring costs.
2. **Improved Crop Yields**
   - Healthy plants with reduced pest pressure produce higher yields, increasing profitability for farmers.
   - Crops grown using biological control often fetch premium prices in markets that value organic and sustainable practices.

4.1.2.2 Enhanced Market Opportunities

The global demand for organic and sustainably grown produce has created lucrative opportunities for farmers adopting biological control strategies.

1. **Organic Certifications**
   - Many biocontrol agents comply with organic farming standards, enabling farmers to access certified organic markets.
   - Certified organic produce commands higher prices, boosting farm incomes.
2. **Consumer Preferences**
   - Consumers increasingly favor pesticide-free food due to health and environmental concerns. Farms using biological control are better positioned to meet this demand.

4.1.2.3 Support for Smallholder Farmers

Biological control systems can be tailored to the needs of smallholder farmers, particularly in developing countries where access to chemical inputs is limited.

1. **Locally Sourced Bioagents**
   - Smallholder farmers can often utilize native bioagents, such as predatory insects or beneficial fungi, reducing dependence on imported products.
   - Community-based initiatives for breeding and distributing bioagents further lower costs.
2. **Capacity Building**
   - Training programs on biological control empower smallholder farmers with the knowledge and skills needed to adopt sustainable practices.

4.1.3 Safety and Public Health

The adoption of biological control agents in agriculture brings transformative benefits for safety and public health, addressing critical concerns for both consumers and farming communities. Biological control represents a sustainable alternative to synthetic chemical pesticides, which often leave harmful residues on crops, contaminate the environment, and pose significant health risks to those involved in food production. The use of biological agents, such as beneficial microorganisms, predators, and parasitoids, significantly reduces or eliminates chemical residues on crops, enhancing food safety and protecting the health of consumers. This transition aligns with global efforts to ensure that agricultural produce meets stringent food safety regulations, which are particularly critical for accessing high-value export markets. Crops grown using biological control methods frequently comply with the Maximum Residue Limits (MRLs) set by regulatory bodies such as the European Union, making them competitive in international markets where consumers demand high-quality, residue-free products. For instance, crops managed with agents like *Trichoderma spp.* for disease suppression or *Bacillus thuringiensis* for pest control often surpass the safety expectations of both local and global markets, providing farmers with economic opportunities while safeguarding public health.

4.1.3.1 Minimal Chemical Residues

One of the most significant contributions of biological control to public health is its ability to minimize or completely eliminate harmful chemical residues on agricultural produce. By replacing or reducing the use of synthetic pesticides, biocontrol ensures that food products are safer for consumption and less likely to expose consumers to toxic chemicals. This advantage is especially crucial in meeting the growing demand for clean, sustainably produced food, as well as ensuring compliance with stringent food safety standards in export markets. For example, countries with strict pesticide residue regulations, such as those in the European Union and North America, increasingly favor crops grown with biological control methods. This compliance facilitates exports to these lucrative markets, providing farmers with enhanced income opportunities. Moreover, consumers benefit from the reduced presence of harmful residues in their diets, which directly addresses public health concerns related to pesticide exposure. Chemical residues on food are often linked to chronic health conditions such as cancer, endocrine disruption, and neurological disorders. The adoption of biological control agents mitigates these risks, promoting healthier diets and improving overall well-being.

The use of biocontrol agents also contributes to healthier farming communities by lowering the dependency on synthetic pesticides, which are associated with significant health risks during their application. Farmers and agricultural workers in conventional systems often face direct exposure to hazardous chemicals due to inadequate protective equipment, improper handling, or repeated application over time. This exposure can lead to acute health problems such as respiratory distress, skin irritation, and poisoning, as well as chronic conditions like cancer and reproductive disorders. By transitioning to biological control methods, farmers significantly reduce these risks, creating safer working environments. For instance, using biocontrol agents such as *Metarhizium anisopliae* (a fungal pathogen of insects) or *Encarsia formosa* (a parasitoid of whiteflies) eliminates the need for repeated applications of harmful synthetic pesticides, fostering the health and safety of farming communities. Furthermore, the reduced environmental contamination associated with biological control indirectly benefits public health by ensuring cleaner air, water, and soil, essential components for the well-being of both rural and urban populations.

The broader implications of minimizing chemical residues extend to global food systems and the sustainability of agricultural practices. Biological control plays a pivotal role in reducing pesticide runoff into water bodies, which is a major cause of environmental pollution and a threat to public health. Contaminated water sources not only harm aquatic ecosystems but also pose risks to human populations relying on these resources for drinking water and irrigation. By adopting biocontrol strategies, farmers contribute to the preservation of water quality and the overall integrity of natural ecosystems, creating healthier environments that support public health on multiple fronts. Additionally, the preservation of beneficial organisms such as pollinators and natural predators, which are often harmed by chemical pesticide use, enhances ecosystem services critical to sustainable agriculture. This ecological balance further supports the production of safe and nutritious food, fostering resilient agricultural systems that benefit society as a whole.

4.1.3.2 Worker Safety

Farmworkers are frequently exposed to harmful chemicals during pesticide applications. Biological control eliminates this risk, improving occupational health.

1. **Reduction in Toxic Exposure**
   - The absence of synthetic chemicals protects workers from acute and chronic health issues, such as respiratory problems and skin disorders.

2. **Safer Working Conditions**
   - The application of bioagents, such as releasing beneficial insects or spraying microbial formulations, poses minimal risk to human health.

4.1.3.3 Protection of Water Resources

Biological control minimizes pesticide runoff, protecting water bodies from contamination.

1. **Safeguarding Aquatic Ecosystems**
   - Reduced pesticide usage prevents harm to aquatic life, including fish, amphibians, and invertebrates.
   - Cleaner water bodies also benefit communities reliant on these resources for drinking and irrigation.

4.1.4 Long-Term Efficacy

Biological control systems provide sustainable solutions for pest and disease management, ensuring long-term agricultural productivity.

4.1.4.1 Self-Sustaining Populations

Many introduced biocontrol agents establish themselves in the ecosystem, providing ongoing pest suppression without additional inputs.

1. **Case Study**:
   - The vedalia beetle (*Rodolia cardinalis*) introduced in California for cottony cushion scale control continues to regulate pest populations decades after its release.

4.1.4.2 Reduced Risk of Secondary Pests

By maintaining ecological balance, biological control prevents outbreaks of secondary pests, a common issue with chemical pesticide use.

1. **Ecological Harmony**
   - Natural predators and parasitoids keep multiple pest species in check, reducing the likelihood of pest resurgence.
   - This holistic approach enhances overall farm resilience.

4.2 Demerits of Biological Control

Despite its advantages, biological control has certain limitations that hinder its broader adoption and consistent efficacy. These challenges, which range from environmental and operational issues to economic and research-related hurdles, must be addressed to fully realize the potential of biocontrol agents in sustainable agriculture.

4.2.1 Environmental and Ecological Challenges

The integration of biocontrol agents into natural ecosystems comes with inherent ecological complexities. While these agents aim to restore balance, their introduction may occasionally lead to unintended consequences.

4.2.1.1 Unintended Ecological Impacts

1. **Invasiveness of Introduced Agents**
   - In some cases, introduced biocontrol agents have become invasive, disrupting native ecosystems instead of controlling pests.
   - For example, the cane toad (*Rhinella marina*), introduced in Australia to control sugarcane pests, failed to suppress the intended targets but caused widespread ecological damage by preying on or outcompeting native species.
2. **Altered Food Webs**
   - Introducing new species into an ecosystem can disrupt food webs, affecting non-target organisms and altering predator-prey dynamics.
   - Such disruptions may lead to unintended population declines or surges in other species.
3. **Case Study:**
   - In Hawaii, the introduction of mongoose to control rats in sugarcane plantations inadvertently affected native bird populations, as mongoose preyed on bird eggs instead.

Mitigation Strategies:

- Rigorous pre-release testing and risk assessments to evaluate potential ecological impacts.
- Continuous monitoring of introduced agents to detect and mitigate unintended consequences.

4.2.1.2 Narrow Environmental Range

The performance of many biocontrol agents is highly dependent on environmental conditions such as temperature, humidity, and soil pH.

1. **Sensitivity to Climatic Conditions**
   - Some biocontrol agents, such as *Trichoderma spp.*, struggle to survive or remain effective in extreme climates, including arid or saline conditions.
   - This limits their applicability in regions with variable or harsh climates.

2. **Seasonal Variability**
   - Seasonal changes can affect the availability and activity of biocontrol agents. For example, certain parasitoids may only be active during specific times of the year, reducing their efficacy in managing year-round pest populations.

Mitigation Strategies:

- Developing bioagents with broader environmental tolerances through genetic engineering or selective breeding.
- Using controlled environments, such as greenhouses, to optimize conditions for bioagent activity.

4.2.1.3 Non-Target Effects

While rare, biocontrol agents may occasionally affect non-target organisms, leading to unintended ecological consequences.

1. **Predator or Parasitoid Impact**
   - Predatory insects or parasitoids introduced to control pests may prey on or parasitize native species, reducing their populations.
   - For instance, certain introduced ladybird beetles have outcompeted native species in some ecosystems.
2. **Pathogen Spillover**
   - Pathogens used in microbial biological control may infect non-target organisms, potentially causing unforeseen damage to beneficial species.

Mitigation Strategies:

- Selecting highly specific bioagents that target only the intended pests or pathogens.
- Implementing stringent regulatory guidelines for the approval and monitoring of biocontrol agents.

4.2.2 Operational Limitations

Operational challenges, including slow action and variable performance, can limit the practicality of biological control in commercial farming systems.

4.2.2.1 Slow Action

Biological control agents often take longer to suppress pest populations compared to chemical pesticides, which provide immediate results.

1. **Gradual Population Control**
   - Predatory insects and parasitoids require time to establish themselves and reduce pest populations, which may not be ideal in acute pest outbreaks.

- For example, the effectiveness of *Cotesia flavipes* in controlling sugarcane borers can take several weeks to manifest.

2. **Delayed Crop Protection**
   - Slow action may expose crops to prolonged pest damage, reducing yields and quality.

Mitigation Strategies:

» Combining biocontrol agents with fast-acting, low-toxicity chemical treatments for immediate pest suppression.

» Employing predictive models to release bioagents preemptively, before pest populations reach damaging levels.

4.2.2.2 Labor-Intensive Application

Deploying biological control agents can be labor-intensive, particularly for methods involving the manual release of predators or parasitoids.

1. **High Labor Requirements**
   - Releasing bioagents like *Cryptolaemus montrouzieri* or *Cotesia spp.* in large-scale fields requires significant manpower and time.
   - This may increase operational costs, especially in labor-scarce regions.
2. **Complex Application Protocols**
   - Some microbial biopesticides require specific preparation and application methods, adding to the complexity of their use.

Mitigation Strategies:

» Developing automated delivery systems, such as drones or robotic dispensers, to streamline bioagent application.

» Simplifying formulation and application techniques to make biocontrol more user-friendly.

4.2.2.3 Variable Field Performance

Biocontrol agents may exhibit inconsistent performance due to environmental variability, poor establishment, or competition with native organisms.

1. **Field Conditions vs. Laboratory Conditions**
   - Agents that perform well under controlled conditions often face challenges in unpredictable field environments.
   - Competition with native species may also reduce the establishment of introduced agents.
2. **Impact on Farmer Confidence**
   - Variability in performance can discourage farmers from adopting biological control, especially in regions where consistency is crucial.

Mitigation Strategies:

» Conducting localized field trials to tailor biocontrol strategies to specific conditions.

» Combining multiple biocontrol agents to enhance reliability and performance.

4.2.3 Economic and Market Challenges

Economic and market-related challenges hinder the widespread adoption of biological control, particularly in developing regions.

4.2.3.1 High Initial Costs

1. **Investment in Infrastructure**
   - Establishing facilities for the production and distribution of biocontrol agents requires substantial investment.
   - Costs are particularly high for large-scale operations targeting commercial agriculture.
2. **Farmer Adoption**
   - High upfront costs may deter smallholder farmers from adopting biological control practices, even if long-term savings are significant.

4.2.3.2 Limited Commercial Availability

1. **Access Issues**
   - Many effective biocontrol agents are not widely available, especially in remote or developing regions.
   - Farmers in these areas often lack access to reliable suppliers or affordable biocontrol products.
2. **Dependency on Imports**
   - Dependence on imported biocontrol agents increases costs and complicates logistics, particularly in regions with weak supply chains.

Mitigation Strategies:

» Promoting local production of bioagents to reduce costs and improve accessibility.

» Establishing public-private partnerships to support the commercialization of biocontrol products.

4.2.3.3 Market and Policy Barriers

1. **Lengthy Regulatory Processes**
   - Approvals for new biocontrol agents often involve lengthy and expensive regulatory procedures, delaying their availability.
2. **Market Awareness**
   - Limited awareness of the benefits and efficacy of biocontrol products among farmers and consumers reduces demand.

Mitigation Strategies:

» Streamlining regulatory frameworks to expedite approvals while ensuring safety.

» Conducting awareness campaigns to educate stakeholders about the advantages of biological control.

4.2.4 Research and Development Challenges

Continuous research and innovation are essential for addressing the limitations of biological control.

4.2.4.1 Need for Continuous Innovation

Pathogens and pests may develop resistance to biocontrol agents, necessitating the ongoing development of new strategies.

1. **Evolution of Resistance**
   - Similar to chemical pesticides, prolonged use of specific bioagents can lead to resistance in target pests.
2. **Emergence of New Pests**
   - Global trade and climate change contribute to the emergence of new pests, requiring novel biocontrol solutions.

Mitigation Strategies:

» Investing in synthetic biology and gene editing to create next-generation bioagents.

» Establishing global research networks to accelerate innovation.

4.2.4.2 Scalability Issues

1. **Production Bottlenecks**
   - Scaling up the production of biocontrol agents to meet the demands of commercial agriculture is a persistent challenge.
2. **Logistical Constraints**
   - Ensuring timely delivery and application of bioagents in large-scale operations requires efficient logistics and infrastructure.

Mitigation Strategies:

» Developing cost-effective mass-production techniques for bioagents.

» Leveraging automation and digital tools to streamline supply chains.

4.3 Balancing Merits and Demerits

The balance between the merits and demerits of biological control lies in its thoughtful implementation, supported by advancements in science, infrastructure, and policy. While biological control provides unparalleled benefits in terms of environmental sustainability, economic viability, and safety, its limitations can be addressed through innovative solutions, collaborative efforts, and a commitment to long-term agricultural resilience.

The Importance of Balance

Biological control exemplifies an approach that integrates natural processes into agricultural systems. Its merits such as reduced chemical usage, enhanced ecosystem health, and economic advantages often outweigh its demerits when implemented correctly. However, ignoring its limitations can lead to inefficiencies, reduced efficacy, or unintended consequences. Achieving balance requires:

1. **Innovation**: Developing new technologies and bioagents to address the challenges of resistance, environmental variability, and scalability.
2. **Collaboration**: Encouraging partnerships among farmers, researchers, policymakers, and international organizations to share knowledge and resources.
3. **Education and Awareness**: Informing stakeholders about the benefits and proper use of biological control to promote adoption and long-term success.

Key Strategies for Improvement

Implementing biological control on a broader scale and with greater efficacy necessitates targeted strategies that address its limitations while enhancing its strengths.

4.3.1 Developing Bioagents with Broader Environmental Tolerances and Faster Action

The effectiveness of biological control agents often depends on environmental factors, limiting their performance in diverse agricultural contexts. Innovations in bioagent development can address this issue.

1. **Broader Environmental Tolerances**
   - Advances in genetic engineering and synthetic biology can create bioagents capable of thriving in extreme climates or varying soil conditions.
   - For example, modifying *Pseudomonas fluorescens* to enhance its survival in saline soils expands its applicability in arid regions.
2. **Faster Action**
   - Developing bioagents that act more quickly can make biological control a viable option in emergency situations where immediate pest suppression is required.
   - Research on microbial bioagents, such as *Beauveria bassiana*, focuses on optimizing their infection cycles to deliver faster results.

Implementation:

- Increased funding for research on genetic modifications and microbial interactions.
- Field trials to assess the performance of enhanced bioagents under diverse conditions.

4.3.2 Investing in Infrastructure for Large-Scale Production and Distribution

Scaling up the production and distribution of biocontrol agents is essential for meeting the demands of modern agriculture.

1. **Mass Production Techniques**
   - Developing cost-effective and efficient production systems for microbial and insect-based bioagents.
   - Innovations in fermentation technology and bioreactor design can enhance the scalability of microbial bioagent production.
2. **Distribution Networks**
   - Establishing local production units and distribution hubs to ensure timely and affordable access to bioagents, particularly in remote and resource-limited areas.
   - Collaborating with cooperatives and extension services to deliver bioagents directly to farmers.
3. **Cold Chain Management**
   - Certain bioagents require specific storage conditions to maintain their viability. Investing in cold chain infrastructure ensures that bioagents remain effective during transportation and storage.

Implementation:

- Public-private partnerships to fund infrastructure projects.
- Incentives for companies investing in biocontrol production and distribution facilities.

4.3.3 Promoting Public Awareness and Education

A lack of awareness about the benefits and proper use of biological control remains a significant barrier to its adoption.

1. **Farmer Training Programs**
   - Conducting workshops and training sessions to educate farmers on the application, monitoring, and benefits of biocontrol agents.
   - Providing region-specific guidance to ensure that farmers can effectively integrate biological control into their practices.
2. **Consumer Awareness Campaigns**
   - Educating consumers about the health and environmental benefits of produce grown using biological control can increase market demand for such products.
   - Highlighting success stories in organic and sustainable farming to build public trust in biocontrol methods.
3. **Demonstration Farms**
   - Establishing demonstration farms to showcase the effectiveness of biological control in real-world conditions.
   - These farms can serve as centers for knowledge exchange and community engagement.

Implementation:

- » Collaboration with agricultural extension services and non-governmental organizations.
- » Development of educational materials in local languages to reach diverse audiences.

4.3.4 Enhancing International Collaboration

Biological control is a global endeavour, requiring cross-border collaboration to address shared challenges and promote best practices.

1. **Streamlining Regulatory Processes**
   - Harmonizing international regulations for the approval and use of biocontrol agents can facilitate their adoption in multiple countries.
   - Establishing global standards for safety, efficacy, and environmental impact ensures consistent quality.
2. **Knowledge Sharing and Research Networks**
   - Creating platforms for researchers and practitioners to share findings, innovations, and success stories.
   - Initiatives like the International Organization for Biological Control (IOBC) play a critical role in fostering global collaboration.
3. **Resource Mobilization**
   - Pooling resources and expertise among nations can accelerate the development and deployment of biocontrol agents, particularly in regions with limited funding and infrastructure.

Implementation:

- Hosting international conferences and workshops focused on biological control.
- Establishing regional biocontrol centres to support local implementation while maintaining global connectivity.

Broader Implications

Balancing the merits and demerits of biological control is not only critical for its success but also for advancing global agricultural sustainability. By addressing its limitations through innovation, collaboration, and education, biological control can evolve into a universally reliable and scalable solution.

1. **Contribution to Climate Resilience**
   - Developing resilient bioagents and sustainable farming systems enhances the ability of agriculture to withstand climate-induced challenges, such as pest outbreaks and extreme weather events.

2. **Advancing Food Security**
   - By reducing dependence on chemical pesticides and enhancing crop productivity, biological control supports global food security goals.
3. **Strengthening Agroecological Systems**
   - Biological control fosters biodiversity, soil health, and ecological balance, contributing to the long-term viability of agricultural landscapes.

Chapter - 2

Biological Interactions in Plant-Pathogen Control

Biological interactions are at the heart of plant-pathogen control, where the dynamic relationships between plants, pathogens, and other organisms define the outcomes of disease suppression and plant health. These interactions encompass a wide range of ecological and molecular processes, including competition, predation, parasitism, mutualism, and antibiosis, all of which contribute to the regulation of pathogen populations and the promotion of plant resistance. In natural and managed ecosystems, plants are constantly exposed to a multitude of microbial organisms, some of which are pathogenic, while others serve as protective allies. The ability of plants to harness beneficial organisms for their defense against pathogens is a fundamental aspect of biological control. For instance, many plant-associated microbes, such as rhizosphere-dwelling bacteria and endophytic fungi, play a pivotal role in suppressing pathogens through mechanisms like competition for nutrients and space, production of antimicrobial compounds, and induction of systemic resistance. A classic example is the role of *Pseudomonas fluorescens*, a well-known biocontrol bacterium, in controlling soil-borne pathogens such as *Fusarium oxysporum* and *Pythium spp.* by producing antibiotics like phenazines and pyrrolnitrin, as well as competing for essential nutrients like iron through siderophore production.

Competition is one of the most prominent biological interactions in plant-pathogen control. Beneficial microbes and pathogens often vie for limited resources such as nutrients, water, and physical space in the plant rhizosphere or phyllosphere. By occupying niches and consuming available resources, beneficial organisms can outcompete pathogens, reducing their ability to colonize and infect plant tissues. For example, Trichoderma species, widely used as biocontrol agents, are effective competitors that colonize plant roots and inhibit pathogen growth through direct competition and mycoparasitism. In addition to competition, parasitism and predation are crucial interactions that suppress pathogen populations. Mycoparasitic fungi such as *Trichoderma harzianum* directly attack and degrade the cell walls of fungal pathogens like *Rhizoctonia solani* and *Sclerotium rolfsii*, effectively reducing their virulence and spread. Similarly, predatory organisms such as nematodes and predatory mites contribute to plant-pathogen control by preying on pathogen propagules or vectors, thereby indirectly reducing disease pressure.

Antibiosis, another critical mechanism of interaction, involves the production of antimicrobial compounds by beneficial organisms that inhibit or kill pathogens. Many bacterial and fungal biocontrol agents produce a diverse array of secondary metabolites with strong antimicrobial activity, targeting pathogens at various stages of their life cycle. For instance, *Bacillus subtilis* and *Bacillus amyloliquefaciens* produce lipopeptides such as surfactin, iturin, and fengycin, which disrupt pathogen cell membranes and inhibit spore germination. These compounds not only suppress the growth of pathogens but also create a hostile environment that deters their establishment. In addition to direct antagonism, these biocontrol agents often enhance plant defense mechanisms, inducing systemic resistance that primes plants to defend themselves against future pathogen attacks. This phenomenon, known as Induced Systemic Resistance (ISR), is mediated by signaling molecules like jasmonic acid and ethylene, which activate a cascade of defensive responses in plants.

Mutualistic interactions also play a significant role in plant-pathogen control, particularly in the context of symbiotic relationships between plants and beneficial microbes. Arbuscular mycorrhizal fungi (AMF), for example, form symbiotic associations with plant roots, enhancing nutrient uptake and providing protection against pathogens. By improving plant vigor and altering root exudation patterns, AMF reduce the susceptibility of plants to root pathogens such as *Phytophthora spp.* and *Verticillium spp.* Similarly, nitrogen-fixing bacteria like *Rhizobium spp.* not only supply plants with essential nutrients but also modulate the microbial community in the rhizosphere, creating an environment less favorable for pathogenic organisms. These mutualistic interactions highlight the importance of maintaining diverse and functional microbial communities in agricultural systems, as they contribute to the natural regulation of pathogens and the overall health of crops.

Biological interactions in plant-pathogen control also extend to higher trophic levels, involving interactions between plants, herbivores, and their natural enemies. Many herbivorous insects serve as vectors for plant pathogens, transmitting diseases such as bacterial wilt and viral mosaics. Natural enemies of these vectors, including predatory insects and parasitic wasps, play an indirect but vital role in plant-pathogen control by reducing vector populations and limiting the spread of diseases. For instance, the parasitoid *Encarsia formosa* effectively suppresses whitefly populations, thereby reducing the transmission of plant viruses they carry. This multi-trophic interplay underscores the interconnectedness of biological interactions in disease management, where the suppression of one organism indirectly benefits plant health by mitigating the impact of another.

The effectiveness of biological interactions in plant-pathogen control is influenced by various environmental and management factors, including soil health, crop diversity, and cultural practices. Soil health, in particular, plays a critical role in determining the composition and activity of microbial communities that drive these interactions. Practices such as crop rotation, cover cropping, and reduced tillage enhance soil biodiversity and structure, creating a favorable environment for beneficial microbes while suppressing pathogen populations. Crop diversity, through intercropping or polyculture systems, further enhances these interactions by providing a variety of niches and resources that support the proliferation of biocontrol agents. These practices not only improve the effectiveness of biological interactions but also contribute to the resilience and sustainability of agricultural systems.

1. Types of Biological Interactions

Biological interactions are foundational to understanding ecological and agricultural systems, particularly in the context of biological control of plant pathogens. These interactions encompass various relationships among organisms, including mutualism, parasitism, competition, and antibiosis, each influencing plant health, microbial dynamics, and pathogen suppression.

1.1 Mutualistic Interactions

Mutualistic interactions are fundamental ecological relationships where both participating organisms derive benefits, contributing to the stability and productivity of ecosystems. In agriculture, mutualistic interactions involving plants, microbes, and fungi are pivotal in promoting plant health, enhancing nutrient uptake, and suppressing diseases. These relationships are crucial for sustainable farming practices, reducing reliance on chemical inputs and fostering biodiversity.

1.1.1 Mycorrhizal Associations

Mycorrhizal fungi establish mutualistic relationships with plant roots, facilitating the exchange of nutrients and water while bolstering plant resilience against environmental and biological stresses. These interactions are among terrestrial ecosystems' most widespread and ecologically significant symbioses.

1.1.1.1 Types of Mycorrhizae

Mycorrhizal fungi are categorized based on their structural and functional relationships with plant roots.

1. **Arbuscular Mycorrhizal Fungi (AMF)**
   - AMF, primarily from the phylum Glomeromycota, penetrate plant root cells to form specialized structures called arbuscules. These structures maximize the surface area for nutrient exchange between the fungi and the plant.
   - **Nutrient Uptake**: AMF enhance the uptake of phosphorus, nitrogen, and other micronutrients, which are often limited in soil.
   - **Stress Tolerance**: AMF colonization improves drought tolerance by enhancing water uptake and reducing oxidative stress in plants.
2. **Ectomycorrhizae**
   - Found in many tree species, ectomycorrhizae form a sheath around plant roots and extend fungal hyphae into the surrounding soil.
   - **Nutrient Uptake**: Ectomycorrhizal fungi aid in acquiring nitrogen, potassium, and other essential elements, particularly in nutrient-poor soils.
   - **Carbon Exchange**: The host plant supplies carbohydrates to the fungi, while the fungi enhance mineral nutrient availability for the plant.
3. **Other Types**
   - **Ericoid Mycorrhizae**: Found in plants like blueberries, these mycorrhizae specialize in acidic, nutrient-deficient soils.
   - **Orchid Mycorrhizae**: Essential for the germination and growth of orchids, these fungi provide necessary nutrients to the developing seedlings.

1.1.1.2 Role in Plant Pathogen Suppression

Mycorrhizal associations indirectly suppress plant pathogens by improving root health and altering rhizosphere microbial communities.

1. **Nutrient Competition**
   - AMF outcompete pathogenic fungi for nutrients in the rhizosphere, limiting pathogen proliferation.
   - This competition reduces the incidence of root diseases caused by pathogens like *Fusarium* and *Rhizoctonia*.
2. **Activation of Plant Defenses**
   - Mycorrhizal colonization triggers systemic resistance in plants, activating defense pathways and enhancing the production of phytoalexins and pathogenesis-related proteins.
   - For example, tomato plants colonized by AMF exhibit reduced severity of *Fusarium* wilt due to enhanced defensive responses.
3. **Physical Barriers**
   - The fungal hyphae form a protective network around plant roots, creating a physical barrier against pathogen invasion.

Case Study:

In controlled trials, AMF inoculation reduced the incidence of *Fusarium* wilt in cucumber plants by over 40%, demonstrating its potential as a biocontrol strategy.

1.1.1.3 Applications in Agriculture

Mycorrhizal fungi have significant applications in sustainable agriculture, contributing to soil health, nutrient cycling, and disease management.

1. **Mycorrhizal Inoculants**
   - Commercial formulations of AMF are increasingly used as soil amendments to enhance crop growth and resilience. These inoculants reduce the need for chemical fertilizers and improve plant health in degraded soils.
2. **Crop Management Practices**
   - Integrating mycorrhizal inoculants into cropping systems enhances nutrient use efficiency and reduces environmental impacts.
   - For example, in maize farming, AMF application has increased phosphorus uptake by 30%, leading to higher yields and reduced fertilizer inputs.
3. **Future Potential**
   - Research into site-specific mycorrhizal strains tailored for particular crops and environments promises to further optimize agricultural productivity and sustainability.

1.1.2 Nitrogen-Fixing Symbiosis

Nitrogen-fixing bacteria form mutualistic relationships with leguminous plants, converting atmospheric nitrogen into bioavailable forms. This process is critical for maintaining soil fertility and supporting plant growth, especially in nitrogen-deficient soils.

1.1.2.1 Mechanism of Interaction

Nitrogen fixation involves a highly specialized interaction between plants and nitrogen-fixing bacteria, such as *Rhizobium* species.

1. **Nodule Formation**
   - *Rhizobium* bacteria enter plant roots through root hairs, triggering the formation of nodules where the bacteria reside.
   - Inside the nodules, bacteria convert nitrogen gas (N_2) from the atmosphere into ammonia (NH_3) through the enzyme nitrogenase.
2. **Nutrient Exchange**
   - Ammonia produced by the bacteria is assimilated by the plant into amino acids and proteins, promoting growth and development.
   - In return, plants provide carbohydrates derived from photosynthesis to fuel bacterial metabolism.
3. **Environmental Significance**
   - Biological nitrogen fixation reduces the need for synthetic nitrogen fertilizers, which are energy-intensive to produce and environmentally damaging when over applied.

1.1.2.2 Impact on Plant Health

The symbiosis between nitrogen-fixing bacteria and legumes has profound effects on plant health and productivity.

1. **Improved Growth and Yield**
   - The availability of bioavailable nitrogen enhances vegetative growth, flowering, and fruiting in leguminous crops.
   - Non-leguminous crops grown in rotation with legumes benefit from residual nitrogen, boosting their productivity.
2. **Pathogen Suppression**
   - Nitrogen-fixing bacteria enhance microbial diversity in the rhizosphere, creating competitive environments that suppress soil-borne pathogens.
   - For instance, *Rhizobium* inoculation has been shown to reduce root rot severity in chickpea crops.
3. **Drought Tolerance**
   - The symbiosis enhances root development and water uptake, improving plant resilience under water-limited conditions.

1.1.2.3 Examples in Cropping Systems

Nitrogen-fixing symbiosis is integral to sustainable cropping systems, particularly in low-input and organic farming practices.

1. **Legume-Based Rotations**
   - Rotating legumes such as soybeans, peas, or clover with non-leguminous crops replenishes soil nitrogen and reduces fertilizer dependency.
   - For example, incorporating alfalfa into crop rotations has increased wheat yields by 20% due to enhanced nitrogen availability.
2. **Intercropping Systems**
   - Intercropping legumes with cereals creates a synergistic effect, where nitrogen fixed by legumes benefits adjacent cereal crops.
   - In maize-legume intercropping systems, nitrogen fixation by legumes reduces the incidence of nitrogen deficiency, improving overall productivity.
3. **Agroforestry Practices**
   - Nitrogen-fixing trees, such as *Gliricidia sepium* and *Leucaena leucocephala*, are used in agroforestry systems to improve soil fertility and support sustainable land management.

1.2 Parasitic and Pathogenic Interactions

Parasitic and pathogenic interactions are fundamental to plant pathology, defining the relationship where one organism (the parasite or pathogen) benefits at the expense of another (the host). These interactions are often detrimental to plants, leading to reduced growth, lower yields, and even death.

1.2.1 Pathogen-Host Dynamics

The dynamics between plant pathogens and their hosts involve complex biochemical and molecular interactions. Pathogens employ various strategies to invade plant tissues, evade host defenses, and establish infections, often resulting in significant damage to the host plant.

1.2.1.1 Mechanisms of Pathogenesis

The pathogenicity of an organism refers to its ability to cause disease, which is achieved through specific mechanisms designed to overcome the structural and biochemical barriers of the host.

1. **Invasion and Colonization**
   - Pathogens invade plant tissues through natural openings (stomata, lenticels), wounds, or direct penetration using specialized structures like appressoria.
   - Once inside, they colonize intercellular spaces or vascular tissues, depending on their trophic lifestyle.
2. **Secretion of Enzymes**
   - Pathogens produce cell wall-degrading enzymes, such as cellulases, pectinases, and ligninases, to break down the plant cell wall and facilitate penetration.
   - For example, *Botrytis cinerea* secretes pectinases to degrade pectin in cell walls, leading to tissue maceration.

3. **Toxin Production**
    - Toxins are produced to kill host cells, suppress immune responses, or create a favourable environment for pathogen proliferation.
    - *Alternaria spp.* produce host-specific toxins that cause necrotic lesions in susceptible plants.
4. **Effector Proteins**
    - Many pathogens secrete effector proteins that manipulate host cell processes, suppress immune responses, or promote nutrient availability.
    - For example, Phytophthora infestans delivers effectors into potato cells to suppress programmed cell death, facilitating infection.
5. **Host Manipulation**
    - Pathogens alter host metabolism to create nutrient-rich environments. For instance, *Puccinia spp.* redirect plant resources to form nutrient-absorbing structures called haustoria.

1.2.1.2 Types of Pathogens

Pathogens are categorized based on their nutritional strategies and the nature of their interaction with the host.

1. **Biotrophs**
    - Biotrophic pathogens require living host cells for survival and nutrition. They maintain host cell viability while extracting nutrients.
    - **Examples**: Rust fungi (*Puccinia spp.*) and powdery mildews (*Erysiphe spp.*).
2. **Necrotrophs**
    - Necrotrophic pathogens kill host cells to derive nutrients from the dead tissue. They produce toxins and hydrolytic enzymes to destroy plant tissues.
    - **Examples**: *Botrytis cinerea* (gray mold) and *Sclerotinia sclerotiorum* (white mold).
3. **Hemi-biotrophs**
    - Hemi-biotrophs exhibit an initial biotrophic phase, during which they maintain living host cells, followed by a necrotrophic phase, where they kill host cells.
    - **Examples**: *Colletotrichum spp.* and *Magnaporthe oryzae* (rice blast fungus).

1.2.1.3 Implications for Plant Pathology

Understanding pathogen-host interactions is critical for developing effective disease management strategies.

1. **Breeding Disease-Resistant Varieties**
    - Knowledge of pathogen mechanisms, such as effector proteins, guides the development of resistant crop varieties through conventional breeding or genetic engineering.

- For example, resistant rice varieties have been developed against *Magnaporthe oryzae* using resistance genes that target specific effectors.

2. **Targeted Biocontrol Strategies**
   - Pathogen-specific biocontrol agents are designed to inhibit key processes, such as enzyme activity or toxin production, disrupting the infection cycle.
3. **Integrated Disease Management (IDM)**
   - Understanding pathogen biology allows for the integration of cultural practices, chemical controls, and biocontrol agents to create holistic disease management systems.

1.2.2 Hyper-parasitism

Hyper-parasitism, a unique interaction where one parasite attacks another, indirectly benefits the host by suppressing the primary parasite. This interaction is a critical mechanism in biological control, particularly against fungal pathogens.

1.2.2.1 Examples of Hyperparasites

Hyperparasitic organisms include fungi, bacteria, and viruses that target plant pathogens, reducing their virulence and population density.

1. **Fungal Hyperparasites**
   - *Trichoderma spp.* are well-known hyperparasitic fungi that attack and degrade pathogenic fungi such as *Rhizoctonia solani* and *Sclerotinia sclerotiorum*.
   - *Gliocladium spp.* are also effective against soil-borne pathogens like *Pythium* and *Phytophthora*.
2. **Viral Hyperparasites**
   - Hypoviruses infect fungal pathogens, such as *Cryphonectria parasitica* (chestnut blight fungus), reducing their ability to cause disease.

1.2.2.2 Mechanisms of Hyperparasitism

Hyperparasitic organisms employ several mechanisms to suppress primary parasites.

1. **Enzyme Production**
   - Hyperparasites produce hydrolytic enzymes like chitinases, glucanases, and proteases to degrade the cell walls of pathogenic fungi.
   - For example, *Trichoderma spp.* produce chitinases that break down the chitin in fungal cell walls, inhibiting pathogen growth.
2. **Nutrient and Space Competition**
   - Hyperparasites compete with pathogens for limited resources, such as nutrients and colonization sites, effectively suppressing pathogen proliferation.

3. **Direct Inhibition**
    - Hyperparasites produce secondary metabolites, such as antibiotics and antifungal compounds, that inhibit pathogen activity.

1.2.2.3 Applications in Biocontrol

Hyperparasitic organisms are widely used in biological control programs to manage soil-borne and foliar diseases.

1. **Formulated Biocontrol Agents**
    - Products containing *Trichoderma spp.* and *Gliocladium spp.* are commercially available for managing diseases like damping-off, root rots, and wilt.
2. **Integrated Use in IDM**
    - Hyperparasites are integrated into IDM programs, complementing cultural and chemical controls to achieve sustainable disease suppression.
    - For instance, *Trichoderma* formulations are used alongside crop rotation and soil solarization to manage *Rhizoctonia solani* in vegetable crops.
3. **Advantages in Agriculture**
    - Hyperparasites provide eco-friendly solutions with minimal environmental impact. Their specificity reduces the risk of harming non-target organisms, making them ideal for sustainable farming systems.

1.3 Competitive Interactions

Competition is a fundamental ecological interaction where organisms vie for limited resources such as nutrients, space, or light. In the context of plant pathology, competitive interactions play a pivotal role in suppressing pathogens and promoting plant health. Beneficial microorganisms exploit competitive exclusion mechanisms to outcompete pathogens, ensuring their dominance in the rhizosphere and other plant-associated environments. This dynamic interaction is critical in biological control strategies, where biocontrol agents are deployed to outcompete harmful pathogens.

1.3.1 Rhizosphere Competition

The rhizosphere, the narrow zone of soil surrounding plant roots, is a hotspot of microbial activity and competition. Beneficial microbes in this zone employ various strategies to dominate the ecological niche, suppressing pathogens and enhancing plant health.

1.3.1.1 Microbial Competition

Microbial competition in the rhizosphere involves direct and indirect interactions that suppress pathogens by depriving them of essential resources.

1. **Mechanisms of Microbial Competition**
    - Beneficial microbes like *Pseudomonas fluorescens*, *Bacillus subtilis*, and *Trichoderma spp.* rapidly colonize root surfaces, occupying niches that pathogens would otherwise exploit.

- These microbes outcompete pathogens for nutrients such as sugars, amino acids, and organic acids exuded by plant roots.
- Pathogens are also deprived of suitable attachment sites, reducing their ability to establish infections.

2. **Case Study: *Pseudomonas fluorescens***
   - *Pseudomonas fluorescens* is a well-known biocontrol agent that effectively competes with soil-borne pathogens like *Fusarium oxysporum*.
   - Its rapid growth rate and efficient utilization of root exudates allow it to establish dominance in the rhizosphere.
   - In tomato crops, *P. fluorescens* reduces root rot caused by *Rhizoctonia solani* by over 50%, showcasing its competitive efficacy.
3. **Role of Exopolysaccharides (EPS)**
   - Certain beneficial microbes produce EPS, which enhance their ability to adhere to root surfaces and form biofilms.
   - These biofilms act as protective barriers, preventing pathogen colonization while facilitating nutrient absorption by the plant.

1.3.1.2 Role of Siderophores

Siderophores are iron-chelating molecules produced by beneficial microbes to outcompete pathogens for this critical nutrient.

1. **Iron Sequestration**
   - Iron is an essential micronutrient for both beneficial microbes and pathogens. However, its bioavailability in soil is often limited.
   - Beneficial microbes like *Pseudomonas spp.* and *Bacillus spp.* secrete siderophores that bind to iron with high affinity, rendering it unavailable to pathogens.
2. **Pathogen Suppression**
   - By depriving pathogens of iron, siderophores inhibit their growth and virulence.
   - For example, the production of pyoverdine by *Pseudomonas fluorescens* suppresses *Phytophthora* species in cucumber crops.
3. **Synergistic Effects**
   - Siderophores not only suppress pathogens but also enhance plant growth by increasing iron availability to the plant through microbial associations.
4. **Application in Biocontrol**
   - Siderophore-producing strains are incorporated into biopesticides to manage soil-borne diseases in crops like rice, wheat, and maize.

1.3.1.3 Impact on Disease Suppression

Competitive exclusion in the rhizosphere serves as a natural defense mechanism against soil-borne pathogens.

1. **Reduction in Pathogen Populations**
   - Beneficial microbes suppress pathogen populations by monopolizing critical resources and colonization sites.
   - This effect is amplified when multiple biocontrol agents with complementary mechanisms are applied.
2. **Enhanced Microbial Diversity**
   - Competition among microbes promotes microbial diversity in the rhizosphere, creating a more resilient ecosystem less prone to pathogen outbreaks.
3. **Examples of Disease Suppression**
   - In rice fields, the application of *Pseudomonas fluorescens* reduces sheath blight caused by *Rhizoctonia solani* by over 40%.
   - In apple orchards, rhizosphere competition involving *Bacillus subtilis* suppresses fire blight caused by *Erwinia amylovora*.

1.3.2 Saprophytic Competition

Saprophytic competition refers to the ability of microbes to colonize and utilize dead organic matter in the soil, effectively outcompeting pathogens during their saprophytic phase. This interaction is crucial for reducing the survival and inoculum levels of soil-borne pathogens.

1.3.2.1 Competitive Saprophytic Ability (CSA)

Competitive saprophytic ability determines the capacity of beneficial microbes to dominate the decomposition process in the soil, thereby suppressing pathogen survival.

1. **Mechanisms of CSA**
   - Beneficial microbes utilize carbon and nitrogen sources in decomposing organic matter more efficiently than pathogens.
   - These microbes produce extracellular enzymes, such as cellulases and hemi-cellulases, to degrade complex organic matter and gain a competitive edge.
2. **Role of *Trichoderma spp.* in CSA**
   - *Trichoderma spp.* are among the most effective saprophytic competitors, capable of colonizing decaying plant residues and suppressing pathogens like *Fusarium oxysporum*.
   - Their ability to secrete a wide range of hydrolytic enzymes enhances their competitive edge in nutrient-limited soils.

3. **Impact on Pathogen Persistence**
   - By dominating the saprophytic phase, beneficial microbes reduce the survival of pathogens in crop residues, minimizing disease inoculum levels in subsequent growing seasons.

Case Study:

The application of *Trichoderma harzianum* in sugarcane fields reduced the inoculum of *Colletotrichum falcatum* by 60%, lowering disease incidence in the following season.

1.3.2.2 Applications in Disease Management

Enhancing saprophytic competition through agricultural practices and microbial inoculants is a key strategy in disease management.

1. **Use of Organic Amendments**
   - Incorporating organic matter, such as compost or green manure, into soils enhances the activity of saprophytic competitors.
   - For example, adding farmyard manure increases the population of *Trichoderma spp.* in the soil, improving its CSA and reducing pathogen persistence.
2. **Bioaugmentation with Inoculants**
   - Commercially available inoculants containing *Trichoderma spp.* and other saprophytic competitors are applied to suppress soil-borne pathogens.
   - In greenhouse vegetable production, bioaugmentation with *Gliocladium spp.* reduces damping-off caused by *Pythium* and *Phytophthora*.
3. **Integrated Disease Management (IDM)**
   - Combining saprophytic competition with cultural practices, such as crop rotation and residue management, enhances disease suppression.
   - For instance, incorporating wheat residues inoculated with *Trichoderma harzianum* into soil effectively suppresses take-all disease in subsequent wheat crops.

1.4 Antagonistic Interactions

Antagonistic interactions occur when one organism inhibits or suppresses the growth and activity of another. These interactions are pivotal in biological control, where antagonistic microbes suppress pathogens through mechanisms such as antibiosis, competition, and resource deprivation. The production of antimicrobial compounds is central to these interactions, making antagonistic microbes valuable tools in plant disease management.

1.4.1 Antibiosis

Antibiosis is a specific form of antagonism where microorganisms produce secondary metabolites that inhibit or kill pathogens. These metabolites, which include antibiotics, lytic enzymes, and volatile organic compounds, disrupt the growth or function of pathogens, offering a natural and targeted method for controlling plant diseases.

1.4.1.1 Definition and Mechanisms

Antibiosis is characterized by the production of bioactive compounds that interfere with pathogen physiology.

1. **Production of Secondary Metabolites**
   - Secondary metabolites are organic compounds not directly involved in the growth, development, or reproduction of the producing organism but play a crucial role in microbial interactions.
   - These compounds exhibit antimicrobial properties, disrupting cell membranes, inhibiting enzyme activity, or interfering with genetic material of pathogens.
2. **Key Examples of Antimicrobial Compounds**
   - **Iturin and Surfactin**: Produced by *Bacillus subtilis*, these lipopeptides disrupt fungal cell membranes, leading to cell lysis.
   - **Phenazines and Pyrrolnitrin**: Produced by *Pseudomonas fluorescens*, these compounds inhibit fungal respiration and growth.
   - **Fusaricidins**: Secreted by certain strains of *Paenibacillus*, these compounds specifically target Fusarium species.
3. **Mechanism of Action**
   - **Membrane Disruption**: Compounds like iturin and surfactin integrate into the lipid bilayer of pathogen membranes, creating pores that lead to ion leakage and cell death.
   - **Inhibition of DNA and Protein Synthesis**: Antibiotics like actinomycin-D inhibit transcription processes in pathogens, preventing their replication and growth.

Example:

In greenhouse trials, *Bacillus subtilis* producing iturin significantly reduced *Rhizoctonia solani* infection in cucumber plants by over 70%.

1.4.1.2 Examples of Antagonistic Microbes

Several microbes exhibit strong antagonistic properties through antibiosis, making them effective biocontrol agents.

1. **Bacillus spp.**
   - *Bacillus subtilis* and *Bacillus amyloliquefaciens* are among the most studied antagonistic bacteria.
   - These species produce a wide range of antimicrobial compounds, including lipopeptides, polyketides, and volatile organic compounds (VOCs).
2. **Pseudomonas spp.**
   - *Pseudomonas fluorescens* produces phenazines, pyrrolnitrin, and siderophores, which inhibit the growth of fungal pathogens like *Fusarium* and *Pythium*.
   - It also produces VOCs that suppress bacterial pathogens such as *Ralstonia solanacearum*.

3. **Streptomyces spp.**
   - Soil-dwelling actinomycetes like *Streptomyces* produce a diverse array of antifungal and antibacterial compounds, including streptomycin and nystatin.
   - These compounds target soil-borne pathogens such as *Phytophthora infestans* and *Verticillium dahliae*.
4. **Trichoderma spp.**
   - Though primarily known for mycoparasitism, *Trichoderma* species also produce antibiotics like gliotoxin, which suppress pathogens like *Sclerotinia sclerotiorum*.
5. **Case Study: *Streptomyces lydicus***
   - *S. lydicus* produces antifungal metabolites that suppress *Fusarium oxysporum*. Its commercial formulation, Actinovate®, has been widely adopted in greenhouse and field applications.

1.4.1.3 Applications in Biocontrol

Antagonistic microbes are integral to the development of biopesticides and sustainable disease management systems.

1. **Biopesticide**s
   - Commercial formulations based on antagonistic microbes are used to manage a wide range of plant diseases.
   - Examples include Serenade® (based on *Bacillus subtilis*), Actinovate® (based on *Streptomyces lydicus*), and Agrilife® (based on *Pseudomonas fluorescens*).
2. **Crop-Specific Applications**
   - **Rice**: *Pseudomonas fluorescens* suppresses sheath blight caused by *Rhizoctonia solani*.
   - **Tomatoes**: *Bacillus subtilis* reduces early blight caused by *Alternaria solani*.
   - **Wheat**: *Streptomyces spp.* manage root rot caused by *Fusarium*.
3. **Integrated Disease Management (IDM)**
   - Antagonistic microbes are integrated with other practices, such as crop rotation and resistant varieties, to enhance disease suppression.
   - For example, the combined use of *Bacillus subtilis* and soil solarization reduces *Verticillium* wilt in eggplants.
4. **Reduction in Chemical Inputs**
   - The use of antagonistic microbes minimizes the reliance on chemical fungicides, reducing environmental contamination and pesticide resistance.
5. **Case Study: Biopesticide Adoption in Cucurbit Cultivation**
   - In cucurbit crops, biopesticides containing *Pseudomonas fluorescens* reduced powdery mildew incidence by 65%, demonstrating their efficacy as a sustainable alternative.

1.5 Induced Resistance

Induced resistance is a critical mechanism by which plants enhance their innate defense responses against a wide range of pathogens and pests. This phenomenon is often triggered by beneficial organisms such as rhizosphere microbes, signaling molecules, or even prior exposure to pathogens. Induced resistance functions as a priming mechanism, enabling plants to mount faster and more robust defense responses when challenged by biotic or abiotic stressors. The two primary forms of induced resistance are **Systemic Acquired Resistance (SAR)** and **Induced Systemic Resistance (ISR)**, each regulated by distinct signaling pathways and mechanisms.

1.5.1 Systemic Acquired Resistance (SAR)

SAR is a plant-wide immune response activated by localized pathogen attack. This long-lasting resistance mechanism is mediated by salicylic acid (SA) and involves the upregulation of pathogenesis-related (PR) proteins and other defense molecules.

1.5.1.1 Role of Salicylic Acid

Salicylic acid (SA) is the central signaling molecule in SAR, orchestrating the activation of defense-related genes and the accumulation of antimicrobial compounds.

1. **Mechanism of SA in SAR Activation**
   - Upon pathogen attack, SA levels increase at the infection site and in distal tissues.
   - SA acts as a signaling molecule, activating PR proteins such as β-1,3-glucanases and chitinases, which inhibit pathogen growth.
   - The transcription factor NPR1 (Non-Expresser of PR genes) plays a pivotal role in translating SA signals into the expression of defense-related genes.
2. **Defense Molecules Induced by SA**
   - **Phytoalexins**: Secondary metabolites with antimicrobial properties that accumulate in infected tissues.
   - **PR Proteins**: Enzymes such as glucanases and chitinases that degrade pathogen cell walls.
3. **Spread of SAR Signals**
   - SAR signals, including methyl salicylate and azelaic acid, are transported systemically to uninfected tissues, priming them for future attacks.

Example:

In tobacco plants, SAR triggered by TMV (Tobacco Mosaic Virus) infection results in the accumulation of PR proteins, preventing subsequent infections.

1.5.1.2 Examples of SAR-Inducing Agents

Certain beneficial microbes and chemicals are known to activate SAR, providing broad-spectrum resistance against pathogens.

1. **Microbial Inducers**
   - *Bacillus subtilis*: Triggers SAR in crops like rice and wheat, enhancing resistance against bacterial blight and rust diseases.
   - *Trichoderma harzianum*: Induces SAR in tomatoes, reducing infection by *Phytophthora infestans*.
2. **Chemical Inducers**
   - Acibenzolar-S-methyl (ASM): A synthetic SA analog that primes SAR in crops like cucumber and wheat.
   - Chitosan: A natural compound derived from chitin that activates SAR and other defense pathways.
3. **Field Applications**
   - In rice, *Bacillus subtilis* application reduces the incidence of bacterial leaf blight by up to 40%, highlighting its effectiveness in SAR activation.
   - ASM-treated wheat fields exhibit a 30% reduction in stripe rust infection, demonstrating its utility in large-scale agriculture.

1.5.2 Induced Systemic Resistance (ISR)

ISR is a plant defense mechanism mediated by jasmonic acid (JA) and ethylene (ET) signaling pathways, activated by non-pathogenic rhizosphere microbes. Unlike SAR, ISR does not involve PR protein accumulation but relies on priming plant defenses to respond more effectively upon pathogen attack.

1.5.2.1 Role of Jasmonic Acid and Ethylene

JA and ET are central signaling molecules in ISR, coordinating the plant's response to biotic and abiotic stressors.

1. **JA-Dependent Responses**
   - JA signaling promotes the synthesis of defense-related proteins, such as proteinase inhibitors and secondary metabolites, which deter herbivores and pathogens.
   - JA also activates genes involved in cell wall reinforcement, restricting pathogen entry.
2. **ET-Dependent Responses**
   - Ethylene signaling regulates stress-responsive genes, enhancing the plant's tolerance to environmental challenges.
   - ET and JA pathways often function synergistically, amplifying the defense response.
3. **Priming Effect**
   - ISR does not directly activate defense responses but primes the plant to react more rapidly and robustly upon pathogen attack.
   - For example, ISR-primed plants exhibit accelerated production of reactive oxygen species (ROS) and callose deposition upon infection.

Example:

In Arabidopsis, ISR induced by *Pseudomonas fluorescens* results in enhanced resistance against *Pseudomonas syringae*, mediated by JA and ET signaling.

1.5.2.2 Examples of ISR-Inducing Agents

Beneficial microbes, particularly plant growth-promoting rhizobacteria (PGPR), are the primary inducers of ISR.

1. **Bacterial Inducers**
   - *Pseudomonas fluorescens*: Reduces the severity of diseases like damping-off caused by *Rhizoctonia solani*.
   - *Bacillus amyloliquefaciens*: Induces ISR in soybean, protecting against *Phytophthora sojae*.
2. **Fungal Inducers**
   - *Trichoderma virens*: Enhances ISR in maize, reducing infections by *Fusarium verticillioides*.
   - *Piriformospora indica*: A mycorrhizal-like fungus that induces ISR in barley, increasing tolerance to abiotic stresses.
3. **Applications in Biocontrol**
   - *Pseudomonas fluorescens* formulations are used in horticultural crops to manage root and foliar diseases.
   - ISR-inducing microbes are integrated into integrated disease management (IDM) programs to reduce dependency on chemical inputs.

2. Competition and Mycoparasitism in Biological Control

Competition and mycoparasitism are two critical mechanisms employed by biological control agents (BCAs) to suppress plant pathogens. These processes operate in diverse ecological settings and play a significant role in disease management by reducing pathogen populations and minimizing their impact on crops.

2.1 Competition

Competition is a core ecological process where organisms contend for scarce resources such as nutrients, space, and energy. In biological control, competition serves as a primary mechanism by which beneficial microbes suppress pathogenic organisms. By dominating ecological niches and depriving pathogens of critical resources, beneficial microbes play an essential role in sustainable plant disease management strategies.

2.1.1 Mechanisms of Competition

The mechanisms of competition employed by beneficial microbes are diverse and multifaceted. These strategies allow them to outcompete pathogens, thereby limiting pathogen growth and activity.

2.1.1.1 Resource Deprivation

Resource competition is one of the most direct forms of competitive exclusion, where beneficial microbes utilize essential nutrients more efficiently than pathogens.

1. **Nutrient Competition**
   - Beneficial microbes rapidly utilize root exudates, which are rich in sugars, amino acids, and organic acids. This deprives pathogens of access to these critical resources.
   - For example, *Pseudomonas fluorescens* thrives in the rhizosphere due to its ability to metabolize a wide range of carbon compounds exuded by plant roots, leaving pathogens like *Fusarium oxysporum* nutrient-starved.
2. **Nitrogen and Phosphorus Utilization**
   - Nitrogen and phosphorus are vital macronutrients in soil ecosystems. Competitive microbes, such as *Trichoderma spp.*, absorb these nutrients at a faster rate, reducing their availability to pathogens.
3. **Microbial Enzyme Activity**
   - Some beneficial microbes secrete enzymes that degrade organic matter into simpler forms, enabling quicker nutrient uptake. This enzymatic advantage further enhances their competitive edge.

2.1.1.2 Siderophore Production

Siderophores are small, iron-chelating molecules produced by microbes to capture iron from the environment. Iron is a critical micronutrient for both pathogens and beneficial microbes, but its bioavailability in the soil is often limited.

1. **Iron Sequestration**
   - Siderophores bind to iron with high affinity, forming complexes that are inaccessible to pathogens.
   - Beneficial microbes such as *Pseudomonas spp.* and *Bacillus subtilis* produce siderophores like pyoverdine and bacillibactin, which suppress pathogens dependent on iron for growth and metabolism.
2. **Impact on Pathogen Virulence**
   - Many pathogens, including *Fusarium* and *Phytophthora* species, rely on iron for the production of virulence factors. By depriving pathogens of iron, siderophores indirectly weaken their ability to infect plants.
3. **Field Applications**
   - Siderophore-producing strains of *Pseudomonas fluorescens* are widely used in agricultural systems to manage soil-borne diseases in crops like wheat, rice, and maize.

Example:

In cucumber fields, the application of a siderophore-producing strain of *Pseudomonas fluorescens* reduced the incidence of *Pythium* root rot by 45%, demonstrating the effectiveness of iron competition.

2.1.1.3 Space Occupation

Rapid colonization of available surfaces by beneficial microbes physically excludes pathogens, preventing them from establishing infections.

1. **Root Colonization**
   - Beneficial microbes such as *Trichoderma spp.* and *Pseudomonas fluorescens* form a protective layer around plant roots, making it difficult for pathogens to access these tissues.
   - The ability of *Trichoderma spp.* to grow rapidly and form dense hyphal networks further enhances its competitive advantage.
2. **Leaf Surface Occupation**
   - On aboveground plant parts, beneficial microbes like *Bacillus subtilis* colonize leaf surfaces, inhibiting the attachment and growth of foliar pathogens.
3. **Implications for Pathogen Suppression**
   - Physical exclusion limits the pathogen's ability to penetrate plant tissues, reducing infection rates and disease severity.

Example:

In tomato crops, *Trichoderma harzianum* colonization of root surfaces reduced the infection potential of *Rhizoctonia solani* by over 60%.

2.1.1.4 Biofilm Formation

Biofilms are structured microbial communities encased in extracellular polymeric substances (EPS), providing stability and protection to the microbes.

1. **Biofilm Advantages**
   - Biofilms enhance microbial survival under adverse environmental conditions.
   - They create a competitive edge by monopolizing space and resources, preventing pathogens from gaining a foothold.
2. **Role in Biocontrol**
   - Beneficial microbes like *Pseudomonas fluorescens* and *Bacillus subtilis* form biofilms on root surfaces, creating a physical and chemical barrier against pathogens.
   - Biofilms also facilitate the sustained release of antimicrobial compounds, enhancing long-term pathogen suppression.

Example:

In wheat fields, *Bacillus subtilis* biofilms reduced the rhizosphere population of *Rhizoctonia solani* by 60%, leading to healthier root systems and improved crop yield.

2.1.2 Applications of Competition in Biological Control

The competitive traits of BCAs are exploited in various agricultural practices to suppress pathogens and improve crop health.

2.1.2.1 Biocontrol Formulations

1. **Commercial Products**
    - Biocontrol products containing competitive microbes, such as *Pseudomonas fluorescens*, *Bacillus subtilis*, and *Trichoderma harzianum*, are widely used in agriculture.
    - These formulations are effective against soil-borne pathogens like *Fusarium oxysporum*, *Pythium ultimum*, and *Rhizoctonia solani*.
2. **Customized Formulations**
    - Tailored formulations are developed for specific crops and regions, considering soil conditions and pathogen profiles.
    - For example, rice paddies in Southeast Asia benefit from formulations of *Pseudomonas fluorescens* optimized for suppressing sheath blight.

2.1.2.2 Soil Health Management

1. **Organic Amendments**
    - Adding organic matter such as compost, manure, or biochar enhances the activity of competitive microbes in the soil.
    - These amendments improve microbial diversity, creating conditions that favor beneficial microbes over pathogens.
2. **Suppressive Soils**
    - Soils naturally rich in competitive microbes exhibit suppressive properties, reducing the prevalence of diseases like take-all in wheat or clubroot in brassicas.
3. **Case Study:**
    - In organic farming systems, compost amendments inoculated with *Trichoderma spp.* increased pathogen suppression, reducing *Fusarium* wilt in tomatoes by 50%.

2.1.2.3 Integrated Disease Management (IDM)

1. **Cultural Practices**
    - Combining BCAs with cultural practices like crop rotation, intercropping, and reduced tillage enhances their efficacy.
    - For instance, crop rotation with legumes promotes siderophore-producing microbes, reducing soil-borne pathogen populations.
2. **Integration with Biopesticides**
    - Competitive BCAs can be used alongside chemical treatments to reduce pesticide

applications. Low-toxicity chemicals paired with BCAs maintain effective pathogen suppression while preserving environmental health.

Example:

In maize fields, the integration of *Pseudomonas fluorescens* with reduced fungicide applications resulted in a 40% reduction in *Pythium* root rot compared to chemical treatment alone.

2.2 Mycoparasitism

Mycoparasitism is a sophisticated antagonistic interaction where one fungus (the mycoparasite) parasitizes another fungus (the host-pathogen), leading to the suppression or complete destruction of the pathogen. This interaction plays a critical role in biological control, particularly in managing fungal diseases in crops. By employing diverse mechanisms such as enzymatic degradation, nutrient extraction, and secondary metabolite production, mycoparasitic fungi establish themselves as key tools in sustainable agriculture.

2.2.1 Mechanisms of Mycoparasitism

The mechanisms underlying mycoparasitism are intricate and involve multiple biochemical and molecular steps. These processes enable the mycoparasite to identify, attach to, and degrade the host pathogen effectively.

2.2.1.1 Recognition and Attachment

Recognition and attachment are the initial stages of the mycoparasitic process, where the mycoparasite identifies specific molecular signals on the host pathogen.

1. **Molecular Recognition**
    - Mycoparasites recognize specific surface molecules on the host pathogen, such as glycoproteins, lectins, and other cell wall components.
    - This recognition process involves signaling pathways that guide the mycoparasite toward the pathogen.
2. **Adhesion to the Host**
    - Adhesion is facilitated by the production of extracellular polysaccharides and adhesive proteins on the mycoparasite's surface.
    - This ensures that the mycoparasite remains in close contact with the host pathogen during subsequent stages of parasitism.

Example:

Trichoderma harzianum exhibits strong adhesion to the hyphae of *Rhizoctonia solani*, initiating mycoparasitic activity.

2.2.1.2 Enzyme Secretion

Enzymatic degradation is a hallmark of mycoparasitism, where the mycoparasite secretes hydrolytic enzymes to break down the host pathogen's cell wall.

1. **Key Enzymes Produced**
   - **Chitinases**: Degrade chitin, a primary structural component of fungal cell walls.
   - **β-Glucanases**: Break down β-glucans, another major component of fungal cell walls.
   - **Proteases**: Target proteinaceous structures within the pathogen's cell wall.
2. **Synergistic Action**
   - These enzymes often act synergistically, ensuring rapid and efficient degradation of the host cell wall.
   - For example, the combined activity of chitinases and β-glucanases enhances the breakdown of fungal cell walls, leaving the pathogen vulnerable.
3. **Pathogen-Specific Targeting**
   - The composition of the pathogen's cell wall determines the specificity and effectiveness of enzymatic degradation.

Example:

Trichoderma spp. secrete high levels of chitinases to degrade *Fusarium oxysporum* cell walls, significantly reducing pathogen viability.

2.2.1.3 Penetration and Nutrient Extraction

After degrading the host pathogen's cell wall, the mycoparasite penetrates the pathogen and extracts nutrients.

1. **Hyphal Penetration**
   - The mycoparasite penetrates the pathogen's hyphae using specialized infection structures.
   - This step often involves the release of additional enzymes to facilitate penetration.
2. **Nutrient Assimilation**
   - The mycoparasite absorbs nutrients from the host pathogen, including carbohydrates, amino acids, and lipids.
   - This extraction weakens the pathogen further and reduces its ability to infect plants.
3. **Impact on Pathogen Viability**
   - Nutrient extraction not only kills the pathogen but also prevents the development of inoculum in the soil.

Example:

Trichoderma virens effectively penetrates the hyphae of *Sclerotinia sclerotiorum* and extracts nutrients, reducing pathogen survival in the soil.

2.2.1.4 Production of Secondary Metabolites

Secondary metabolite production is a critical aspect of mycoparasitism, enhancing the suppression of the host pathogen.

1. **Types of Secondary Metabolites**
   - **Gliotoxin**: Produced by *Gliocladium spp.*, gliotoxin disrupts fungal metabolism and inhibits spore germination.
   - **Viridin**: An antifungal compound produced by *Trichoderma spp.*, effective against a broad spectrum of pathogens.
   - **6-Pentyl-α-pyrone (6-PAP)**: A volatile organic compound with strong antifungal properties.
2. **Dual Role**
   - These metabolites not only inhibit pathogen growth but also enhance the competitive fitness of the mycoparasite in the soil ecosystem.

Example:

Gliocladium virens produces gliotoxin to suppress *Pythium ultimum*, preventing root rot in cucumber crops.

2.2.2 Examples of Mycoparasitic Fungi

Several fungi exhibit mycoparasitic behavior and have been effectively utilized in biological control.

2.2.2.1 *Trichoderma spp.*

1. **Target Pathogens**
   - *Trichoderma spp.* are effective against a wide range of soil-borne pathogens, including *Fusarium oxysporum*, *Rhizoctonia solani*, and *Sclerotinia sclerotiorum*.
2. **Mechanisms**
   - These fungi employ enzymatic degradation, secondary metabolite production, and competition for resources to suppress pathogens.
3. **Applications**
   - *Trichoderma harzianum* is widely used in seed treatments and soil applications to manage damping-off and root rots.

Example:

In controlled trials, *Trichoderma harzianum* reduced *Rhizoctonia solani* infections in lettuce crops by 75%.

2.2.2.2 *Gliocladium spp.*

1. **Target Pathogens**
   - Effective against *Pythium spp.* and *Phytophthora spp.*, which cause damping-off and root rots.
2. **Secondary Metabolites**
   - Gliotoxin production makes *Gliocladium spp.* highly effective in pathogen suppression.

3. **Applications**
    - Often used in compost inoculation to enhance soil health and reduce pathogen populations.

Example:

The application of *Gliocladium spp.* in cucumber fields reduced *Pythium ultimum* incidence by 60%.

2.2.2.3 *Coniothyrium minitans*

1. **Specialization**
    - Specifically targets the sclerotia of *Sclerotinia sclerotiorum*, reducing its survival and infection potential.
2. **Applications**
    - Applied as a soil amendment to manage diseases caused by sclerotial pathogens in crops like canola and beans.

Example:

In canola fields, *Coniothyrium minitans* reduced sclerotial inoculum of *Sclerotinia sclerotiorum* by 80%.

2.2.2.4 Hypoviruses

1. **Pathogen Targeting**
    - Hypoviruses infect fungal pathogens like *Cryphonectria parasitica*, reducing their virulence and fitness.
2. **Applications**
    - Widely used in the management of chestnut blight, aiding in the recovery of chestnut populations in North America.

Example:

Hypoviruses have reduced chestnut blight severity by over 50% in affected forests.

2.2.3 Applications of Mycoparasitism in Biological Control

Mycoparasitic fungi are integral to sustainable agricultural practices and biological control strategies.

2.2.3.1 Seed and Soil Treatments

1. **Seed Coating**
    - Seeds coated with *Trichoderma harzianum* or *Gliocladium spp.* are protected from soil-borne pathogens during germination.
2. **Soil Drenches**
    - Soil drenches with mycoparasitic fungi reduce pathogen populations in the rhizosphere, promoting healthier root systems.

Example:

In tomato production, seed treatment with *Trichoderma harzianum* reduced damping-off by 70%.

2.2.3.2 Compost and Organic Amendments

1. **Compost Inoculation**
   - Compost inoculated with mycoparasitic fungi enhances microbial diversity and suppresses pathogen activity.
2. **Improved Soil Health**
   - These amendments contribute to soil fertility, structure, and resilience against disease outbreaks.

Example:

In strawberry cultivation, compost inoculated with *Trichoderma spp.* reduced *Verticillium dahliae* incidence by 50%.

2.2.3.3 Biopesticide Development

1. **Commercial Products**
   - Products like Rootshield® (based on *Trichoderma harzianum*) are widely used in horticulture and greenhouse production.
2. **Field Applications**
   - Biopesticides are applied to manage soil-borne diseases in crops such as wheat, maize, and vegetables.

Example:

The use of Rootshield® in cucumber crops reduced *Fusarium wilt* incidence by 65%.

2.2.3.4 Integration with IDM Practices

1. **Complementary Control Measures**
   - Mycoparasitic fungi are integrated with resistant crop varieties, crop rotation, and cultural practices for effective disease management.
2. **Sustainability Benefits**
   - This approach reduces dependency on chemical pesticides, lowering environmental impact while maintaining crop health.

Example:

In soybean fields, the integration of *Trichoderma harzianum* with crop rotation reduced *Sclerotinia stem rot* by 45%.

3. Exploitation for Hypovirulence and Rhizosphere Colonization

Biological control strategies leverage natural mechanisms like hypovirulence and rhizosphere colonization to suppress plant pathogens and enhance crop health. These processes involve the attenuation of pathogen virulence through biological agents and the strategic colonization of the rhizosphere by beneficial microbes to outcompete pathogens, promote plant growth, and establish a robust plant-microbe interaction network.

3.1 Exploitation for Hypovirulence

Hypovirulence represents a significant natural mechanism for biological control, leveraging reduced pathogen virulence to manage plant diseases effectively. This phenomenon, mediated by biological agents such as viruses, fungi, or bacteria, diminishes a pathogen's ability to cause severe disease without eliminating it. The strategic use of hypovirulence promotes ecological balance while allowing the coexistence of pathogens and their hosts in a controlled manner.

3.1.1 Mechanisms of Hypovirulence

Hypovirulence occurs through intricate interactions between biotic agents and the pathogen. These mechanisms include viral infections, genetic alterations, and disruption of pathogen signaling pathways.

3.1.1.1 Viral Infections in Pathogens

Viruses play a pivotal role in inducing hypovirulence in fungal pathogens.

1. **Mycoviruses**
   - Mycoviruses infect fungal pathogens and replicate within their cells, interfering with critical metabolic functions.
   - They reduce the pathogen's ability to produce virulence factors, such as enzymes and toxins, which are essential for host colonization.
2. **Disruption of Pathogen Metabolism**
   - Mycoviruses alter the metabolic pathways of fungal pathogens, reducing their ability to acquire nutrients or propagate effectively.

Example:

In chestnut blight caused by *Cryphonectria parasitica*, hypoviruses suppress the pathogen's ability to cause severe lesions, enabling chestnut trees to recover. This case stands as a landmark example of hypovirulence in action.

3.1.1.2 Genetic Alterations

Hypovirulence can also result from genetic mutations within the pathogen, often induced by external agents like hypoviruses or chemical mutagens.

1. **Mutational Changes**
   - Mutations in genes associated with toxin production or enzyme secretion weaken the pathogen's pathogenicity.

- These mutations may occur naturally or be induced by targeted interventions.

2. **Retention of Ecological Competitiveness**
   - Hypovirulent strains retain their ability to compete in their ecological niche, ensuring they can coexist with beneficial microorganisms and hosts.

Example:

Genetically altered strains of *Sclerotinia sclerotiorum* have shown reduced virulence while maintaining soil compatibility, making them effective biocontrol tools.

3.1.1.3 Interference with Signaling Pathways

Pathogens rely on complex signaling pathways to produce virulence factors critical for host colonization. Hypovirulent agents disrupt these pathways, rendering the pathogen less effective.

1. **Inhibition of Toxin Production**
   - Hypovirulence-inducing agents suppress the expression of genes responsible for toxin synthesis, reducing the pathogen's ability to damage host tissues.
2. **Disruption of Enzymatic Activity**
   - Enzymes that degrade plant cell walls, such as cellulases and pectinases, are downregulated in hypovirulent strains.

Example:

In rice, hypovirulent strains of *Magnaporthe oryzae* exhibit reduced production of melanin, a critical virulence factor, leading to less severe infections.

3.1.2 Applications of Hypovirulence

The practical application of hypovirulence in agriculture has proven effective in controlling several devastating plant diseases.

3.1.2.1 Chestnut Blight Management

The control of chestnut blight caused by *Cryphonectria parasitica* stands as the most notable example of hypovirulence in action.

1. **Role of Hypoviruses**
   - Hypoviruses infect *Cryphonectria parasitica*, reducing its ability to cause lethal lesions on chestnut trees.
   - The infection spreads horizontally among pathogen populations, amplifying the effect of hypovirulence.
2. **Recovery of Chestnut Populations**
   - In North America and Europe, the deliberate introduction of hypovirulent strains has significantly reduced blight incidence.

3.1.2.2 Sclerotinia Stem Rot

Hypovirulent strains of *Sclerotinia sclerotiorum* have been utilized to manage stem rot in canola and other crops.

1. **Reduction in Disease Severity**
   - Hypovirulent strains outcompete virulent pathogens in the soil, reducing disease spread and severity.
   - These strains have minimal impact on beneficial soil microbes, preserving soil health.
2. **Integration into Field Practices**
   - Combining hypovirulent strains with cultural practices such as crop rotation enhances their efficacy.

Example:

In canola fields, the use of hypovirulent *Sclerotinia sclerotiorum* strains has reduced stem rot incidence by 50%, increasing yields significantly.

3.1.2.3 Integration with Other Practices

Hypovirulence is most effective when combined with other disease management practices.

1. **Crop Rotation and Resistant Varieties**
   - Rotating crops and using disease-resistant varieties reduce pathogen populations, complementing the effects of hypovirulent strains.
2. **Biopesticide Applications**
   - Hypovirulent strains can be integrated with biopesticides to provide a multi-faceted approach to disease suppression.

Case Study:

In soybean fields, integrating hypovirulent *Fusarium oxysporum* strains with biopesticides reduced root rot incidence by 60%.

3.1.3 Advantages of Hypovirulence

Hypovirulence offers several advantages that make it a sustainable and practical tool in agricultural disease management.

3.1.3.1 Eco-Friendly

Hypovirulence relies on natural biological agents, eliminating the need for synthetic chemicals and reducing environmental impact.

1. **Reduction in Pesticide Use**
   - Hypovirulent agents lower the reliance on fungicides and other chemical treatments, promoting sustainable farming practices.

2. **Preservation of Ecosystems**
   - The specificity of hypovirulent agents ensures minimal disruption to non-target organisms, preserving biodiversity.

3.1.3.2 Specificity

The targeted action of hypovirulent agents ensures effective pathogen suppression without affecting beneficial microbes or plant health.

1. **Pathogen-Specific Action**
   - Hypovirulent agents selectively suppress the virulence of specific pathogens, leaving beneficial microbes unaffected.
2. **Compatibility with Agroecosystems**
   - Hypovirulent strains integrate seamlessly into existing soil microbial communities, maintaining ecological balance.

3.1.3.3 Long-Term Control

Once established, hypovirulent strains can persist in the environment, providing durable and ongoing disease suppression.

1. **Persistence in Soil**
   - Hypovirulent strains remain viable in the soil for extended periods, reducing the need for repeated applications.
2. **Horizontal Transmission**
   - In pathogens like *Cryphonectria parasitica*, hypoviruses spread naturally among pathogen populations, amplifying their impact over time.

3.1.3.4 Compatibility

Hypovirulence integrates effectively with other biological, cultural, and chemical control practices.

1. **Integration with IDM**
   - Hypovirulent agents complement integrated disease management (IDM) strategies, enhancing overall efficacy.
2. **Synergy with Biocontrol Agents**
   - Hypovirulent strains can be used alongside antagonistic microbes such as *Trichoderma spp.* for comprehensive disease management.

3.2 Rhizosphere Colonization

The rhizosphere, a dynamic interface between plant roots and the surrounding soil, is a hub of complex biological interactions that significantly influence plant health and productivity. This region is teeming with microbial activity, as it provides nutrients and signals that attract a diverse range of organisms, including beneficial microbes. Rhizosphere colonization by these beneficial microbes is a cornerstone of biological control, as it not only establishes a protective barrier against pathogens but also facilitates nutrient cycling, stress tolerance, and overall plant growth.

3.2.1 Mechanisms of Rhizosphere Colonization

Beneficial microbes employ a variety of adaptive strategies to establish themselves in the rhizosphere. These strategies enable them to outcompete pathogens, optimize resource utilization, and form synergistic relationships with plant roots.

3.2.1.1 Root Surface Adhesion

Adhesion to the root surface is the first step in effective rhizosphere colonization.

1. **Mechanisms of Adhesion**
   - Beneficial microbes produce extracellular polysaccharides (EPS) and adhesive proteins that facilitate attachment to root surfaces.
   - These structures enhance microbial persistence in the rhizosphere and provide protection from desiccation and environmental stressors.
2. Niche Occupation
   - By adhering to the root surface, beneficial microbes occupy ecological niches that pathogens might exploit, creating a protective shield around the plant.
   - For example, *Pseudomonas fluorescens* forms a dense layer on wheat roots, preventing colonization by *Fusarium oxysporum*.

Example:

In soybean crops, *Bacillus subtilis* adheres to root surfaces and significantly reduces the incidence of root rot caused by *Phytophthora sojae*.

3.2.1.2 Biofilm Formation

Biofilm formation is a critical adaptation that enhances the ability of beneficial microbes to colonize and persist in the rhizosphere.

1. **Structure and Function**
   - Biofilms are structured microbial communities encased in a matrix of EPS, proteins, and DNA.
   - This matrix provides physical protection and retains moisture and nutrients, ensuring microbial survival under fluctuating environmental conditions.
2. **Synergistic Interactions**
   - Biofilms often harbor multiple microbial species that interact synergistically, enhancing their collective ability to suppress pathogens and promote plant growth.
   - For instance, biofilms formed by *Bacillus subtilis* and *Pseudomonas fluorescens* show increased antifungal activity against *Rhizoctonia solani*.

Example:

In rice paddies, biofilms formed by *Trichoderma harzianum* on root surfaces have been shown to reduce sheath blight caused by *Rhizoctonia solani* by 60%.

3.2.1.3 Chemotaxis and Root Exudate Utilization

Root exudates are a rich source of nutrients and chemical signals that attract beneficial microbes to the rhizosphere.

1. **Attraction to Root Exudates**
   - Beneficial microbes exhibit chemotaxis, moving toward root exudates that contain sugars, amino acids, organic acids, and secondary metabolites.
   - This targeted movement ensures that beneficial microbes colonize areas rich in resources and microbial signals.
2. **Efficient Nutrient Utilization**
   - Once in proximity to the roots, microbes utilize exudates to grow and proliferate, establishing dominance over competing organisms.
   - For example, *Pseudomonas fluorescens* metabolizes amino acids and organic acids from root exudates, outcompeting soil-borne pathogens.

Example:

In maize, root exudates attract *Pseudomonas putida*, which colonizes the rhizosphere and protects plants from *Fusarium verticillioides*.

3.2.1.4 Production of Antimicrobial Compounds

Beneficial microbes produce a wide array of antimicrobial compounds that suppress pathogen growth in the rhizosphere.

1. **Types of Compounds**
   - **Antibiotics**: Compounds like phenazines, pyrrolnitrin, and iturin directly inhibit pathogen growth.
   - **Lytic Enzymes**: Enzymes such as chitinases and glucanases degrade the cell walls of fungal pathogens.
   - **Volatile Organic Compounds (VOCs)**: Substances like 2,3-butanediol and acetoin inhibit pathogens and stimulate plant defenses.
2. **Targeted Pathogen Suppression**
   - These compounds act specifically against pathogens without affecting beneficial microbes, maintaining the balance of the rhizosphere microbiome.

Example:

Bacillus subtilis produces iturin, which disrupts fungal cell membranes, suppressing root rot caused by *Pythium ultimum* in cucumber plants.

3.2.2 Examples of Rhizosphere Colonizers

A diverse array of microbes, including bacteria, fungi, and endophytes, effectively colonize the rhizosphere, providing protection against pathogens and enhancing plant growth.

3.2.2.1 Bacterial Colonizers

Bacteria are among the most prolific rhizosphere colonizers, known for their versatility and wide-ranging benefits.

1. **Pseudomonas spp.**
   - These bacteria produce siderophores, antibiotics, and biofilms that suppress pathogens like *Fusarium* and *Pythium*.
   - They also induce systemic resistance in plants, enhancing overall defense mechanisms.

Example:

In wheat, *Pseudomonas fluorescens* colonization reduced infections by *Fusarium oxysporum* by 45%.

2. **Bacillus spp.**
   - Known for producing lipopeptides like surfactin and fengycin, *Bacillus spp.* enhance plant health and suppress soil-borne pathogens.
   - They also solubilize phosphorus, making it available to plants.

Example:

Bacillus amyloliquefaciens reduced *Rhizoctonia solani* infections in soybean by 50% in field trials.

3.2.2.2 Fungal Colonizers

Fungi, particularly those in the genera *Trichoderma* and *Gliocladium*, are effective rhizosphere colonizers.

1. **Trichoderma spp.**
   - These fungi suppress pathogens through mycoparasitism, competition, and enzyme production.
   - They also produce growth-promoting hormones that enhance root development.

Example:

In tomato plants, *Trichoderma harzianum* colonization suppressed root rot caused by *Rhizoctonia solani* by 70%.

2. **Gliocladium spp.**
   - These fungi are particularly effective against damping-off pathogens, such as *Pythium* and *Phytophthora*.

Example:

In cucumber fields, *Gliocladium spp.* reduced *Pythium ultimum* infections by 60%.

3.2.2.3 Endophytes

Endophytes colonize the internal tissues of plants while maintaining interactions with the rhizosphere.

1. **Bacterial Endophytes**
   - These include species like *Azospirillum* and *Burkholderia*, which fix nitrogen and promote plant growth.
2. **Fungal Endophytes**

» Fungi such as *Piriformospora indica* enhance drought tolerance and nutrient uptake in plants.

Example:

In barley, colonization by *Piriformospora indica* improved phosphorus uptake and reduced stress-related damage.

3.2.3 Applications of Rhizosphere Colonization

Rhizosphere colonization has diverse applications in agriculture, offering solutions for sustainable disease management, nutrient optimization, and plant growth enhancement.

3.2.3.1 Biocontrol Agents

1. **Soil Inoculants and Seed Treatments**
   - Beneficial microbes are applied as inoculants or seed coatings to establish dominance in the rhizosphere.
2. **Pathogen Suppression**
   - Colonizing microbes suppress soil-borne pathogens like *Fusarium oxysporum* and *Rhizoctonia solani*.

Example:

In rice, inoculation with *Pseudomonas fluorescens* reduced sheath blight incidence by 50%.

3.2.3.2 Nutrient Cycling and Plant Growth Promotion

1. **Nitrogen Fixation**
   - Rhizosphere colonizers like *Rhizobium* and *Azotobacter* fix atmospheric nitrogen, making it available to plants.
2. **Phosphorus Solubilization**
   - Fungi like *Trichoderma spp.* release organic acids that solubilize phosphate, enhancing nutrient uptake.

Example:

In maize, colonization by *Bacillus megaterium* improved phosphorus availability and boosted yields.

3.2.3.3 Disease Suppression

1. **Direct Pathogen Inhibition**
   - Antimicrobial compounds produced by colonizers reduce pathogen viability.
2. **Enhanced Plant Immunity**
   - Rhizosphere colonizers induce systemic resistance, enabling plants to better withstand infections.

Example:

In sugarcane, colonization by *Pseudomonas fluorescens* reduced red rot incidence by 40%.

3.2.4 Integration with Sustainable Practices

Rhizosphere colonization aligns with sustainable farming practices, offering eco-friendly solutions for crop management.

3.2.4.1 Organic Farming Systems

1. **Compost and Green Manure**
   - Adding organic amendments enhances microbial activity and supports rhizosphere colonization.
2. **Chemical-Free Disease Management**
   - Beneficial microbes replace chemical fungicides in organic farming systems.

Example:

In strawberry cultivation, compost inoculated with *Trichoderma spp.* reduced *Verticillium dahliae* infections by 50%.

3.2.4.2 Integrated Disease Management (IDM)

1. **Complementary Strategies**
   - Combining rhizosphere colonizers with crop rotation, resistant varieties, and reduced tillage enhances disease suppression.
2. **Long-Term Soil Health**
   - Colonizers improve soil structure and microbial diversity, ensuring sustainable productivity.

Example:

In maize systems, integrating *Trichoderma spp.* with crop residue management reduced *Pythium ultimum* infections by 40%.

4. Competitive Saprophytic Ability and Antibiosis

Biological control agents (BCAs) employ multiple mechanisms to suppress plant pathogens and maintain ecological balance. Among these, **competitive saprophytic ability (CSA)** and **antibiosis** are two critical processes that play pivotal roles in disease suppression. CSA enables beneficial microbes to outcompete pathogens for resources during their saprophytic phase, while antibiosis involves the production of bioactive compounds that directly inhibit or kill pathogens. Together, these mechanisms form an integral part of biological control strategies in sustainable agriculture.

4.1 Competitive Saprophytic Ability (CSA)

Competitive saprophytic ability (CSA) is a crucial mechanism employed by biological control agents (BCAs) to suppress plant pathogens by outcompeting them for resources within the soil. This capacity enables beneficial microbes to colonize and utilize dead organic matter in the soil ecosystem more effectively than pathogens, reducing the latter's survival and infection potential. By dominating the saprophytic niche, BCAs decrease the inoculum density of harmful pathogens, thereby limiting the risk of plant disease outbreaks. CSA is particularly effective in managing soil-borne diseases, where pathogens rely on decomposing organic residues as a survival strategy between cropping cycles.

Mechanisms of CSA

The competitive saprophytic ability of beneficial microbes hinges on various biological and ecological processes that ensure they dominate organic substrates in the soil. These mechanisms collectively allow BCAs to suppress pathogen populations during their saprophytic phase.

One primary mechanism is rapid resource utilization, where beneficial microbes such as *Trichoderma spp.* and *Bacillus spp.* exhibit an extraordinary ability to degrade organic matter swiftly. These microbes secrete a suite of extracellular enzymes, including cellulases, ligninases, and proteases, to break down complex organic compounds into simpler, bioavailable forms. This enzymatic advantage ensures that BCAs access nutrients from organic residues before pathogens can utilize them, effectively starving the pathogens. For instance, in wheat fields, *Trichoderma harzianum* has demonstrated exceptional CSA by quickly colonizing decomposing plant residues, significantly reducing the population of *Rhizoctonia solani.*

Another critical aspect of CSA is physical exclusion. Beneficial microbes colonize organic substrates densely, forming robust microbial networks that physically block pathogens from accessing these resources. This colonization creates an ecological barrier, preventing pathogens from establishing themselves on the same substrate. *Trichoderma spp.* are well-known for rapidly occupying crop residues, outcompeting pathogens such as *Fusarium oxysporum* and inhibiting their ability to survive and proliferate.

Microbial antagonism further strengthens CSA. Beneficial microbes produce antimicrobial compounds that directly inhibit the growth and activity of pathogens. These compounds include secondary metabolites with antifungal or antibacterial properties, such as antibiotics and lytic enzymes, which enhance the competitive edge of BCAs in the saprophytic environment. For example, the production of antifungal metabolites by *Trichoderma spp.* ensures that pathogens cannot establish dominance in their ecological niche. Additionally, biochemical competition plays a pivotal role in

CSA. Beneficial microbes release siderophores—iron-chelating compounds that bind to and sequester iron in the soil. Iron is an essential micronutrient for microbial growth and is often a limiting resource in soil ecosystems. By capturing iron more efficiently than pathogens, BCAs deprive pathogens of a critical nutrient, curbing their growth and survival. This mechanism is particularly effective against pathogens like *Fusarium* and *Pythium*, which depend heavily on iron for metabolic activity.

Factors Influencing CSA

The effectiveness of competitive saprophytic ability is influenced by various environmental, biological, and ecological factors. These factors determine the success of BCAs in suppressing pathogens through CSA. Environmental conditions, such as soil pH, temperature, and moisture content, play a crucial role in shaping the CSA potential of beneficial microbes. For instance, *Trichoderma spp.* exhibit optimal CSA activity in well-drained, slightly acidic soils. These conditions promote their growth and metabolic activity, enabling them to outcompete pathogens effectively. Conversely, in waterlogged or highly alkaline soils, the activity of *Trichoderma spp.* may be inhibited, reducing their competitive advantage.

Nutrient availability is another significant factor influencing CSA. The presence of organic amendments such as compost or biochar enhances the growth and activity of CSA-active microbes by providing additional substrates for colonization. These amendments not only enrich the soil with organic matter but also create favorable conditions for microbial proliferation. For example, in organic tomato farming systems, compost inoculated with *Trichoderma spp.* significantly improved CSA, resulting in a 45% reduction in *Fusarium* wilt incidence.

Microbial diversity within the soil ecosystem also impacts the CSA of beneficial microbes. Diverse microbial communities foster synergistic interactions, where multiple microbial species collaborate to suppress pathogens more effectively. This collective action enhances the overall CSA of the microbial community, ensuring that pathogens are excluded from the saprophytic niche. High microbial diversity also promotes resilience against environmental fluctuations, maintaining the effectiveness of CSA under varying conditions.

Applications of CSA

Competitive saprophytic ability has widespread applications in sustainable agriculture, contributing to soil health, disease management, and crop productivity. One of the most significant applications of CSA is in soil health management. CSA-active microbes play a vital role in decomposing organic matter, converting it into bioavailable nutrients that enrich soil fertility. This decomposition process not only reduces the inoculum potential of pathogens but also enhances the nutrient profile of the soil, promoting healthier plant growth. By incorporating CSA-active microbes into soil amendments or crop rotations, farmers can maintain soil fertility while suppressing soil-borne diseases.

Residue management is another critical area where CSA is applied. Crop residues left in the field after harvest serve as a breeding ground for many soil-borne pathogens. By applying CSA-capable microbes to these residues, farmers can minimize pathogen survival during the off-season. For example, the application of *Trichoderma harzianum* to straw residues in rice paddies has been shown to reduce *Sclerotium rolfsii* inoculum levels, leading to improved seedling emergence rates in the subsequent cropping cycle.

Seed and root treatments with CSA-active microbes also offer a proactive approach to disease prevention. Treating seeds or roots with beneficial microbes such as *Bacillus subtilis* ensures early colonization of the rhizosphere, providing a competitive advantage over soil-borne pathogens. This early establishment reduces the likelihood of infection and promotes robust plant growth from the outset. In addition to these applications, CSA is integral to integrated disease management (IDM) strategies, where it complements other biological, cultural, and physical control methods. By integrating CSA with practices such as crop rotation, resistant varieties, and organic amendments, farmers can achieve more sustainable and effective disease management outcomes.

4.2 Antibiosis

Antibiosis is a vital biological control mechanism wherein beneficial microorganisms suppress pathogens through the production of bioactive compounds. These compounds—ranging from antibiotics to lytic enzymes and volatile organic compounds (VOCs)—target and inhibit the growth or survival of plant pathogens. Antibiosis plays a critical role in the soil microbiome, not only providing direct pathogen suppression but also maintaining ecological balance and promoting plant health. Its specificity and versatility make it an integral strategy in sustainable agricultural practices.

Mechanisms of Antibiosis

Antibiosis relies on a variety of biochemical pathways and metabolites to suppress pathogens effectively. These mechanisms provide direct inhibition, ensuring pathogens cannot colonize plant tissues or thrive in the rhizosphere.

One of the primary mechanisms is the production of antibiotics, which are specialized metabolites that interfere with the growth, reproduction, or metabolism of pathogens. Antibiotics are highly specific and can target bacterial, fungal, or viral pathogens. For instance, *Bacillus subtilis* produces iturin, a lipopeptide that disrupts fungal cell membranes, causing leakage and eventual cell death. Similarly, *Pseudomonas fluorescens* produces phenazines, which inhibit the respiration of fungal pathogens by targeting their electron transport chains. These antibiotics not only reduce pathogen populations but also enhance the resilience of beneficial microbes in the soil.

Another important mechanism is the secretion of lytic enzymes such as chitinases, glucanases, and proteases. These enzymes degrade the structural components of fungal cell walls, effectively lysing the pathogen. Chitinases, for example, target chitin—a key component of fungal cell walls—weakening the pathogen's structural integrity and rendering it non-viable. This enzymatic action is particularly effective against soil-borne fungi such as *Fusarium* and *Pythium*.

Volatile organic compounds (VOCs) add another dimension to antibiosis. These gaseous compounds, such as hydrogen cyanide and 2,3-butanediol, diffuse through the soil and suppress pathogens at a distance. VOCs are particularly effective in suppressing pathogens in the rhizosphere, where gaseous exchange is high. For example, *Bacillus subtilis* produces VOCs that inhibit the growth of root-infecting pathogens like *Pythium ultimum*. Finally, some beneficial microbes produce specific toxins that selectively target pathogens. These toxins do not harm plants but are highly effective in inhibiting or killing pathogenic microbes. For instance, surfactin, a toxin produced by *Bacillus subtilis*, disrupts the lipid membranes of fungal pathogens, further enhancing antibiosis-driven suppression.

Example:

In cucumber cultivation, *Bacillus subtilis* reduced root rot caused by *Pythium ultimum* by producing iturin and surfactin, leading to healthier root systems and improved crop yield.

Examples of Antibiosis-Producing Microbes

Several genera of beneficial microbes are renowned for their ability to produce antibiotics and other bioactive compounds, making them critical agents in biological control.

Bacillus spp. are prolific producers of antibiotics such as iturin, fengycin, and surfactin. These compounds are highly effective against a range of fungal pathogens. In addition, *Bacillus spp.* produce VOCs that further enhance their antagonistic activity in the rhizosphere.

Pseudomonas spp. are equally notable for their production of phenazines, pyrrolnitrin, and siderophores. Phenazines, in particular, are potent antifungal agents that disrupt pathogen respiration. This genus is also known for its ability to produce hydrogen cyanide, a VOC that inhibits pathogen growth while stimulating plant defenses.

Streptomyces spp., soil-dwelling actinomycetes, produce a wide range of antibiotics, including streptomycin and nystatin. These compounds are effective against bacterial and fungal pathogens, making *Streptomyces spp.* valuable in managing diseases in high-value crops.

Trichoderma spp. are well-known for their production of secondary metabolites such as gliotoxin. Gliotoxin disrupts the metabolic pathways of fungal pathogens, effectively inhibiting their growth. These fungi also produce lytic enzymes that enhance their antibiosis-driven antagonistic activity.

Case Study:

In greenhouse trials, *Pseudomonas fluorescens* reduced infections of *Rhizoctonia solani* in tomatoes by 70%, primarily through the production of phenazines and hydrogen cyanide.

Applications of Antibiosis

Antibiosis has widespread applications in agriculture, offering solutions for disease management in both conventional and organic farming systems. Its versatility and effectiveness make it a cornerstone of biological control strategies.

One of the most significant applications of antibiosis is in the development of biopesticides. Antibiosis-producing microbes such as *Bacillus subtilis* and *Pseudomonas fluorescens* are widely used in commercial biopesticides. Products like Serenade® and Agrilife® utilize the antibiotic and VOC-producing capabilities of these microbes to suppress pathogens effectively. These biopesticides are safe for non-target organisms and contribute to the sustainability of agricultural practices.

In integrated disease management (IDM) programs, antibiosis-producing microbes are combined with resistant crop varieties, cultural practices, and other biological control strategies to achieve comprehensive disease suppression. This integrated approach enhances the overall efficacy of disease management systems while reducing dependency on chemical fungicides. Antibiosis has also been successfully applied to control diseases in high-value crops such as fruits, vegetables, and cereals. For example, in strawberry cultivation, *Streptomyces lydicus* has been used to reduce *Verticillium dahliae* populations by 60%, leading to improved fruit yield and quality.

Example:

In potato fields, the application of *Bacillus subtilis* reduced black scurf caused by *Rhizoctonia solani*, increasing marketable tuber yield by 40%.

4.3 Integration of Competitive Saprophytic Ability (CSA) and Antibiosis

The integration of competitive saprophytic ability (CSA) and antibiosis represents a highly effective and synergistic approach in biological control strategies. Both mechanisms address pathogen suppression through complementary pathways: CSA deprives pathogens of essential resources by dominating the saprophytic niche, while antibiosis provides direct inhibition through the production of bioactive compounds. Together, these strategies form a comprehensive defense system against a wide range of plant pathogens, contributing significantly to sustainable disease management and enhanced crop productivity.

Complementary Mechanisms

The synergy between CSA and antibiosis lies in their distinct yet interdependent modes of action. CSA primarily targets the survival and proliferation of pathogens in the soil by outcompeting them for organic matter, essential nutrients, and ecological niches. Beneficial microbes exhibiting strong CSA, such as *Trichoderma spp.*, colonize organic substrates rapidly, leaving little opportunity for pathogens to establish themselves. This dominance is further reinforced by microbial networks that physically exclude pathogens, creating a protective environment around plant roots.

On the other hand, antibiosis operates through the production of secondary metabolites, such as antibiotics, lytic enzymes, and volatile organic compounds (VOCs), which directly inhibit or kill pathogens. These compounds disrupt the structural integrity, metabolic pathways, or reproduction of the pathogens, providing immediate and targeted suppression. For example, *Pseudomonas fluorescens* produces phenazines and hydrogen cyanide, which suppress soil-borne pathogens like *Fusarium oxysporum* and *Rhizoctonia solani*.

When combined, CSA and antibiosis address both the competitive and biochemical aspects of pathogen suppression. CSA reduces the availability of resources that pathogens need for survival, weakening them further and making them more susceptible to the inhibitory effects of antibiotics and other bioactive compounds produced by antibiosis-active microbes. This dual action ensures comprehensive control over pathogen populations, reducing their ability to infect plants and spread within the agricultural ecosystem.

Example:

In maize fields, a combined formulation of *Trichoderma harzianum* (CSA-active) and *Pseudomonas fluorescens* (antibiosis-active) reduced root rot incidence by 55%, highlighting the effectiveness of integrating these mechanisms.

Applications in Sustainable Farming

The integration of CSA and antibiosis aligns perfectly with the principles of sustainable and organic farming. By leveraging these natural mechanisms, farmers can minimize the use of synthetic chemicals, promote soil health, and enhance crop resilience to diseases.

In organic farming systems, where chemical inputs are restricted, CSA and antibiosis provide eco-friendly alternatives for managing soil-borne and foliar pathogens. CSA-active microbes, such as *Trichoderma spp.*, improve soil structure and nutrient cycling by decomposing organic matter, while antibiosis-active microbes suppress pathogens without harming non-target organisms. This combination ensures that crops remain protected from diseases while maintaining the ecological balance of the farming system.

The use of integrated CSA-antibiosis strategies is particularly effective in managing persistent soil-borne pathogens in sustainable farming systems. For example, in organic tomato production, the co-application of compost inoculated with *Trichoderma spp.* and biopesticides containing *Bacillus subtilis* significantly reduced *Fusarium* wilt incidence and improved overall plant health. Such approaches not only suppress pathogens but also enhance the microbial diversity and functionality of the soil, leading to long-term benefits for agricultural productivity.

Case Study:

In rice cultivation, integrating *Trichoderma harzianum* for CSA with *Bacillus subtilis* for antibiosis reduced sheath blight severity by 40% while improving nitrogen uptake and yield in organic systems.

Enhanced Biocontrol Formulations

Modern advancements in biopesticide development have focused on combining CSA- and antibiosis-active strains to create multi-faceted formulations that address a broader spectrum of plant pathogens. These formulations capitalize on the complementary strengths of both mechanisms, providing robust and reliable pathogen suppression under diverse environmental conditions.

1. **Multi-Component Biopesticides**
   - Biopesticides that integrate CSA-active fungi like *Trichoderma spp.* with antibiosis-producing bacteria like *Pseudomonas fluorescens* or *Bacillus subtilis* offer synergistic effects.
   - Such formulations are designed to suppress multiple pathogens simultaneously, reducing the need for separate applications of different control agents.
2. **Broad-Spectrum Disease Management**
   - The integration of CSA and antibiosis ensures that both resource competition and direct pathogen inhibition are addressed. This dual action is particularly effective in managing complex diseases involving multiple pathogens.
3. **Enhanced Field Performance**
   - Combined formulations are more resilient to environmental variability, as the CSA mechanism provides a stable foundation for pathogen suppression, while antibiosis offers rapid and targeted control.

Example:

In greenhouse cucumber production, a biopesticide combining *Trichoderma viride* (CSA) and *Pseudomonas fluorescens* (antibiosis) reduced damping-off caused by *Pythium ultimum* by 65%, outperforming single-strain formulations.

Long-Term Benefits of Integration

The integration of CSA and antibiosis not only improves immediate disease control but also contributes to long-term agricultural sustainability. By reducing the reliance on chemical fungicides, these mechanisms help prevent the development of pathogen resistance, preserve beneficial soil microbiota, and minimize environmental pollution. Moreover, the dual action of CSA and antibiosis promotes healthier and more resilient cropping systems, enabling farmers to achieve higher yields and better-quality produce.

5. Induced Resistance and Mycorrhizal Associations

The complex interactions between plants and beneficial organisms underpin many strategies in sustainable agriculture, particularly in the context of biological control. Two critical mechanisms that contribute significantly to plant health and disease management are **induced resistance** and **mycorrhizal associations**. These mechanisms leverage natural processes to enhance plant immunity, suppress pathogens, and improve overall plant growth, creating a balanced and productive agricultural ecosystem.

5.1 Induced Resistance

Induced resistance refers to the enhancement of a plant's innate defense mechanisms in response to specific stimuli, such as interactions with beneficial microbes or environmental signals. This heightened state of defense allows plants to better resist pathogens, pests, and even abiotic stressors. The process is often categorized into two major types: systemic acquired resistance (SAR) and induced systemic resistance (ISR).

Mechanisms of Induced Resistance

Induced resistance involves a cascade of molecular and biochemical events that prime the plant's immune system to respond more effectively to potential threats.

Systemic Acquired Resistance (SAR)

SAR is triggered by localized pathogen attacks and involves the systemic activation of plant defense pathways. The accumulation of **salicylic acid (SA)** plays a pivotal role in SAR, signaling the production of pathogenesis-related (PR) proteins, phytoalexins, and other antimicrobial compounds. These defense molecules spread throughout the plant, preparing distant tissues to fend off future attacks.

For example, plants exposed to a mild infection by *Pseudomonas syringae* exhibit SAR, enabling them to resist subsequent infections more effectively. The presence of PR proteins in distal tissues has been shown to reduce the severity of diseases such as powdery mildew and rusts in crops like wheat and rice.

Induced Systemic Resistance (ISR)

ISR, on the other hand, is typically mediated by beneficial microbes such as *Pseudomonas fluorescens* and *Bacillus subtilis*. These microbes activate the plant's defense pathways via **jasmonic acid (JA)** and **ethylene (ET)** signaling, enhancing the plant's resistance to both biotic and abiotic stresses. Unlike SAR, ISR does not rely on PR protein production but instead primes the plant to respond faster and more robustly to pathogen attacks.

For instance, *Pseudomonas fluorescens* colonizing the rhizosphere of tomato plants has been shown to induce ISR, reducing the incidence of *Fusarium* wilt by 50%.

Cross-Talk Between SAR and ISR

While SAR and ISR are distinct pathways, they often interact, creating a robust and multilayered defense system. Cross-talk between SA, JA, and ET signaling pathways enables plants to tailor their responses to specific pathogens, ensuring optimal protection without unnecessary metabolic costs.

Applications of Induced Resistance

The use of beneficial microbes to trigger induced resistance has broad applications in agriculture.

1. **Seed and Root Treatments**
   - Treating seeds with ISR-inducing microbes like *Bacillus subtilis* enhances early-stage resistance, protecting seedlings from soil-borne pathogens.
2. **Biocontrol Formulations**
   - Commercial products such as Serenade® (based on *Bacillus subtilis*) leverage ISR to manage foliar and root diseases in a wide range of crops.
3. **Integrated Disease Management (IDM)**
   - Induced resistance is a key component of IDM strategies, complementing other biological, cultural, and chemical controls.

Case Study: In cucumber production, inoculation with *Pseudomonas fluorescens* reduced bacterial wilt by 40%, primarily through ISR activation.

5.2 Mycorrhizal Associations

Mycorrhizal fungi form mutualistic relationships with plant roots, significantly enhancing plant health and resilience. These associations involve the exchange of nutrients, signaling molecules, and protective factors between the fungus and the plant, creating a symbiotic interaction that benefits both partners.

Types of Mycorrhizal Associations

The two primary types of mycorrhizal associations are **arbuscular mycorrhizae (AM)** and **ectomycorrhizae (ECM)**, each with unique roles in supporting plant health.

Arbuscular Mycorrhizae (AM)

AM fungi, such as those in the genus *Glomus*, penetrate plant root cells, forming specialized structures called arbuscules. These structures facilitate the exchange of nutrients and signaling molecules.

» **Nutrient Uptake**: AM fungi significantly enhance the uptake of phosphorus, nitrogen, and micronutrients like zinc and copper, which are often limited in agricultural soils.

» **Pathogen Suppression**: By improving root health and altering rhizosphere microbial communities, AM fungi reduce the incidence of root diseases caused by pathogens like *Fusarium* and *Pythium*.

Ectomycorrhizae (ECM)

ECM fungi, commonly associated with trees, form a sheath around the root surface and extend their hyphae into the soil.

- » **Nutrient Foraging**: ECM fungi enhance the absorption of nitrogen and potassium, particularly in forest and perennial systems.
- » **Drought Tolerance**: ECM associations improve water uptake during dry conditions, enhancing plant resilience to drought stress.

Case Study:

In maize cultivation, inoculation with *Glomus intraradices* increased phosphorus uptake by 30% while reducing the incidence of *Fusarium* root rot by 50%.

Mechanisms of Mycorrhizal Benefits

1. **Nutrient Exchange:**Mycorrhizal fungi extend their hyphal networks far beyond the root zone, accessing nutrient pools unavailable to plant roots. In return, plants supply the fungi with carbohydrates derived from photosynthesis.
2. **Enhanced Root Development**: Mycorrhizal associations stimulate root branching and elongation, increasing the plant's absorptive surface area.
3. **Induction of Plant Defense**: Mycorrhizal fungi activate plant defense pathways, similar to ISR, enhancing resistance to both biotic and abiotic stressors.
4. **Modification of Rhizosphere Microbial Communities**

By altering the composition of the rhizosphere microbiome, mycorrhizal fungi create a hostile environment for pathogens, further reducing disease incidence.

Applications of Mycorrhizal Associations

1. **Sustainable Fertilizer Use:** Mycorrhizal inoculants are increasingly used as biofertilizers to reduce dependency on chemical phosphorus and nitrogen inputs.
2. **Disease Management:** The ability of mycorrhizal fungi to suppress root pathogens makes them valuable in managing soil-borne diseases in crops like tomatoes, cucumbers, and wheat.
3. **Stress Tolerance**: Mycorrhizal associations enhance plant tolerance to drought, salinity, and heavy metal toxicity, broadening their applicability in marginal agricultural lands.

Case Study:

In organic tomato farming, AM fungi reduced the severity of *Verticillium* wilt by 60% while improving fruit yield by 20%.

Integration of Induced Resistance and Mycorrhizal Associations

The integration of induced resistance and mycorrhizal associations offers a multi-layered approach to plant health management. While induced resistance primes the plant's immune system to respond effectively to pathogen attacks, mycorrhizal fungi enhance nutrient availability, stress tolerance, and pathogen suppression. Together, these mechanisms create a robust and resilient cropping system capable of withstanding biotic and abiotic challenges.

Example:

In wheat cropping systems, the combination of *Pseudomonas fluorescens* (induced resistance) and *Glomus mosseae* (mycorrhizal association) reduced root rot incidence by 45% and increased yield by 25%.

6. Operational Mechanisms in Biological Control and Their Relevance

Biological control represents a sustainable alternative to chemical interventions in agriculture, utilizing living organisms or their derivatives to suppress pathogens and promote plant health. The success of biological control relies heavily on the operational mechanisms employed by biocontrol agents (BCAs), which include direct interactions such as mycoparasitism, antibiosis, and competition, as well as indirect mechanisms like induced resistance and nutrient cycling. Each mechanism plays a distinct role in managing plant diseases and enhancing crop productivity, making them integral to the development of robust agricultural systems.

6.1 Direct Mechanisms

Direct mechanisms are at the forefront of biological control, involving active suppression of pathogens through physical, biochemical, or enzymatic interactions. These mechanisms target the pathogens directly, inhibiting their growth, reproduction, or ability to infect plant hosts.

6.1.1 Mycoparasitism

Mycoparasitism is a specialized antagonistic interaction in which one fungus parasitizes another. It is an important operational mechanism employed by fungal BCAs such as *Trichoderma spp.*, which are known for their ability to suppress a wide range of soil-borne pathogens. Mycoparasitism involves a series of steps, including recognition, attachment, enzymatic degradation of the pathogen's cell wall, and nutrient extraction. During mycoparasitism, *Trichoderma spp.* recognize specific surface molecules on the target pathogen and attach to its hyphae. This is followed by the secretion of hydrolytic enzymes such as chitinases, glucanases, and proteases, which break down the structural components of the pathogen's cell walls. The enzymatic degradation not only kills the pathogen but also releases nutrients that the mycoparasitic fungus can absorb, further strengthening its competitive advantage.

The ecological benefits of mycoparasitism extend beyond immediate pathogen suppression. By reducing the population of harmful fungi in the soil, mycoparasitic BCAs limit the availability of inoculum, thus decreasing the likelihood of future infections. For example, *Trichoderma harzianum* has been shown to parasitize *Rhizoctonia solani*, significantly reducing the incidence of root rot in tomatoes by up to 70% in field conditions. The dual role of mycoparasitism in pathogen suppression and nutrient recycling makes it a cornerstone of fungal biological control strategies.

6.1.2 Antibiosis

Antibiosis is another critical direct mechanism of biological control, characterized by the production of bioactive compounds that inhibit or kill pathogens. These compounds include antibiotics, lytic enzymes, toxins, and volatile organic compounds (VOCs). Antibiosis is particularly effective in the rhizosphere, where microbial interactions are dense, and competition for resources is high.

Beneficial bacteria such as *Bacillus subtilis* and *Pseudomonas fluorescens* are prolific producers of antibiotics. For instance, *Bacillus subtilis* synthesizes iturin, a lipopeptide that disrupts fungal membranes, causing leakage of cellular contents and subsequent cell death. Similarly, *Pseudomonas fluorescens* produces phenazines, which interfere with fungal respiration by targeting the electron transport chain. These antibiotics not only suppress pathogen populations but also confer an ecological advantage to the producing microbes, allowing them to dominate the microbial community in the rhizosphere.

Lytic enzymes play a complementary role in antibiosis by degrading the structural components of pathogens. Chitinases and glucanases, for example, break down chitin and glucans in fungal cell walls, weakening the pathogen's defenses and rendering it susceptible to further microbial attacks.

VOCs represent a unique dimension of antibiosis, as they can act at a distance, diffusing through soil pores to inhibit pathogens that are not in direct contact with the BCA. Hydrogen cyanide, produced by certain strains of *Pseudomonas spp.*, is a well-documented VOC that suppresses root pathogens like *Fusarium oxysporum*. These gaseous compounds add versatility to antibiosis, enabling BCAs to influence pathogen populations beyond their immediate vicinity. Antibiosis is widely applied in agricultural systems through commercial biopesticides. Products like Serenade® (based on *Bacillus subtilis*) and Agrilife® (based on *Pseudomonas fluorescens*) utilize antibiosis to control a range of diseases in high-value crops. In cucumber cultivation, for instance, *Bacillus subtilis* reduced damping-off caused by *Pythium ultimum* by producing iturin and surfactin, leading to healthier plants and improved yields.

6.1.3 Competition

Competition for resources is a fundamental ecological interaction that plays a significant role in biological control. BCAs compete with pathogens for nutrients, space, and ecological niches, effectively suppressing their growth and reproduction. This mechanism is particularly prominent in the rhizosphere, where microbial communities interact intensively with plant roots and each other. One of the primary strategies employed by BCAs in competitive interactions is the production of siderophores. These iron-chelating molecules bind to iron in the soil, making it unavailable to pathogens that rely on this essential micronutrient for their metabolic processes. Beneficial microbes such as *Pseudomonas fluorescens* and *Bacillus spp.* are known for their ability to sequester iron efficiently, thereby depriving pathogens like *Fusarium oxysporum* of a critical resource.

In addition to nutrient competition, BCAs also compete for physical space on plant roots and in the soil. Rapid colonization of root surfaces by BCAs such as *Trichoderma spp.* creates a physical barrier that prevents pathogens from accessing infection sites. This competitive exclusion is further reinforced by the formation of biofilms—structured microbial communities that provide a stable environment for beneficial microbes while excluding pathogens. The competitive saprophytic ability (CSA) of certain BCAs, such as *Trichoderma spp.*, adds another layer of effectiveness. These microbes efficiently colonize and utilize organic matter in the soil, outcompeting pathogens during their saprophytic phase. By dominating the decomposition of crop residues, CSA-active BCAs reduce the inoculum potential of pathogens, limiting their ability to infect plants in subsequent cropping cycles.

6.2 Indirect Mechanisms

Indirect mechanisms enhance plant health and resilience, enabling plants to withstand pathogen attacks and environmental stresses. These mechanisms often involve interactions between BCAs and plants that activate or reinforce the plant's innate defense systems.

6.2.1 Induced Resistance

Induced resistance is a process in which BCAs stimulate the plant's immune system, priming it to respond more effectively to pathogen attacks. There are two main types of induced resistance: systemic acquired resistance (SAR) and induced systemic resistance (ISR). SAR is typically triggered by localized pathogen attacks and involves the accumulation of salicylic acid, which signals the production of pathogenesis-related (PR) proteins and other defense compounds. ISR, on the other hand, is mediated by beneficial microbes like *Pseudomonas fluorescens* and *Bacillus subtilis*. These microbes activate defense pathways via jasmonic acid and ethylene signaling, enhancing the plant's resistance to both biotic and abiotic stressors. The integration of SAR and ISR creates a multilayered defense system, offering broad-spectrum protection against pathogens. For example, in cucumber plants, inoculation with *Pseudomonas fluorescens* induced ISR, reducing bacterial wilt by 40%. These mechanisms are particularly valuable in integrated disease management (IDM) programs, where they complement other control strategies to provide comprehensive disease suppression.

6.2.2 Nutrient Cycling and Plant Growth Promotion

BCAs contribute to nutrient cycling by decomposing organic matter and making nutrients available to plants. Fungi such as *Trichoderma spp.* and bacteria like *Azospirillum spp.* solubilize phosphorus, fix atmospheric nitrogen, and produce growth-promoting hormones like auxins and gibberellins. These activities not only enhance plant growth but also improve overall soil health, creating an environment that is less conducive to pathogen proliferation. Mycorrhizal associations also play a significant role in nutrient cycling and plant growth promotion. Arbuscular mycorrhizal fungi (AMF) form symbiotic relationships with plant roots, facilitating the uptake of phosphorus, nitrogen, and other nutrients. By improving root health and altering rhizosphere microbial communities, AMF indirectly suppress pathogens and enhance the resilience of plants to environmental stresses.

6.3 Relevance of Operational Mechanisms

The operational mechanisms of biological control are highly relevant to modern agriculture, particularly in the context of sustainable farming systems. By reducing reliance on chemical inputs, these mechanisms contribute to environmental conservation, food safety, and long-term soil health.

The integration of direct and indirect mechanisms into biocontrol strategies ensures comprehensive pathogen suppression while promoting plant health. For example, combining mycoparasitic fungi like *Trichoderma spp.* with ISR-inducing bacteria like *Bacillus subtilis* provides multi-faceted disease management, addressing both pathogen inhibition and plant resilience. The relevance of these mechanisms extends beyond disease control. By enhancing soil fertility and microbial diversity, biological control agents contribute to the overall sustainability of agricultural systems. This holistic approach aligns with global efforts to promote eco-friendly farming practices and ensure food security in the face of growing environmental challenges.

Chapter - 3

Factors Governing Biological Control

The success of biological control is influenced by a multitude of factors that govern the interactions between biological control agents, target pests or pathogens, and their surrounding environment. These factors encompass biological, ecological, environmental, and operational dimensions, each of which plays a critical role in determining the effectiveness, sustainability, and scalability of biocontrol interventions. At the biological level, the characteristics of the biocontrol agent itself are paramount. The agent must exhibit a high degree of specificity toward the target pest or pathogen to avoid non-target effects, while also demonstrating efficacy under diverse environmental conditions. For instance, the predatory insect *Cryptolaemus montrouzieri* (commonly known as the mealybug destroyer) is highly effective against mealybug infestations due to its strong predation specificity. Similarly, microbial agents like *Trichoderma spp.* and *Bacillus subtilis* are valued for their dual functionality—suppressing pathogens and promoting plant growth—making them versatile components in biological control programs. The reproductive capacity, life cycle synchronization, and adaptability of the biocontrol agent are also critical factors. Agents with rapid reproductive rates and life cycles closely aligned with those of the target pest are more likely to establish and exert sustained control over pest populations.

Ecological factors further shape the outcomes of biological control by influencing the interactions between biocontrol agents, target organisms, and the broader community of organisms within the ecosystem. The availability of resources, such as food, habitat, and environmental niches, determines the establishment and persistence of biocontrol agents. For example, the presence of flowering plants that provide nectar and pollen can enhance the survival and efficacy of parasitoids like *Encarsia formosa* and predators like ladybird beetles, which rely on these resources during their life cycles. The composition and diversity of the microbial or insect community also affect the efficacy of biocontrol. High biodiversity in the rhizosphere or crop canopy often creates competitive environments that suppress pathogens or pests, amplifying the effects of biocontrol agents. However, the presence of competing organisms or secondary predators can sometimes hinder the success of biocontrol programs by preying on or outcompeting the biocontrol agents themselves. For instance, intraguild predation, where natural enemies feed on one another, can reduce the overall effectiveness of biological control, necessitating careful management of ecological interactions.

Environmental factors, including abiotic conditions such as temperature, humidity, and soil pH, play a crucial role in determining the viability and performance of biocontrol agents. Many biocontrol agents have specific environmental requirements for optimal activity. For instance, fungal pathogens like *Beauveria bassiana* and *Metarhizium anisopliae*, which are used to control insect pests, thrive in moderate temperatures and high humidity but may be less effective under arid conditions. Similarly, the effectiveness of microbial agents like *Pseudomonas fluorescens* depends on soil pH and moisture levels, as these factors influence their colonization and antagonistic activity. Climatic variability, including extreme weather events, can disrupt biocontrol programs by altering the dynamics of pest populations and the survival of biocontrol agents. For example, prolonged droughts or heavy rains can adversely impact the population densities of natural enemies, reducing their capacity to regulate pests effectively. As a result, understanding and managing the environmental requirements of biocontrol agents is essential for ensuring their success across diverse agricultural landscapes.

Operational factors, such as the methods of application, timing, and scale of deployment, are equally critical in governing the success of biological control. The effectiveness of biocontrol agents is often contingent on their timely introduction and appropriate density in the field. Releasing insufficient numbers of agents or applying them at suboptimal times can lead to poor establishment and limited impact on target pests. Conversely, excessive release rates may result in resource competition among agents, reducing their effectiveness. For example, augmentative biological control programs involving the periodic release of mass-reared parasitoids like *Trichogramma spp.* must carefully balance release rates to ensure that the agents can effectively parasitize pest eggs without depleting available resources. The method of delivery also influences the success of biocontrol agents. Techniques such as seed coating, soil drenching, or aerial spraying must be tailored to the specific characteristics of the biocontrol agent and the target pest or pathogen. For microbial agents like *Trichoderma spp.* or *Rhizobium spp.*, formulations that enhance their shelf life, ease of application, and compatibility with existing farming practices are crucial for widespread adoption.

Another key operational factor is the integration of biological control into broader pest management frameworks, such as Integrated Pest Management (IPM). Biocontrol agents are most effective when used in conjunction with other pest management strategies, including cultural, physical, and chemical methods. For example, combining crop rotation and intercropping with the release of biocontrol agents can create synergistic effects, improving the overall efficacy of pest suppression. However, the compatibility of biocontrol agents with chemical pesticides must be carefully assessed, as some pesticides may harm natural enemies or disrupt their interactions with target pests. The use of selective or reduced-risk chemicals, which minimize non-target effects, is often recommended to preserve the integrity of biocontrol programs within IPM frameworks.

Lastly, socioeconomic factors, including farmer knowledge, access to resources, and market incentives, play a pivotal role in determining the adoption and success of biological control. Farmer training and awareness programs are essential for ensuring that biocontrol practices are correctly implemented and maintained over time. For example, educating farmers about the identification and preservation of natural enemies, as well as the appropriate timing and methods of releasing biocontrol agents, enhances the likelihood of success. Access to affordable and high-quality biocontrol products is another critical factor, particularly for smallholder farmers in developing regions. Supportive policies and market incentives, such as subsidies for biocontrol products or premium prices for sustainably grown crops, further encourage the adoption of biological control as a viable alternative to chemical-intensive farming.

3.1 Role of Physical Environment in Biocontrol Success

The physical environment is a critical determinant in the success or failure of biological control strategies. It encompasses climatic conditions, soil properties, and other abiotic factors that influence the survival, activity, and efficiency of biological control agents (BCAs). Since biological control relies on living organisms, environmental conditions play a significant role in shaping their growth, metabolism, and interactions with both pathogens and host plants. Understanding how these factors affect BCAs is crucial for optimizing their application and ensuring consistent results across diverse agricultural systems. Environmental variables such as temperature, humidity, soil characteristics, and light intensity directly affect the behaviour and performance of BCAs. Suboptimal conditions can weaken the efficacy of biocontrol agents, while favourable environments enhance their ability to suppress pathogens and promote plant health. Therefore, tailoring biocontrol strategies to align with the prevailing environmental conditions is a key component of successful agricultural disease management.

3.1.1 Temperature

Temperature is one of the most influential factors affecting the performance of BCAs. As microbial activity is temperature-dependent, fluctuations in temperature can alter the growth, metabolic rate, and persistence of beneficial organisms. Each biocontrol agent operates within an optimal temperature range, and deviations from this range can significantly reduce its effectiveness. Understanding the interactions between temperature and microbial activity is essential for maximizing the efficacy of biological control.

Impact on Biocontrol Agents

Temperature directly influences the physiological processes of BCAs, including their ability to grow, reproduce, and produce secondary metabolites. Beneficial microbes such as *Trichoderma spp.* and *Pseudomonas fluorescens* thrive within specific temperature ranges, where their antagonistic activities are most pronounced. For instance, *Trichoderma harzianum* demonstrates peak activity against pathogens like *Fusarium oxysporum* when temperatures are maintained between 25°C and 30°C. Outside this range, the growth and metabolic functions of the fungus decline, weakening its ability to suppress pathogens effectively. In colder conditions, microbial activity slows down, reducing the efficacy of BCAs. Conversely, excessively high temperatures can denature enzymes and disrupt microbial cell functions, leading to reduced survival rates. For example, *Bacillus subtilis* exhibits better resilience in cooler climates compared to heat-sensitive species like *Pseudomonas fluorescens*. As a result, selecting temperature-adapted BCAs is vital for specific climatic zones.

Pathogen-Temperature Interactions

Temperature also modulates the behaviour of pathogens, influencing their aggressiveness and ability to infect plants. In warmer climates, pathogens such as *Rhizoctonia solani* and *Sclerotium rolfsii* become more virulent, often outpacing the suppressive capacity of non-thermotolerant BCAs. In such cases, the deployment of thermotolerant BCAs like certain strains of *Trichoderma spp.* or *Bacillus spp.* is necessary to counteract the increased aggressiveness of pathogens. In cooler climates, pathogens such as *Pythium ultimum* thrive, necessitating the use of BCAs adapted to low-temperature environments.

Pseudomonas spp. are particularly effective in these conditions, as they maintain their activity and colonization ability even under cooler soil temperatures. The dynamic interplay between temperature and pathogen activity highlights the importance of selecting BCAs that are compatible with the thermal conditions of the target environment.

Strategies to Address Temperature Sensitivity

To overcome temperature-related challenges, several strategies have been developed to improve the resilience and efficacy of BCAs. One approach involves the development of temperature-tolerant strains through genetic selection or microbial engineering. These strains are capable of maintaining their suppressive activities under a broader range of temperatures, making them suitable for diverse agroecological zones.

Adjusting the timing of BCA applications is another practical strategy. By aligning application schedules with periods of optimal temperature, farmers can enhance the colonization and activity of BCAs. For example, applying *Pseudomonas fluorescens* during cooler evening hours in hot climates minimizes heat stress on the bacteria, improving their persistence and efficacy. Formulation improvements also play a significant role in addressing temperature sensitivity. Encapsulation techniques and protective carriers help shield BCAs from temperature extremes, ensuring their viability and performance during storage and field application. These formulations extend the shelf life of BCAs and enhance their resilience against fluctuating environmental conditions.

Example

In wheat cultivation, the application of *Pseudomonas fluorescens* at optimal soil temperatures of 20–25°C resulted in successful suppression of *Fusarium* root rot. The bacteria exhibited robust colonization of the rhizosphere and high antagonistic activity against the pathogen. However, when applied at lower temperatures, the colonization ability and pathogen suppression efficacy of *Pseudomonas fluorescens* were significantly reduced, underscoring the importance of temperature optimization in biocontrol strategies.

Significance of Temperature in Biocontrol

Temperature serves as a foundational parameter in biological control, influencing not only the survival and activity of BCAs but also their interactions with pathogens and host plants. By understanding the temperature preferences of BCAs and pathogens, farmers and researchers can design more effective application strategies, tailored to the specific climatic conditions of their agricultural systems. Furthermore, advances in strain development and formulation technologies are expanding the temperature adaptability of BCAs, paving the way for more versatile and resilient biological control solutions. Temperature remains one of the most critical factors for ensuring the success of biological control in diverse agroecosystems. Its careful management can significantly enhance the consistency and reliability of biocontrol interventions, making them an indispensable tool in sustainable agriculture.

3.1.2 Soil Properties

Soil properties are among the most critical factors influencing the success of biological control agents (BCAs). Characteristics such as pH, texture, organic matter content, and moisture levels play significant roles in determining the survival, activity, and efficacy of these beneficial microbes. Since soil is the primary habitat for most BCAs, understanding and optimizing these properties is crucial for enhancing their performance in agricultural systems.

Soil pH

Soil pH is a fundamental factor that affects the growth and metabolic activity of BCAs. Most beneficial microbes, including *Trichoderma spp.* and *Pseudomonas fluorescens*, exhibit optimal activity in slightly acidic to neutral soils, with a pH range of 6.0–7.5. In this range, microbial enzymes function efficiently, facilitating pathogen suppression through mechanisms such as mycoparasitism, antibiosis, and competition. However, extreme pH levels can adversely affect the viability and antagonistic potential of BCAs. In acidic soils (pH < 6), the activity of microbial enzymes such as chitinases and glucanases is often inhibited, reducing the ability of BCAs to degrade the cell walls of fungal pathogens. Conversely, in alkaline soils (pH > 7.5), the production of secondary metabolites, including antibiotics and siderophores, may decline. For example, *Pseudomonas fluorescens* shows reduced siderophore production in high-pH soils, limiting its ability to compete with pathogens for essential nutrients like iron. Adjusting soil pH through amendments such as lime (for acidic soils) or sulphur (for alkaline soils) can improve the efficacy of BCAs in suboptimal conditions.

Organic Matter

The organic matter content of soil is another critical determinant of BCA activity. Organic matter serves as a nutrient reservoir, providing carbon, nitrogen, and other essential elements that support the proliferation of beneficial microbes. High organic matter content enhances the competitive saprophytic ability (CSA) of BCAs, such as *Trichoderma spp.*, enabling them to dominate the saprophytic niche and suppress soil-borne pathogens effectively. In addition to serving as a nutrient source, organic matter improves soil structure and water-holding capacity, creating a more favourable environment for microbial colonization. Crop residues, compost, and manure are common sources of organic matter that can be incorporated into soils to promote BCA activity. For instance, the addition of compost inoculated with *Trichoderma spp.* has been shown to reduce the inoculum potential of *Fusarium oxysporum* in vegetable cropping systems, highlighting the importance of organic matter in enhancing biocontrol efficacy.

Soil Moisture and Aeration

Soil moisture and aeration are critical abiotic factors influencing the mobility, activity, and persistence of BCAs. Proper soil moisture ensures that beneficial microbes can access the rhizosphere, where they establish symbiotic relationships with plant roots and compete with pathogens. However, excessive moisture can lead to oxygen depletion, creating anaerobic conditions that are unfavourable for aerobic BCAs such as *Pseudomonas fluorescens*. Under such conditions, the survival and efficacy of these microbe's decline, reducing their ability to suppress pathogens. Conversely, drought conditions can desiccate microbial populations, particularly those that lack protective structures or formulations. To mitigate these challenges, researchers have developed moisture-retaining formulations and carriers, such as hydrogels, that protect BCAs from desiccation. Adjusting irrigation schedules to maintain optimal soil moisture levels further enhances the activity of BCAs in arid and semi-arid environments.

Example

In maize fields with high organic matter content, the application of *Trichoderma harzianum* reduced infections by *Pythium ultimum* by 50%. This success highlights the importance of favorable soil conditions, including optimal pH, organic matter, and moisture levels, in achieving effective biological control.

3.1.3 Humidity

Humidity is another vital environmental factor influencing the efficacy of biological control, particularly in the context of foliar applications. The interaction between BCAs and their target pathogens is significantly affected by the moisture levels in the environment, as high or low humidity can alter microbial survival, pathogen dynamics, and the effectiveness of biocontrol formulations.

Impact on Foliar BCAs

Foliar BCAs such as *Bacillus subtilis* and *Pseudomonas fluorescens* depend on optimal humidity levels to maintain their activity on leaf surfaces. Moderate humidity conditions are ideal for these microbes, as they provide sufficient moisture to support microbial colonization and activity without washing off the formulations. In excessively humid environments, however, BCAs may struggle to adhere to leaf surfaces, leading to reduced persistence and efficacy. Additionally, high humidity levels often favor the proliferation of foliar pathogens such as *Alternaria* and *Phytophthora*, further complicating disease management efforts. On the other hand, low humidity levels can desiccate foliar BCAs, reducing their viability and effectiveness. This challenge is particularly relevant in arid regions or during dry seasons, where supplemental irrigation or misting may be required to create favorable conditions for biocontrol applications.

Humidity-Based Timing

The timing of BCA applications is critical for optimizing their performance under varying humidity conditions. In greenhouse environments, where humidity levels can be controlled, maintaining moderate humidity enhances the persistence of foliar BCAs and improves their ability to suppress pathogens. For example, *Bacillus subtilis* sprays are more effective against *Botrytis cinerea* when applied during periods of optimal humidity, as this minimizes desiccation and maximizes microbial activity. Field applications require more strategic timing to align with favorable environmental conditions. Early morning or late evening applications are often recommended, as these periods typically exhibit higher humidity levels and reduced evaporation rates, ensuring better adherence and survival of BCAs on plant surfaces.

Example

In strawberry cultivation, applications of *Bacillus subtilis* under moderate humidity conditions reduced the incidence of gray mold caused by *Botrytis cinerea* by 60%. In contrast, applications during high humidity conditions led to inconsistent results, underscoring the importance of optimizing humidity levels for effective biocontrol.

3.1.4 Light Intensity

Light intensity is a significant environmental factor that influences the activity and persistence of biological control agents (BCAs), particularly those applied as foliar sprays. While light is essential for plant growth and microbial photosynthesis, certain components of light, such as ultraviolet (UV) radiation, can negatively impact the viability and effectiveness of BCAs. Understanding how light intensity affects microbial formulations is critical for optimizing their application and ensuring consistent efficacy in field conditions.

UV Sensitivity

UV radiation, particularly UV-B and UV-C wavelengths, can damage microbial cells by disrupting their DNA and cellular structures. Many BCAs, including *Bacillus subtilis* and *Pseudomonas fluorescens*, are highly sensitive to UV light, which can degrade their formulations and reduce their persistence on leaf surfaces. As a result, the efficacy of these BCAs in suppressing pathogens declines when applied under intense sunlight or during peak daylight hours. The vulnerability of BCAs to UV light poses a challenge for foliar applications, where exposure to sun*light is inevitable. The degradation of microbial populations reduces their ability to coloniz*e leaf surfaces and produce the antimicrobial compounds needed to suppress pathogens effectively. For example, *Bacillus subtilis* loses significant viability when exposed to prolonged UV radiation, limiting its capacity to control diseases such as *Botrytis cinerea* in greenhouse and field environments.

Photoprotection Strategies

To mitigate the impact of UV radiation on BCAs, several photoprotection strategies have been developed. One of the most effective approaches is the use of UV-protective carriers in microbial formulations. These carriers include natural and synthetic materials that shield BCAs from harmful UV rays, enhancing their stability and persistence in field conditions. For instance, alginate and starch-based carriers have been successfully incorporated into biocontrol formulations to protect microbial cells from UV-induced damage.

Timing the application of BCAs to avoid peak sunlight hours is another practical strategy. Early morning or late evening applications minimize UV exposure, allowing BCAs to establish themselves on leaf surfaces before the intensity of sunlight increases. This approach has been particularly effective in improving the persistence of *Pseudomonas fluorescens*, which thrive under low-light conditions. Advances in formulation technology have also introduced encapsulation techniques that embed BCAs in protective microcapsules. These encapsulations not only shield microbes from UV radiation but also provide controlled release of the active agents, ensuring prolonged activity against pathogens.

Example:

In tomato fields, early morning applications of *Pseudomonas fluorescens* enhanced its persistence on leaf surfaces and improved its efficacy in controlling bacterial spot caused by *Xanthomonas campestris*. The reduced UV exposure during early hours allowed the bacterium to establish itself and produce antimicrobial compounds before sunlight intensity increased.

3.1.5 Wind and Rainfall

Wind and rainfall are critical environmental factors that influence the distribution, persistence, and overall effectiveness of BCAs in the field. These abiotic factors can disrupt the uniform application of microbial agents and reduce their concentration in target areas, posing challenges for the consistent implementation of biological control strategies.

Impact of Wind

High wind speeds can disperse BCAs unevenly during application, reducing their effectiveness in suppressing pathogens. Wind not only affects the distribution of foliar-applied BCAs but also increases the likelihood of their drift to non-target areas, potentially leading to reduced pathogen suppression in the intended zones. For instance, during aerial or mechanized spray applications, wind can cause significant loss of microbial formulations, decreasing their field efficacy. In soil applications, wind-induced drying can desiccate surface-applied BCAs, especially in arid or semi-arid environments. This desiccation reduces the survival and activity of microbes such as *Trichoderma spp.* and *Bacillus subtilis*, limiting their ability to colonize the rhizosphere and suppress soil-borne pathogens.

Impact of Rainfall

Rainfall poses another significant challenge to the effectiveness of BCAs, particularly those applied to foliar surfaces. Heavy rain can wash away microbial formulations, reducing their concentration on leaves and stems. This runoff not only decreases the efficacy of biocontrol agents but also necessitates repeated applications, increasing labor and input costs. In soil applications, excessive rainfall can lead to waterlogging, creating anaerobic conditions that are unfavorable for aerobic BCAs like *Pseudomonas fluorescens*. Conversely, light rainfall or irrigation can enhance the mobility and distribution of BCAs in the soil, improving their colonization of plant roots and interaction with pathogens.

Mitigation Strategies

To counteract the challenges posed by wind and rainfall, several mitigation strategies have been developed:

1. **Adhesive Agents in Foliar Applications**: Incorporating adhesive agents into biocontrol formulations helps BCAs adhere to leaf surfaces, reducing the impact of wind and rainfall. These adhesives create a protective film around the microbial cells, ensuring their persistence even under adverse weather conditions.
2. **Soil Incorporation**: For soil-applied BCAs, incorporating them into the soil rather than applying them on the surface minimizes the risk of runoff and desiccation. This approach enhances the survival of BCAs and their ability to colonize the rhizosphere.
3. **Precision Application Techniques**: Advances in precision agriculture, such as drone-based application, allow for more accurate and uniform distribution of BCAs in the field. These technologies account for wind speed and direction, ensuring that microbial formulations reach the intended target areas with minimal loss.
4. **Timing Adjustments**: Timing applications to coincide with favorable weather conditions, such as low wind speeds and light rainfall, improves the effectiveness of BCAs. For instance, applying microbial agents just before predicted light rainfall can enhance their soil mobility without the risk of runoff.

Example:

In rice cultivation, incorporating *Trichoderma harzianum* into soil compost reduced the impact of heavy rainfall, maintaining its effectiveness in suppressing sheath blight by 40%. Similarly, the use of adhesive agents in foliar applications of *Bacillus subtilis* improved its persistence during moderate rain events, reducing disease incidence in vegetable crops.

3.2 Influence of Agroecosystem on Pathogen and Antagonist Dynamics

Agroecosystems, the managed ecosystems designed for agricultural production, are complex environments where biological, chemical, and physical components interact. These interactions significantly influence the dynamics between pathogens and antagonists, shaping the outcomes of plant disease management strategies. The structure and management of agroecosystems determine the availability of resources, diversity of microbial communities, and resilience of plants, which collectively influence the effectiveness of biological control agents (BCAs) in suppressing pathogens. Understanding how agroecosystem factors affect these interactions is essential for optimizing biocontrol strategies and ensuring sustainable agricultural productivity.

3.2.1 Structure of Agroecosystems and Pathogen Dynamics

The structure of an agroecosystem encompasses its spatial configuration, crop diversity, and resource distribution. These elements create the ecological context within which pathogens thrive or are suppressed.

Monoculture Systems and Pathogen Proliferation

Monoculture, the continuous cultivation of a single crop, is a common agricultural practice that creates favorable conditions for specialized pathogens. These systems provide a consistent host base, allowing pathogens to proliferate without ecological barriers. For example, in continuous wheat cultivation, pathogens such as *Fusarium spp.* accumulate in the soil, leading to severe outbreaks of Fusarium head blight. The lack of crop diversity eliminates natural disruptions to pathogen life cycles, making monoculture systems highly vulnerable to diseases.

The persistence of inoculum in monoculture systems further exacerbates disease problems. Pathogens overwinter in crop residues, soil, or seeds, maintaining a consistent presence in the field. Over time, the absence of crop rotation allows pathogens to adapt to the monoculture environment, increasing their virulence and resistance to control measures.

Crop Diversification and Disease Suppression

Crop diversification strategies, such as intercropping and crop rotation, play a critical role in disrupting pathogen dynamics. By alternating host and non-host crops, these practices break the continuous availability of a susceptible host, thereby reducing pathogen populations. For example, rotating maize with legumes disrupts the life cycle of *Rhizoctonia solani*, reducing its prevalence in the soil. Intercropping systems enhance microbial diversity in the rhizosphere, fostering antagonistic interactions that suppress pathogens. For instance, intercropping wheat with clover has been shown to increase populations of *Pseudomonas spp.*, which produce siderophores and antibiotics that inhibit soil-borne pathogens. Crop diversification also introduces root exudates from different plants, altering the soil microbiome in ways that favor beneficial microbes over pathogens.

3.2.2 Agroecosystem Management and Antagonist Dynamics

The management practices employed in agroecosystems have a profound impact on the survival, activity, and efficacy of antagonists. Factors such as soil amendments, chemical inputs, and cultural practices influence the availability of resources and ecological niches for BCAs.

Organic Amendments and Antagonist Proliferation

The application of organic amendments, such as compost, manure, and biochar, enhances soil fertility and microbial diversity, creating conditions favorable for BCAs. Organic matter provides a nutrient-rich substrate for the growth and reproduction of beneficial microbes like *Trichoderma spp.* and *Pseudomonas fluorescens*. For example, compost-enriched soils support higher populations of *Trichoderma spp.*, which exhibit strong competitive saprophytic ability (CSA) and suppress pathogens by colonizing organic residues. In addition to serving as a nutrient source, organic amendments improve soil structure and water-holding capacity, which are critical for the survival of antagonists. Biochar, for instance, creates microporous habitats that protect BCAs from desiccation and predation, enhancing their persistence in the soil.

Impact of Chemical Inputs

While agrochemicals such as fertilizers and pesticides are integral to conventional farming, their overuse can disrupt the ecological balance of agroecosystems, negatively affecting antagonists. Excessive nitrogen inputs can stimulate the growth of fast-growing, nutrient-demanding pathogens like *Pythium ultimum*, outcompeting slower-growing BCAs. Similarly, broad-spectrum fungicides may inadvertently suppress beneficial fungi such as *Trichoderma spp.*, reducing their efficacy in biological control. Integrated approaches that combine reduced chemical inputs with biocontrol agents can mitigate these negative effects. For example, the use of low-toxicity fungicides in conjunction with *Bacillus subtilis* enhances disease suppression by targeting pathogens without harming BCAs.

Cultural Practices and Antagonist Activity

Cultural practices such as tillage, irrigation, and residue management influence the spatial distribution and activity of antagonists in agroecosystems.

- **Tillage**: Conventional tillage exposes pathogen propagules to environmental stressors like UV radiation and desiccation, reducing their viability. However, it can also disrupt soil microbial communities, negatively impacting BCAs. Reduced tillage systems preserve microbial habitats, promoting the persistence of BCAs such as *Trichoderma spp.*.
- **Irrigation**: Adequate irrigation enhances the mobility and colonization ability of rhizosphere-associated BCAs. Conversely, excessive irrigation can create waterlogged conditions that inhibit aerobic microbes like *Pseudomonas fluorescens*.
- **Residue Management**: Incorporating crop residues into the soil provides a substrate for BCAs, enhancing their competitive advantage over pathogens during the saprophytic phase.

3.2.3 Interactions Between Pathogens and Antagonists

The dynamics between pathogens and antagonists in agroecosystems are governed by their interactions, which are influenced by resource availability, environmental conditions, and microbial diversity.

Resource Competition

Competition for nutrients, space, and ecological niches is a primary mechanism through which antagonists suppress pathogens. Beneficial microbes such as *Pseudomonas fluorescens* and *Trichoderma spp.* outcompete pathogens for critical resources like iron and carbon, reducing their ability to infect plants. Siderophore production by *Pseudomonas spp.* exemplifies this mechanism, as these compounds sequester iron, depriving pathogens of this essential micronutrient.

Antagonistic Activities

Antagonists employ various mechanisms, such as antibiosis and mycoparasitism, to directly inhibit pathogens. For instance, *Trichoderma spp.* parasitize fungal pathogens by secreting enzymes that degrade their cell walls, while *Bacillus subtilis* produces antibiotics like iturin that disrupt fungal membranes. These activities are enhanced in agroecosystems with high microbial diversity, where synergistic interactions between BCAs amplify their suppressive effects.

Pathogen Adaptation and Resistance

Pathogens in agroecosystems may develop resistance to antagonists over time, particularly in systems with prolonged exposure to a single biocontrol agent. This underscores the importance of rotating BCAs and integrating multiple control strategies to maintain their efficacy. For example, combining *Trichoderma spp.* with *Pseudomonas fluorescens* provides complementary mechanisms of action, reducing the likelihood of resistance development.

3.2.4 Agroecosystem Resilience and Disease Suppression

Resilient agroecosystems are characterized by their ability to suppress pathogens and support antagonists through natural ecological processes. Practices such as increasing plant diversity, minimizing chemical inputs, and enhancing soil organic matter contribute to agroecosystem resilience.

Case Study:

In rice-based agroecosystems, the integration of organic amendments with *Trichoderma harzianum* applications reduced sheath blight incidence by 40% while improving soil microbial diversity. Similarly, in maize-legume intercropping systems, the use of *Pseudomonas fluorescens* enhanced pathogen suppression and increased crop yields.

3.3 Cultural Practices Enhancing Biocontrol Efficacy

Cultural practices are agricultural techniques that influence crop health, soil conditions, and the overall agroecosystem. These practices play a significant role in enhancing the efficacy of biological control agents (BCAs) by creating favorable conditions for their establishment, activity, and interaction with pathogens. Cultural practices such as crop rotation, intercropping, tillage, irrigation, and organic amendments not only contribute to pathogen suppression but also support the proliferation and functionality of BCAs. Integrating these practices with biological control strategies can maximize their impact, making them vital components of sustainable agricultural systems.

3.3.1 Crop Rotation

Crop rotation, a foundational agricultural practice, involves systematically alternating the types of crops grown in a specific field over successive seasons. This practice has proven to be an essential strategy for managing soil health, improving nutrient cycling, and most importantly, disrupting the life cycles of plant pathogens. Crop rotation breaks the continuity of host availability for pathogens, thereby reducing their population densities and enhancing the activity of biological control agents (BCAs). By diversifying cropping systems, farmers create environments that support beneficial microbes while suppressing pathogens, making crop rotation a cornerstone of sustainable disease management.

Pathogen Suppression through Host Disruption

The effectiveness of crop rotation in pathogen suppression lies in its ability to interrupt the host-pathogen relationship. Continuous monoculture systems, where the same crop is grown repeatedly in the same field, provide pathogens with a consistent host base, enabling them to proliferate unchecked. These systems often lead to the build-up of pathogen inoculum in the soil, creating severe disease outbreaks over time.

Crop rotation disrupts this cycle by introducing non-host crops that deprive pathogens of their preferred hosts. For example, alternating cereals like wheat with legumes such as soybean breaks the life cycles of soil-borne pathogens like *Rhizoctonia solani* and *Fusarium oxysporum*. Non-host crops do not support the growth or reproduction of these pathogens, effectively reducing their populations in the soil. Over multiple growing seasons, this reduction in pathogen inoculum leads to decreased disease pressure and healthier crops. The introduction of non-host crops can also induce changes in the soil microbial community that favor pathogen suppression. Non-host plants release root exudates that alter the rhizosphere environment, promoting microbial diversity and encouraging the growth of antagonistic microbes. This microbial shift further enhances the suppression of soil-borne pathogens.

Synergy Between Crop Rotation and Biological Control Agents

Crop rotation not only disrupts pathogen life cycles but also creates favorable conditions for the activity of BCAs. By diversifying cropping systems, rotation increases the availability of diverse root exudates, which serve as carbon sources for beneficial microbes. These exudates support the proliferation of antagonistic organisms such as *Pseudomonas fluorescens* and *Trichoderma spp.*, which suppress pathogens through competition, antibiosis, and induced resistance. Legume-based rotation systems offer an additional benefit by improving soil fertility through nitrogen fixation. Nitrogen-fixing bacteria, such as *Rhizobium spp.*, enhance soil nutrient levels, indirectly supporting the growth and activity of BCAs. For example, in a wheat-soybean rotation, the nitrogen fixed by soybean roots enriches the soil, creating a nutrient-rich environment that supports the proliferation of antagonistic microbes. Furthermore, the diversity of root exudates in rotation systems promotes interactions among microbial communities, enhancing their collective ability to suppress pathogens. These interactions often result in synergistic effects, where the combined activity of multiple BCAs is greater than the sum of their individual effects. This synergy maximizes the suppression of residual pathogens and enhances overall crop health.

Benefits to Soil Health and Long-Term Sustainability

Beyond its immediate effects on pathogen suppression and BCA activity, crop rotation contributes to long-term soil health and sustainability. By alternating crops with different root structures and nutrient demands, rotation prevents soil degradation and improves its physical and chemical properties. For instance, deep-rooted crops like maize can break up compacted soil layers, enhancing aeration and water infiltration, while shallow-rooted crops like beans improve topsoil quality. Improved soil structure and nutrient availability create a stable environment for BCAs, ensuring their persistence and activity across growing seasons. These conditions also enhance the resilience of the agroecosystem, reducing the risk of disease outbreaks and promoting sustainable agricultural productivity.

Example

In wheat cultivation, crop rotation with brassicas has been particularly effective in managing take-all disease caused by *Gaeumannomyces graminis*. Brassicas release biofumigant compounds such as glucosinolates during decomposition, which suppress the growth of the pathogen. Additionally, the rotation promotes the activity of *Pseudomonas spp.*, which produce siderophores that sequester iron, depriving the pathogen of this essential nutrient. The combined effects of host disruption, biofumigation, and microbial activity result in significant reductions in disease incidence and improved wheat yields.

3.3.2 Intercropping

Intercropping, the practice of cultivating two or more crops simultaneously in the same field, represents a dynamic agricultural strategy that enhances the efficacy of biological control agents (BCAs). By increasing plant and microbial diversity, intercropping fosters a complex agroecosystem that supports beneficial microbial communities, suppresses pathogens, and promotes overall plant health. This practice aligns with ecological principles, offering a sustainable approach to disease management that complements the activity of BCAs.

Pathogen Suppression Through Microbial Diversity

One of the most significant benefits of intercropping is its ability to enhance microbial diversity in the rhizosphere. Each crop species releases a unique profile of root exudates, including sugars, organic acids, and secondary metabolites, which serve as nutrients for soil microorganisms. When multiple crops are grown together, the diversity of root exudates creates a rich microbial habitat, supporting a wide array of beneficial microbes. Beneficial microbes such as *Bacillus spp.* and *Pseudomonas spp.* thrive in intercropping systems due to the increased availability of resources. These microbes suppress pathogens through mechanisms like antibiosis, competition, and nutrient sequestration. For example, in maize-legume intercropping systems, populations of *Pseudomonas fluorescens* increase, producing antibiotics and siderophores that inhibit soil-borne pathogens like *Fusarium oxysporum*. Similarly, *Bacillus subtilis* exhibits enhanced activity against *Rhizoctonia solani*, benefiting from the diverse exudates provided by the intercropped plants.

The enhanced microbial diversity in intercropping systems not only suppresses pathogens directly but also fosters synergistic interactions among beneficial microbes. These interactions amplify the overall suppressive effect, reducing the need for chemical inputs and enhancing the resilience of the agroecosystem.

Physical and Environmental Benefits

Intercropping systems also provide physical and environmental benefits that indirectly enhance the activity of BCAs and suppress pathogens. The mixed canopy structure of intercropped fields modifies microclimatic conditions, such as light intensity, temperature, and humidity, creating environments less favorable for the proliferation of certain pathogens.

» **Temperature Regulation**: The shading effect of taller crops in an intercropping system can lower soil and canopy temperatures, reducing the activity of heat-sensitive pathogens. For instance, in maize-bean intercropping, the maize canopy protects the beans from excessive heat, minimizing stress and pathogen vulnerability.

» **Humidity Control**: The diverse plant architecture in intercropping systems regulates humidity levels, preventing conditions that favor foliar pathogens like *Alternaria* and *Phytophthora*. By reducing leaf wetness duration, intercropping indirectly supports the persistence and activity of foliar-applied BCAs like *Bacillus subtilis*.

» **Windbreaks and Erosion Prevention**: The dense canopy of intercropped systems acts as a natural windbreak, reducing the spread of airborne pathogens and minimizing soil erosion. This stabilization of the soil environment further supports the survival of soil-active BCAs.

These physical and environmental modifications create a more balanced and resilient agroecosystem, enhancing the efficacy of BCAs and reducing pathogen pressure.

Synergy Between Intercropping and BCAs

The integration of intercropping with biological control strategies creates a synergistic relationship that enhances disease suppression and crop productivity.

» **Enhanced Colonization and Activity**: The increased microbial diversity and nutrient availability in intercropping systems support the colonization and activity of BCAs. For instance, *Trichoderma harzianum*, a widely used fungal BCA, exhibits improved root colonization in intercropped systems due to the complementary root exudates provided by multiple crops.

» **Reduced Pathogen Inoculum**: By suppressing pathogen populations through competition and antibiosis, intercropping indirectly enhances the effectiveness of BCAs. For example, in rice and mung bean intercropping systems, the suppression of *Fusarium spp.* by microbial communities enhances the activity of *Pseudomonas fluorescens*, creating a multi-tiered defense against pathogens.

These synergies not only improve disease management but also contribute to the long-term sustainability of agricultural systems.

3.3.3 Tillage Practices

Tillage practices, which include conventional tillage, reduced tillage, and no-till systems, are integral to agricultural management and play a significant role in influencing the dynamics of soil-borne pathogens and biological control agents (BCAs). These practices directly affect soil structure, moisture, aeration, and the distribution of organic matter, all of which impact the survival, activity, and efficacy of BCAs. Understanding the relationship between tillage systems and biological control is essential for designing integrated disease management strategies that promote sustainable crop production.

Conventional Tillage and Pathogen Exposure

Conventional tillage, characterized by the intensive turning of soil, is widely used to prepare seedbeds, control weeds, and incorporate crop residues. One of the primary effects of conventional tillage on pathogens is the exposure of soil-borne propagules to environmental stressors such as UV radiation, desiccation, and microbial antagonism.

Pathogen Suppression

The mechanical disturbance of soil during conventional tillage disrupts the protective environment of pathogens, bringing their spores or resting structures to the surface. Exposure to sunlight and drying conditions can significantly reduce the viability of propagules from pathogens like *Fusarium oxysporum* and *Sclerotium rolfsii*. This effect can be particularly beneficial in regions with intense sunlight, where UV-induced damage to fungal pathogens is pronounced.

Impact on BCAs

While conventional tillage can suppress pathogens, it may also have negative consequences for BCAs. The disruption of soil structure and microbial habitats can reduce the survival of beneficial microbes like *Trichoderma spp.* and *Pseudomonas fluorescens*. These BCAs thrive in stable microenvironments, which are often disturbed by frequent tillage. Additionally, conventional tillage may deplete soil organic matter, limiting the substrate available for saprophytic BCAs to colonize and compete with pathogens.

Balancing Benefits and Challenges

To maximize the benefits of pathogen suppression while minimizing the negative impact on BCAs, farmers can adopt modified tillage practices. For instance, shallow tillage limits soil disturbance to the upper layers, reducing the impact on microbial communities while exposing pathogen propagules to environmental stressors.

Reduced and No-Tillage Systems

Reduced tillage and no-tillage systems involve minimal soil disturbance, preserving soil structure and microbial habitats. These systems are gaining popularity due to their benefits for soil health, carbon sequestration, and biological control.

Enhancing BCA Persistence

By maintaining soil structure and minimizing disruption, reduced tillage and no-till systems create a stable environment for BCAs. Beneficial fungi such as *Trichoderma spp.* and bacteria like *Bacillus subtilis* can establish themselves in the rhizosphere and organic matter-rich zones, where they suppress pathogens through competition, antibiosis, and mycoparasitism.

Residue Retention

No-till systems often retain crop residues on the soil surface, providing a nutrient-rich substrate for saprophytic BCAs. These residues act as a food source for microbes during decomposition, enhancing their competitive saprophytic ability (CSA). For example, *Trichoderma harzianum* colonizes crop residues effectively, outcompeting pathogens like *Pythium spp.* during the saprophytic phase.

Improved Soil Health

Reduced tillage and no-till systems improve soil health by increasing organic matter content, enhancing water retention, and reducing erosion. These benefits create a more resilient agroecosystem, where BCAs can thrive and suppress pathogens over the long term.

Synergy Between Tillage Practices and Biological Control

The choice of tillage system significantly influences the interaction between pathogens and BCAs. Reduced and no-till systems often synergize with biological control strategies by preserving microbial diversity and promoting the persistence of beneficial microbes.

Combining Tillage Practices with BCAs

Integrating BCAs into reduced or no-till systems can amplify their effectiveness. For example, seed treatments with *Trichoderma spp.* or *Bacillus subtilis* ensure early colonization of plant roots, giving BCAs a competitive advantage in undisturbed soils. In conventional tillage systems, incorporating organic amendments or cover crops can offset the negative impacts on microbial communities, supporting the activity of BCAs.

Case Study

In maize cultivation, reduced tillage combined with the application of *Trichoderma harzianum* decreased root rot caused by *Pythium spp.* by 40%. The no-till system preserved crop residues, providing a substrate for *Trichoderma* colonization and enhancing its suppressive effects on pathogens.

Challenges and Considerations

While reduced and no-tillage systems offer significant benefits for biological control, they also present challenges that must be addressed to optimize their efficacy.

- **Residue Management**: Retaining crop residues can sometimes favor pathogens that use organic matter as a substrate. Careful management, including the use of residue-decomposing BCAs, is essential to mitigate this risk.
- **Adaptation to Local Conditions**: The performance of tillage practices varies with soil type, climate, and cropping system. Tailoring tillage strategies to local conditions is crucial for achieving the desired balance between pathogen suppression and BCA activity.
- **Integration with Other Practices**: Tillage systems should be integrated with other cultural practices, such as crop rotation and organic amendments, to maximize their impact on biological control.

3.3.4 Irrigation Management

Irrigation management is a vital agricultural practice that significantly impacts the efficacy of biological control agents (BCAs). Moisture availability in the soil and on plant surfaces directly affects the survival, activity, and colonization ability of BCAs. Properly managed irrigation creates optimal conditions for BCAs to thrive and suppress pathogens, whereas excessive or insufficient water can undermine biocontrol efforts. Tailoring irrigation strategies to the needs of both crops and BCAs enhances the effectiveness of biological control in integrated disease management systems.

Optimizing Soil Moisture for BCAs

Soil moisture is a critical factor influencing the mobility, nutrient acquisition, and overall activity of soil-active BCAs. Adequate moisture ensures that microbes such as *Pseudomonas fluorescens* and *Bacillus subtilis* can effectively colonize the rhizosphere and interact with pathogens.

Importance of Moisture for Microbial Activity

BCAs rely on soil water to transport nutrients and facilitate movement toward plant roots and pathogen inoculum. Moisture also supports microbial metabolism and the production of secondary metabolites, such as antibiotics and siderophores, which are crucial for pathogen suppression. For instance, *Pseudomonas fluorescens* produces phenazines and pyoluteorin under optimal moisture conditions, inhibiting pathogens like *Fusarium oxysporum*.

Challenges of Over-Irrigation

Excessive irrigation can lead to waterlogging, creating anaerobic conditions that are detrimental to aerobic BCAs. Reduced oxygen availability in waterlogged soils inhibits the respiration and metabolic activity of microbes like *Bacillus subtilis*, compromising their ability to suppress pathogens. Furthermore, waterlogging can promote the growth of facultative anaerobic pathogens such as *Pythium spp.* and *Phytophthora spp.*, increasing disease pressure.

Impact of Drought Conditions

Drought conditions pose an equally significant challenge, as low soil moisture levels can desiccate microbial populations, reducing their survival and efficacy. Desiccation stress particularly affects BCAs that lack protective structures or formulations designed for dry conditions. To address this, researchers have developed microbial formulations with hydrophilic carriers, such as alginate beads or hydrogels, to protect BCAs from desiccation and enhance their resilience in drought-prone environments.

Strategies for Optimizing Soil Moisture

To maintain favorable moisture conditions for BCAs, irrigation schedules must be carefully aligned with crop water requirements and microbial needs. This involves monitoring soil moisture levels using tools such as tensiometers or soil moisture sensors and adjusting irrigation frequency and volume accordingly.

Synergy Between Irrigation and Soil-Active BCAs

Irrigation practices play a critical role in enhancing the interaction between soil-active BCAs and plant roots, as well as their ability to suppress pathogens. By creating localized zones of moisture, irrigation systems can concentrate BCAs around the root zone, where they are most effective.

Localized Irrigation Systems

Drip irrigation is an effective method for delivering water directly to the root zone, minimizing water wastage and reducing the risk of waterlogging. The localized wet zones created by drip irrigation promote the colonization of plant roots by BCAs like *Trichoderma harzianum* and *Pseudomonas fluorescens*. These microbes establish protective biofilms around root surfaces, enhancing their ability to suppress pathogens through competition, antibiosis, and mycoparasitism.

Irrigation Timing and Pathogen Suppression

The timing of irrigation applications also influences the effectiveness of BCAs. For instance, applying water during cooler parts of the day, such as early morning or late evening, minimizes evaporation and ensures that moisture levels remain consistent in the rhizosphere. This supports the survival of BCAs while limiting the proliferation of moisture-dependent pathogens.

Integration with Biological Control

Irrigation systems can be integrated with biocontrol programs by incorporating BCAs into the irrigation water, a technique known as fertigation. This approach ensures uniform distribution of BCAs across the field, enhancing their interaction with plant roots and pathogens. For example, fertigation with *Bacillus subtilis* has been shown to suppress *Ralstonia solanacearum* in tomato crops, reducing bacterial wilt incidence.

Example

In tomato fields, the integration of drip irrigation with *Bacillus subtilis* application improved the control of *Ralstonia solanacearum* by 50%. The localized wet zones created by the drip irrigation system concentrated the bacterial BCA around the root zone, enhancing its colonization and antagonistic activity against the pathogen. This demonstrates the importance of tailored irrigation strategies in maximizing the efficacy of biological control programs.

Considerations and Challenges

While irrigation management offers significant benefits for biological control, it also presents challenges that must be addressed to optimize its impact.

1. **Balancing Water Availability**

Maintaining the right balance between over-irrigation and under-irrigation is critical for supporting BCAs and suppressing pathogens. This requires continuous monitoring and adjustment of irrigation practices based on environmental conditions and crop requirements.

2. **Compatibility with BCAs**

Not all BCAs are compatible with irrigation systems, particularly those that require specific carriers or formulations to maintain viability. Selecting BCAs that can withstand the physical and chemical conditions of irrigation water is essential for successful implementation.

3. **Environmental and Economic Factors**

In regions with limited water availability, optimizing irrigation efficiency is crucial for reducing water usage while maintaining biocontrol efficacy. Advanced technologies such as precision irrigation and automated moisture sensors can help achieve this balance, though their initial cost may be a barrier for smallholder farmers.

3.3.5 Organic Amendments

Organic amendments, including compost, manure, and biochar, are vital tools in enhancing soil health and supporting the activity of biological control agents (BCAs). These amendments improve the physical, chemical, and biological properties of soil, creating a favorable environment for BCAs to thrive and suppress pathogens effectively. By enriching soil organic matter, increasing nutrient availability, and fostering microbial diversity, organic amendments align with the principles of sustainable agriculture and integrated disease management.

Nutrient Supply and Microbial Stimulation

Organic amendments act as a nutrient reservoir, providing essential elements like carbon, nitrogen, phosphorus, and potassium that stimulate microbial activity. These nutrients not only support crop growth but also enhance the proliferation and functionality of BCAs.

Substrate for Microbial Growth

Organic amendments serve as a food source for BCAs, supporting their population growth and metabolic activity. For instance, composted manure provides a rich mix of organic compounds that promote the colonization of beneficial fungi like *Trichoderma spp.* and bacteria like *Bacillus subtilis*. These microbes compete with pathogens for resources, produce antimicrobial compounds, and enhance plant resistance to diseases.

Enhancement of Antagonistic Activities

The nutrients supplied by organic amendments stimulate the production of secondary metabolites, such as antibiotics and lytic enzymes, by BCAs. For example, *Trichoderma harzianum* produces hydrolytic enzymes like chitinases and glucanases more effectively in compost-enriched soils, enabling it to suppress pathogens like *Fusarium oxysporum* and *Rhizoctonia solani*.

Impact on Microbial Communities

Organic amendments increase microbial diversity in the soil, fostering synergistic interactions between BCAs and other beneficial microbes. This diversity strengthens the suppressive potential of the soil, reducing the risk of pathogen dominance. For instance, biochar-amended soils often exhibit enhanced populations of *Pseudomonas fluorescens*, which suppresses pathogens through siderophore production and competitive exclusion.

Soil Resilience

The addition of organic amendments significantly improves soil resilience, allowing BCAs and crops to withstand environmental stresses such as drought, temperature fluctuations, and nutrient depletion.

Improvement of Soil Organic Matter

Organic matter plays a critical role in enhancing soil structure, water-holding capacity, and nutrient retention. These properties create a stable environment for BCAs to survive and function effectively. For example, in soils with high organic matter content, BCAs like *Trichoderma spp.* can maintain their activity even under suboptimal conditions, such as low moisture levels.

Mitigation of Environmental Stress

Amended soils are better equipped to buffer against extreme weather events. Biochar, for instance, improves soil aeration and water infiltration, reducing the likelihood of waterlogging or desiccation. These properties benefit both BCAs and plants, ensuring consistent disease suppression and crop growth.

Reduction of Pathogen Inoculum

Organic amendments can directly reduce pathogen populations by promoting the activity of saprophytic BCAs. Crop residues or compost incorporated into the soil serve as a substrate for antagonistic microbes to colonize and outcompete pathogens during the decomposition phase.

Example

In strawberry cultivation, compost enriched with *Trichoderma harzianum* reduced *Verticillium wilt* incidence by 60%. The organic matter in the compost provided a substrate for *Trichoderma* to establish and proliferate, enhancing its competitive saprophytic ability and mycoparasitic activity. This case demonstrates the synergistic effects of organic amendments and biological control in managing soil-borne diseases.

Application Strategies for Organic Amendments

To maximize the benefits of organic amendments, their application must be tailored to specific cropping systems, soil conditions, and target pathogens.

Selection of Appropriate Amendments

Different organic amendments vary in their nutrient composition, decomposition rates, and effects on microbial communities. Selecting the right amendment based on the soil's needs and the desired microbial response is essential. For instance, composted manure is suitable for enriching nutrient-deficient soils, while biochar is effective in improving soil structure and aeration.

Timing and Method of Application

Incorporating organic amendments at the right time ensures optimal interaction between BCAs, plants, and pathogens. Pre-planting applications allow sufficient time for amendments to decompose and stimulate microbial activity, while post-harvest applications help reduce pathogen inoculum for the next growing season.

Combination with BCAs

Organic amendments can be inoculated with BCAs before application to enhance their suppressive effects. For example, mixing compost with *Trichoderma spp.* or *Bacillus subtilis* ensures that these BCAs establish themselves in the soil, providing long-term disease suppression.

Integration with Other Practices

Organic amendments should be integrated with other cultural practices, such as crop rotation, reduced tillage, and irrigation management, to create a comprehensive approach to disease management. This integration enhances the overall resilience and productivity of the agroecosystem.

3.3.6 Residue Management

Crop residue management is a critical practice in agriculture that significantly impacts the dynamics of soil-borne pathogens and biological control agents (BCAs). Residues, such as crop stubble, straw, and roots, serve as substrates for microbial communities, influencing their composition and activity.

Proper management of residues can promote the proliferation of saprophytic BCAs, suppress pathogen populations, and enhance soil health. The balance between pathogen suppression and BCA activity depends on how residues are managed—whether incorporated into the soil or retained on the surface.

Residue Incorporation

Incorporating crop residues into the soil is a widely practiced method to enhance decomposition and nutrient recycling. This practice creates a nutrient-rich environment that supports the growth of BCAs while suppressing pathogens reliant on the same substrates during their saprophytic phase.

Enhanced Competitive Saprophytic Ability (CSA)

Incorporated residues serve as a carbon source for saprophytic BCAs such as *Trichoderma spp.* and *Bacillus subtilis*. These microbes excel in colonizing decomposing organic matter, outcompeting pathogens for resources. For instance, *Trichoderma harzianum* is known for its ability to rapidly colonize plant residues, suppressing pathogens like *Fusarium oxysporum* by occupying ecological niches and producing antimicrobial compounds.

Reduction of Pathogen Inoculum

Residue incorporation accelerates the decomposition of organic matter, reducing the survival of pathogen propagules such as spores, sclerotia, and mycelium. Decomposing residues create an environment rich in antagonistic microbial activity, breaking down the structural components of pathogen propagules and minimizing their ability to infect the next crop cycle.

Promotion of Nutrient Cycling

Incorporated residues enrich the soil with organic matter and nutrients, improving soil structure and microbial diversity. This nutrient enrichment supports the metabolic activity of BCAs, enhancing their efficacy in suppressing pathogens.

Example:

In rice fields, incorporating straw residues inoculated with *Trichoderma spp.* reduced *Sclerotium rolfsii* populations by 50%, improving seedling emergence rates. This demonstrates how residue incorporation, combined with BCAs, can significantly reduce pathogen pressure and enhance crop establishment.

Surface Retention

Surface retention of crop residues involves leaving plant residues on the soil surface after harvest. This practice has both benefits and challenges for biological control, depending on how it is managed.

Creation of Protective Microbial Habitats

Retained residues create a protective habitat for BCAs by shielding them from environmental stressors such as UV radiation, temperature extremes, and desiccation. This enhances the survival and persistence of BCAs like *Pseudomonas fluorescens* and *Trichoderma spp.* on the soil surface. These microbes can colonize the residue layer, forming a biological barrier that suppresses pathogen activity.

Challenges of Pathogen Proliferation

Surface-retained residues can also serve as substrates for pathogens, particularly if the residues originate from diseased plants. For example, pathogens like *Rhizoctonia solani* and *Fusarium spp.* can colonize surface residues, increasing inoculum potential. Proper management, such as combining surface retention with BCA inoculation or periodic residue turnover, is essential to mitigate this risk.

Moisture Retention and Microclimatic Benefits

Surface residues help retain soil moisture and regulate temperature, creating a favorable microclimate for both BCAs and crops. This is particularly beneficial in arid and semi-arid regions, where moisture availability is a limiting factor for microbial activity.

Example:

In wheat cultivation, surface retention of stubble combined with *Trichoderma harzianum* application reduced *Fusarium* inoculum levels by 40% while improving soil moisture retention.

Balancing Benefits and Challenges

Effective residue management requires a balance between promoting BCA activity and minimizing pathogen proliferation. Strategies include:

1. Inoculating Residues with BCAs

Adding BCAs to crop residues before incorporation or surface retention enhances their ability to suppress pathogens. For example, treating wheat stubble with *Bacillus subtilis* before incorporation improves its suppressive effects against *Rhizoctonia solani*.

2. Monitoring Residue Quality

Residue quality, including its carbon-to-nitrogen (C:N) ratio, affects its decomposition rate and microbial colonization. High C:N ratios may slow decomposition, favoring pathogens. Balancing residue composition through the addition of nitrogen-rich amendments can accelerate decomposition and support BCA activity.

3. Periodic Turnover of Surface Residues

Periodically incorporating surface-retained residues into the soil prevents the accumulation of pathogen inoculum while maintaining the benefits of residue retention.

4. Integration with Other Practices

Combining residue management with practices like crop rotation, reduced tillage, and organic amendments enhances its impact on pathogen suppression and BCA proliferation.

Chapter - 4

Relationships Between Pathogens, Antagonists and Bioagents

The interactions between pathogens, antagonists, and bioagents are central to the success of biological control in managing plant diseases. These relationships are characterized by a complex web of ecological and molecular interactions that govern the dynamics of pathogen suppression and the promotion of plant health. Pathogens, as the primary targets in biocontrol strategies, are organisms—such as fungi, bacteria, viruses, or nematodes—that cause diseases in plants by invading their tissues and disrupting physiological processes. Antagonists, on the other hand, are organisms that inhibit or suppress the growth, survival, or activity of these pathogens through various mechanisms, including competition, antibiosis, parasitism, and predation. Bioagents, which include antagonists as well as other beneficial organisms, are intentionally introduced or encouraged to regulate pathogen populations and enhance plant health. Together, the interactions between these three groups—pathogens, antagonists, and bioagents—form the foundation of biological control systems.

Competition is one of the most critical interactions between pathogens and antagonists. Both groups often compete for limited resources, such as nutrients, water, and physical space within the plant rhizosphere or phyllosphere. Antagonistic bioagents like *Trichoderma spp.*, for example, are known to outcompete pathogens such as *Fusarium oxysporum* by rapidly colonizing the root surface and utilizing available nutrients more efficiently. This competitive exclusion prevents pathogens from establishing themselves and causing infections. Similarly, bacterial antagonists such as *Pseudomonas fluorescens* produce siderophores—iron-chelating molecules—that deprive pathogens of the iron necessary for their growth and survival, thereby suppressing diseases like damping-off and wilt.

Another key interaction is antibiosis, where bioagents produce antimicrobial compounds that directly inhibit or kill pathogens. These compounds include antibiotics, lytic enzymes, and toxins that disrupt pathogen cell walls, membranes, or metabolic pathways. For instance, *Bacillus subtilis* produces lipopeptides like surfactin and iturin, which have potent antifungal properties and are effective against pathogens like *Botrytis cinerea* and *Rhizoctonia solani*. Similarly, *Streptomyces spp.* produce a range of secondary metabolites with antibacterial and antifungal activity, targeting diverse pathogens in the soil and on plant surfaces. Antibiosis is particularly valuable in biological control as it allows bioagents to exert their effects even in the absence of direct physical contact with the pathogens, broadening their scope of activity and enhancing their effectiveness in diverse environmental conditions.

Parasitism and predation are also central to the interactions between pathogens and bioagents.

Parasitic antagonists, such as mycoparasitic fungi, directly attack and degrade the structures of pathogenic fungi. For example, *Trichoderma harzianum* and *Trichoderma viride* secrete enzymes like chitinases and glucanases that break down the cell walls of fungal pathogens, rendering them inactive. This form of biological control is highly specific and effective, as it directly targets the pathogen without harming the plant or non-target organisms. Predatory organisms, such as nematodes and predatory fungi, contribute to pathogen suppression by feeding on pathogen propagules, such as spores or sclerotia. These interactions reduce the reproductive potential of pathogens and limit their capacity to cause disease, creating a natural check on their populations.

In addition to these antagonistic relationships, bioagents also interact with plants to enhance their defensive capacity against pathogens. Many bioagents, such as plant growth-promoting rhizobacteria (PGPR) and mycorrhizal fungi, induce systemic resistance in plants, priming them to respond more effectively to pathogen attacks. This phenomenon, known as Induced Systemic Resistance (ISR), involves the activation of plant defense pathways regulated by signaling molecules like jasmonic acid, salicylic acid, and ethylene. For example, *Pseudomonas fluorescens* triggers ISR in crops like tomato and cucumber, enhancing their resistance to foliar and soil-borne diseases. Similarly, arbuscular mycorrhizal fungi (AMF) improve nutrient uptake and modify root exudates, creating an environment that is less favorable for pathogenic organisms. These mutualistic interactions between bioagents and plants add another layer of protection, complementing the direct antagonistic effects on pathogens.

The relationships between pathogens, antagonists, and bioagents are influenced by various biotic and abiotic factors. The diversity and composition of the microbial community in the rhizosphere play a critical role in shaping these interactions. A high level of microbial diversity often creates a competitive environment that suppresses pathogen populations while supporting the activity of beneficial bioagents. Conversely, monoculture systems with low microbial diversity may favor the proliferation of pathogens, reducing the efficacy of biological control. Abiotic factors such as temperature, moisture, pH, and soil organic matter also affect the interactions between these groups. For example, the activity of fungal bioagents like *Beauveria bassiana* and *Metarhizium anisopliae* is highly dependent on humidity levels, while the effectiveness of bacterial bioagents like *Bacillus spp.* may vary with soil pH and nutrient availability. Understanding and managing these factors are essential for optimizing the performance of biological control agents and ensuring their long-term success.

4.1 Ecological and Biochemical Dynamics of Pathogen-Antagonist Relationships

The relationships between pathogens, antagonists, and bioagents form the foundation of biological control in plant pathology. These interactions are governed by ecological and biochemical dynamics that determine the success of antagonists and bioagents in suppressing plant pathogens. These dynamics are critical for optimizing the efficacy of biological control agents (BCAs) and designing sustainable disease management strategies.

4.1.1 Ecological Dynamics of Pathogen-Antagonist Relationships

Ecological dynamics govern the interactions between pathogens, antagonists, and the environment, forming the core mechanisms by which biological control agents (BCAs) suppress diseases in agricultural systems. These interactions are driven by competition for resources, spatial and temporal colonization dynamics, microbial diversity, and environmental factors.

Competition for Resources

Competition for essential resources like carbon, nitrogen, and micronutrients such as iron is a fundamental ecological mechanism employed by antagonists to suppress pathogens. In agricultural ecosystems, pathogens and antagonists cohabit the rhizosphere and compete for limited resources to proliferate and infect host plants.

1. Siderophore Production

Iron is a critical micronutrient for microbial growth but is often present in insoluble forms in the soil. To compete effectively, many antagonistic microbes secrete siderophores, specialized iron-chelating molecules that scavenge and sequester iron from the environment. By binding iron with high affinity, siderophores deprive pathogens of this essential nutrient, weakening their growth and virulence.

Mechanism in Pathogen Suppression:

» *Pseudomonas fluorescens*, a widely studied antagonist, produces pyoverdines, high-affinity siderophores that outcompete pathogens like *Fusarium oxysporum* in the rhizosphere. This mechanism is particularly effective in suppressing soil-borne pathogens in crops such as tomatoes, cucumbers, and wheat.

Advanced Applications:

» Modern biocontrol strategies include formulating BCAs to enhance siderophore production under varying soil conditions. Research has demonstrated that combining siderophore-producing BCAs with organic amendments further increases their efficacy by improving the availability of nutrients for beneficial microbes while depleting those for pathogens.

2. Carbon Utilization

Antagonists such as *Trichoderma spp.* are highly efficient at metabolizing a wide range of organic compounds, which enables them to dominate in environments with limited carbon availability. Their ability to utilize plant residues and other organic substrates gives them a competitive edge over pathogens during the saprophytic phase.

Case Study:

» In strawberry fields, the application of *Trichoderma harzianum* suppressed *Verticillium dahliae* populations by outcompeting them for carbon-rich substrates in decomposing plant residues. This also reduced the inoculum potential for subsequent cropping cycles.

3. Nitrogen Competition

Nitrogen, a key element for microbial and plant growth, often becomes a limiting factor in soil. BCAs such as nitrogen-fixing bacteria (*Rhizobium spp.*) not only contribute to soil fertility but also suppress pathogens by establishing dominance in nitrogen-limited environments. This dual functionality of nutrient provision and pathogen suppression makes nitrogen competition a valuable strategy in integrated disease management systems.

Spatial and Temporal Interactions

The spatial and temporal distribution of pathogens and antagonists within the soil and plant environment plays a pivotal role in shaping their interactions and outcomes. The ability of antagonists to establish themselves in key niches, such as the rhizosphere, and to colonize these spaces before pathogens arrive often determines the success of biological control strategies.

1. Rhizosphere Colonization

The rhizosphere, a nutrient-rich zone surrounding plant roots, serves as a hotspot for microbial activity. Antagonists that effectively colonize this zone gain a competitive advantage over pathogens.

Biofilm Formation:

» Antagonistic bacteria such as *Bacillus subtilis* and *Pseudomonas fluorescens* form biofilms on root surfaces, creating a protective barrier that prevents pathogens from establishing infection sites. Biofilms also enhance the persistence and activity of BCAs, improving their ability to suppress pathogens over time.

Root-Surface Interactions:

» Colonization of root surfaces by BCAs is influenced by plant exudates, which serve as attractants and nutrient sources. Certain crops, such as legumes, release exudates that favor the proliferation of beneficial microbes like *Rhizobium spp.* and *Pseudomonas spp.*, indirectly suppressing pathogens.

2. Timing of Colonization

The timing of colonization by antagonists is critical for outcompeting pathogens. Early establishment ensures that pathogens encounter an environment already dominated by beneficial microbes.

Seed Treatments:

» Seed inoculation with BCAs like *Trichoderma harzianum* and *Pseudomonas fluorescens* allows these microbes to colonize roots immediately after germination, providing early protection against soil-borne pathogens like *Rhizoctonia solani* and *Fusarium oxysporum*.

Environmental Adaptation:

» BCAs must be adapted to local environmental conditions to colonize effectively. For example, temperature-tolerant strains of *Trichoderma spp.* have been developed for use in tropical climates, where high temperatures often favor aggressive pathogens.

Microbial Diversity and Synergistic Interactions

Diverse microbial communities in the rhizosphere enhance pathogen suppression through synergistic interactions among antagonists. This microbial diversity is crucial for maintaining soil health and resilience against pathogen invasions.

1. Functional Redundancy

Functional redundancy refers to the presence of multiple microbial species that perform similar ecological functions. This redundancy ensures consistent pathogen suppression, even if some microbial populations decline due to environmental changes. For example, both *Trichoderma spp.* and *Pseudomonas spp.* produce hydrolytic enzymes that degrade fungal cell walls, providing overlapping layers of protection.

2. Synergistic Biocontrol

Co-inoculation of complementary BCAs often results in enhanced disease suppression. For instance, *Pseudomonas fluorescens* produces antibiotics and siderophores, while *Trichoderma harzianum* exhibits mycoparasitic activity. Together, they target multiple aspects of pathogen biology, offering robust protection against soil-borne diseases.

Example:

» In rice paddies, the combination of *Pseudomonas fluorescens* and *Trichoderma harzianum* reduced sheath blight incidence caused by *Rhizoctonia solani* by 60%, demonstrating the potential of synergistic interactions in biological control.

3. Diversity in Cropping Systems

Intercropping and crop rotation enhance microbial diversity by introducing varied root exudates and substrates into the soil. These practices indirectly support the activity of BCAs by fostering a dynamic and competitive microbial environment that suppresses pathogens.

Environmental Factors

Abiotic factors such as soil pH, temperature, moisture, and nutrient availability significantly influence the interactions between pathogens, antagonists, and their environment.

1. Soil pH

Soil pH affects the growth and activity of both pathogens and antagonists. Most BCAs, including *Trichoderma spp.* and *Pseudomonas fluorescens*, thrive in slightly acidic to neutral soils (pH 6.0–7.5). Extremes of pH can inhibit the activity of BCAs while favouring certain pathogens, necessitating site-specific adjustments to optimize biocontrol efficacy.

Case Study:

» In acidic soils, lime application improved the performance of *Bacillus subtilis*, enhancing its suppression of *Fusarium spp.* in wheat fields.

2. Temperature

Temperature directly influences microbial metabolism and activity. While pathogens such as *Pythium spp.* thrive in cooler soils, antagonists like *Trichoderma harzianum* are more active at moderate temperatures (25–30°C). Tailoring BCA applications to align with optimal temperature conditions is crucial for maximizing their effectiveness.

3. Moisture

Soil moisture levels affect the mobility and survival of BCAs. Adequate moisture supports rhizosphere colonization and microbial activity, while excessive moisture can lead to waterlogging, favoring anaerobic pathogens like *Phytophthora spp.* Conversely, drought conditions can desiccate microbial populations, necessitating the use of protective formulations for BCAs.

4. Nutrient Availability

Nutrient-rich soils support the proliferation of BCAs, enhancing their ability to suppress pathogens. Organic amendments such as compost and biochar can improve nutrient availability while simultaneously fostering beneficial microbial communities.

Example:

» In strawberry cultivation, compost enriched with *Trichoderma harzianum* reduced *Verticillium wilt* by 60%, showcasing the role of nutrient-rich environments in enhancing biocontrol efficacy.

4.1.2 Biochemical Dynamics of Pathogen-Antagonist Relationships

The biochemical dynamics underlying pathogen-antagonist relationships are critical to understanding how antagonists suppress pathogens and establish themselves as effective biological control agents (BCAs). These dynamics are mediated by the production of antimicrobial compounds, enzymatic activities, plant resistance induction, and interference with pathogen signaling systems. Together, these molecular and metabolic processes provide antagonists with the means to directly inhibit pathogens, modulate plant defense responses, and disrupt pathogen communication.

1. Antibiosis

Antibiosis is a fundamental biochemical mechanism through which antagonists suppress pathogens by producing bioactive compounds. These compounds disrupt vital processes in pathogens, directly inhibiting their growth, reproduction, or survival.

1.1 Antibiotics

Antibiotics are secondary metabolites produced by antagonistic microorganisms that target specific cellular components of pathogens.

Mechanisms of Action:

» Antibiotics disrupt pathogen cell membranes, interfere with protein synthesis, and inhibit essential metabolic pathways. For example, *Bacillus subtilis* produces cyclic lipopeptides such as iturin, surfactin, and fengycin, which integrate into fungal cell membranes, causing leakage and eventual cell death.

Spectrum of Activity:

» While some antibiotics exhibit broad-spectrum activity against a range of pathogens, others are highly specific. *Pseudomonas fluorescens* produces phenazines and pyrrolnitrin, which are particularly effective against fungal pathogens like *Fusarium spp.* and *Rhizoctonia solani.*

1.2 Volatile Organic Compounds (VOCs)

VOCs are gaseous metabolites that antagonists release into their environment, providing an additional layer of pathogen suppression.

- **Hydrogen Cyanide (HCN):** Produced by *Pseudomonas spp.*, HCN inhibits the respiratory enzymes of pathogens, leading to their death.
- **2,3-Butanediol:** This VOC not only suppresses pathogens but also promotes plant growth by modulating hormone pathways.

Applications in Biocontrol:

Antibiotics and VOCs are increasingly utilized in the formulation of biopesticides. For example, biocontrol products based on *Bacillus subtilis* incorporate its antibiotic-producing capabilities to manage soil-borne diseases in crops like tomatoes and cucumbers.

2. Enzymatic Degradation

Enzymatic degradation is another cornerstone of pathogen suppression, involving the production of hydrolytic enzymes that degrade structural components of pathogen cells.

2.1 Chitinases and Glucanases

Fungal cell walls are primarily composed of chitin and glucan, making them ideal targets for enzymatic degradation by BCAs.

- **Chitinases:** These enzymes hydrolyze the chitin polymer into smaller units, weakening fungal cell walls. For instance, *Trichoderma harzianum* secretes high levels of chitinases, enabling it to parasitize and kill *Rhizoctonia solani.*
- **Glucanases:** Glucanases break down glucan, another structural polymer in fungal cell walls. This dual enzymatic action makes *Trichoderma spp.* potent mycoparasites.

2.2 Proteases

Proteases degrade proteins that pathogens use for adhesion, colonization, and infection. By targeting these proteins, proteases reduce pathogen virulence and prevent the establishment of infections.

Example:

- Proteases produced by *Bacillus subtilis* disrupt the colonization of *Phytophthora spp.* on plant roots, significantly reducing disease incidence.

Enhanced Biocontrol Formulations:

Modern formulations of BCAs often include enzyme-boosting agents to enhance the production of hydrolytic enzymes under field conditions, improving their effectiveness against pathogens.

3. Induction of Systemic Resistance

Certain antagonists induce systemic resistance in plants, effectively priming the plant's immune system to respond more effectively to pathogen attacks. This induction occurs through two primary mechanisms: Induced Systemic Resistance (ISR) and Systemic Acquired Resistance (SAR).

3.1 Induced Systemic Resistance (ISR)

ISR is mediated by rhizosphere-associated microbes that activate plant defense pathways involving jasmonic acid (JA) and ethylene (ET).

Mechanism:

» ISR does not involve direct pathogen attack but primes the plant to mount a stronger defense upon encountering a pathogen. For example, *Pseudomonas fluorescens* triggers ISR, enhancing the production of defense-related enzymes like peroxidases and phenylalanine ammonia-lyase in crops such as rice and wheat.

Abiotic Stress Protection:

» In addition to pathogen suppression, ISR also improves plant tolerance to abiotic stresses like drought and salinity, making it a holistic approach to crop health management.

3.2 Systemic Acquired Resistance (SAR)

SAR is a pathogen-induced response involving the accumulation of salicylic acid (SA) and the activation of pathogenesis-related (PR) genes.

Mechanism:

» Antagonists like *Trichoderma spp.* mimic pathogen attacks by releasing elicitors that activate SAR pathways. This leads to the systemic production of antimicrobial compounds, reducing the plant's susceptibility to subsequent infections.

Long-Lasting Immunity:

SAR provides long-lasting protection against a broad spectrum of pathogens, including bacteria, fungi, and viruses.

Applications in Disease Management:

The dual activation of ISR and SAR by BCAs is leveraged in integrated pest management (IPM) programs to provide comprehensive protection against plant diseases.

4. Quorum Sensing and Signal Disruption

Pathogens often rely on quorum sensing, a communication system that uses signaling molecules to coordinate activities such as biofilm formation and toxin production. Antagonists disrupt these signaling pathways, effectively disarming the pathogen's virulence mechanisms.

4.1 Quorum Quenching

Antagonists produce enzymes that degrade quorum-sensing molecules, preventing pathogens from coordinating their activities.

» **N-acyl Homoserine Lactones (AHLs):** Many Gram-negative bacterial pathogens use AHLs for quorum sensing. Enzymes produced by *Bacillus spp.* and *Pseudomonas spp.* degrade AHLs, disrupting processes like biofilm formation and virulence factor production.

4.2 Interference with Biofilms

Biofilms are protective structures formed by pathogens to shield themselves from environmental stressors and antimicrobial agents. By interfering with biofilm formation, antagonists expose pathogens to natural suppression mechanisms.

Example:

» In potato crops, quorum quenching by *Bacillus subtilis* reduced the virulence of *Pectobacterium carotovorum*, a quorum-sensing pathogen responsible for soft rot.

Future Directions in Biocontrol:

Research into quorum-sensing inhibitors is paving the way for next-generation biopesticides that target pathogen communication without disrupting the surrounding microbial ecosystem.

Example

The combined biochemical strategies of *Bacillus subtilis* exemplify the power of antagonists in pathogen suppression. Through the production of antibiotics like iturin, the secretion of lytic enzymes such as proteases, and the disruption of quorum-sensing pathways, *Bacillus subtilis* has proven effective against a wide range of pathogens, including *Fusarium spp.* and *Phytophthora spp.*. In field trials, its application reduced disease incidence by 50–70%, demonstrating the practical relevance of these biochemical mechanisms.

4.1.3 Applications in Biological Control

The application of biological control agents (BCAs) has evolved significantly with a deeper understanding of the ecological and biochemical dynamics governing pathogen-antagonist relationships. This knowledge has enabled the development of precise and innovative strategies to manage plant diseases effectively. Tailored formulations, co-inoculation approaches, and the integration of BCAs with cultural practices represent the cornerstones of modern biological control.

1. Tailored Biocontrol Formulations

Tailored biocontrol formulations are designed to maximize the synergistic interactions between BCAs and their target pathogens. These formulations exploit specific ecological and biochemical dynamics such as resource competition, antibiosis, and systemic resistance induction to suppress pathogens in diverse agroecosystems.

Seed Treatments

Seed treatments are one of the most effective ways to deliver BCAs to the root zone, where many soil-borne pathogens initiate infections. BCAs like *Bacillus subtilis* and *Pseudomonas fluorescens* are applied as coatings on seeds, ensuring early colonization of the rhizosphere.

Mechanisms in Action:

» *Bacillus subtilis* combines multiple modes of action—antibiosis through lipopeptide production, competition for nutrients like iron, and the induction of systemic resistance. These actions protect young seedlings from soil-borne pathogens such as *Fusarium oxysporum* and *Pythium ultimum*.

Case Study:

» In wheat cultivation, seed treatment with *Pseudomonas fluorescens* reduced the incidence of root rot caused by *Rhizoctonia solani* by 60%, highlighting the efficacy of tailored biocontrol formulations in protecting emerging crops.

Advanced Formulations

The development of advanced formulations such as microencapsulation, alginate beads, and liquid suspensions has improved the stability, shelf life, and field performance of BCAs. Encapsulation techniques protect BCAs from environmental stressors like UV radiation and desiccation, ensuring their viability until they reach the target site.

2. Co-Inoculation Strategies

Co-inoculation involves the simultaneous application of multiple BCAs that employ complementary mechanisms of action. By combining agents that target different aspects of pathogen biology, co-inoculation enhances the overall efficacy of biological control programs.

Mechanisms of Synergy

Targeting Different Niches:

» *Trichoderma spp.* colonize the rhizosphere and suppress soil-borne pathogens through mycoparasitism and enzymatic degradation, while *Pseudomonas fluorescens* targets the phyllosphere, producing antibiotics and VOCs to suppress foliar pathogens. Together, these BCAs provide comprehensive protection to the entire plant.

Enhanced Pathogen Suppression:

» Co-inoculation with BCAs that exhibit synergistic effects reduces the likelihood of pathogen resistance and ensures long-term disease suppression. For instance, *Trichoderma harzianum* and *Bacillus subtilis* act synergistically to suppress *Fusarium* species by combining enzymatic degradation with resource competition.

Case Study

In tomato cultivation, the co-inoculation of *Trichoderma harzianum* and *Pseudomonas fluorescens* reduced wilt incidence by 70% while promoting plant growth. This approach demonstrated the value of combining rhizosphere- and phyllosphere-active BCAs for holistic disease management.

Commercial Applications

Several commercial products now include co-inoculation formulations. For example, biopesticides such as RootShield® Plus combine *Trichoderma spp.* and *Gliocladium spp.*, targeting multiple pathogens and improving efficacy across different environmental conditions.

3. Integration with Cultural Practices

Integrating BCAs with cultural practices amplifies the ecological dynamics that favor antagonists over pathogens. By creating favorable environments for BCAs to thrive, these practices improve the reliability and effectiveness of biological control.

Crop Rotation

Crop rotation disrupts the life cycle of host-specific pathogens, reducing their populations in the soil. When combined with BCAs, rotation systems provide a dual benefit: reduced pathogen pressure and enhanced BCA activity. For example, alternating cereal crops with legumes promotes the proliferation of nitrogen-fixing bacteria, which indirectly suppress soil-borne pathogens.

Example:

» Rotating wheat with brassica crops reduced the severity of take-all disease caused by *Gaeumannomyces graminis*, while promoting the activity of *Pseudomonas fluorescens* in the rhizosphere.

Residue Management

Incorporating crop residues into the soil provides a substrate for saprophytic BCAs to colonize, suppressing pathogens during their saprophytic phase. Retaining residues on the soil surface also enhances the activity of BCAs like *Trichoderma spp.* by protecting them from environmental stressors.

Example:

» In rice fields, the incorporation of straw residues inoculated with *Trichoderma harzianum* reduced *Sclerotium rolfsii* populations by 50%, improving seedling emergence rates and overall soil health.

Organic Amendments

The addition of organic amendments such as compost, manure, and biochar enriches soil organic matter, providing nutrients for BCAs while suppressing pathogens. These amendments improve soil structure, water-holding capacity, and microbial diversity, creating an ideal environment for BCAs to thrive.

Example:

» Compost enriched with *Bacillus subtilis* suppressed *Phytophthora spp.* in strawberry cultivation, reducing disease incidence by 40%.

4. Integration with Precision Agriculture

Advancements in precision agriculture have opened new avenues for optimizing the application of BCAs. Technologies such as drones, automated irrigation systems, and remote sensing tools allow for targeted delivery of BCAs to areas with high pathogen pressure.

Drone-Based Applications

Drones equipped with sprayers can evenly distribute BCAs over large fields, ensuring uniform coverage of foliage and soil. This method is particularly useful for foliar pathogens like *Botrytis cinerea* in high-value crops such as grapes and strawberries.

Soil-Moisture Management

Automated irrigation systems can create localized wet zones that enhance the mobility and activity of rhizosphere-active BCAs. For instance, drip irrigation combined with *Pseudomonas fluorescens* applications has been shown to reduce bacterial wilt in tomato fields.

4.2 Biocontrol Agents and Their Comparative Approaches

Biocontrol agents (BCAs) are the cornerstone of sustainable plant disease management, offering environmentally friendly alternatives to chemical pesticides. These agents include a diverse array of beneficial microbes, fungi, and sometimes even viruses, each with unique mechanisms for suppressing plant pathogens. Comparative approaches to using BCAs, whether resident or introduced, provide strategic insights into their deployment across various agroecosystems.

4.2.1 Types of Biocontrol Agents

Biocontrol agents (BCAs) represent a diverse array of microorganisms and biological entities that suppress plant pathogens through various mechanisms, including competition, antibiosis, parasitism, and induced resistance. These agents are classified based on their taxonomic identity, ecological niche, or mode of action. Among the most utilized BCAs in agriculture are bacteria, fungi, and viruses, each with unique capabilities and applications in plant disease management. Their versatility and effectiveness make them indispensable components of sustainable agriculture.

1. Bacterial BCAs

Bacterial BCAs are some of the most extensively studied and widely applied biological control agents. Their ability to colonize diverse plant habitats—such as the rhizosphere, phyllosphere, and endosphere—enables them to interact closely with pathogens and plants. Bacterial BCAs employ multiple mechanisms, including the production of antimicrobial compounds, competition for nutrients, and induction of systemic resistance, making them highly versatile in agricultural systems.

1.1 Key Characteristics of Bacterial BCAs

- **Adaptability**: Bacteria thrive in diverse environmental conditions and can rapidly colonize plant surfaces.
- **Multifunctionality**: Many bacterial BCAs combine several mechanisms of pathogen suppression, such as producing antibiotics while also inducing systemic resistance in plants.
- **Ease of Production**: Bacteria can be mass-produced and formulated into stable products, facilitating their commercial use in agriculture.

1.2 Key Examples

1.2.1 *Bacillus subtilis*

Bacillus subtilis is one of the most widely used bacterial BCAs due to its robust nature and multiple modes of action.

» **Antibiotic Production**: *Bacillus subtilis* produces cyclic lipopeptides, including iturin, surfactin, and fengycin, which disrupt the cell membranes of fungal pathogens, leading to their death. These antibiotics are highly effective against soil-borne pathogens such as *Fusarium oxysporum*, *Rhizoctonia solani*, and *Botrytis cinerea*.

» **Biofilm Formation**: This bacterium forms biofilms on root surfaces, creating a protective barrier against pathogen invasion.

» **Induced Resistance**: *Bacillus subtilis* triggers systemic resistance in plants by activating defense-related genes, enhancing their ability to resist subsequent pathogen attacks.

Applications:

Commercial formulations like Serenade® leverage the properties of *Bacillus subtilis* to control diseases in high-value crops like tomatoes, strawberries, and cucumbers.

1.2.2 *Pseudomonas fluorescens*

Pseudomonas fluorescens is a prominent rhizosphere colonizer and a highly effective BCA due to its multifaceted mechanisms of action.

» **Siderophore Production**: *Pseudomonas fluorescens* produces high-affinity siderophores that sequester iron from the soil, depriving pathogens of this essential nutrient and limiting their growth.

» **Volatile Organic Compounds (VOCs)**: This bacterium produces VOCs like hydrogen cyanide and 2,3-butanediol, which inhibit pathogen growth and promote plant health.

» **Enzymatic Degradation**: The secretion of hydrolytic enzymes such as chitinases and proteases enables *Pseudomonas fluorescens* to degrade fungal cell walls, effectively suppressing pathogens.

Applications:

In rice cultivation, *Pseudomonas fluorescens* has been shown to reduce sheath blight caused by *Rhizoctonia solani* by more than 50%, highlighting its effectiveness in managing soil-borne diseases.

1.2.3 *Streptomyces spp.*

Streptomyces species, a group of filamentous actinomycetes, are known for their ability to produce a wide range of antimicrobial compounds.

» **Antibiotic Production**: These bacteria produce antifungal and antibacterial compounds such as streptomycin, nystatin, and actinomycin, which suppress pathogens like *Fusarium spp.* and *Phytophthora spp.*.

» **Saprophytic Activity**: *Streptomyces* species degrade organic matter in the soil, outcompeting pathogens for resources.

- **Soil Health Improvement**: By contributing to nutrient cycling and organic matter decomposition, these bacteria enhance soil health and indirectly suppress pathogens.

Applications:

Streptomyces lydicus has been used in biopesticide formulations like Actinovate®, which is effective against a variety of root and foliar diseases in crops like vegetables, fruits, and ornamentals.

2. Fungal BCAs

Fungal BCAs are highly effective against a broad spectrum of fungal pathogens due to their ability to parasitize pathogens, compete for resources, and produce antimicrobial compounds. Their versatility and resilience make them valuable tools in managing soil-borne diseases.

2.1 Key Characteristics of Fungal BCAs

- **Mycoparasitism**: Fungi exhibit mycoparasitic behavior, directly attacking and degrading pathogenic fungi.
- **Enzyme Production**: Hydrolytic enzymes such as chitinases and glucanases play a critical role in breaking down the cell walls of fungal pathogens.
- **Ecological Versatility**: Fungal BCAs are effective in various soil types and climatic conditions, making them widely applicable.

2.2 Key Examples

2.2.1 *Trichoderma spp.*

Trichoderma species are among the most studied fungal BCAs and are widely used for their robust mycoparasitic activity.

- **Mycoparasitism**: *Trichoderma spp.* attack pathogenic fungi by producing enzymes such as chitinases and glucanases that degrade fungal cell walls.
- **Antibiosis**: Secondary metabolites produced by *Trichoderma spp.*, such as gliotoxin and viridin, inhibit the growth of fungal pathogens.
- **Induced Resistance**: These fungi also activate plant defense mechanisms, enhancing systemic resistance to diseases.

Applications:

Trichoderma harzianum is extensively used to control diseases like damping-off caused by *Pythium spp.* and root rot caused by *Rhizoctonia solani*. It is a key component of biopesticide products such as RootShield® and PlantShield®.

2.2.2 *Gliocladium spp.*

Gliocladium species are effective against soil-borne pathogens such as *Pythium* and *Phytophthora* due to their ability to colonize the rhizosphere and compete for resources.

- **Nutrient Competition**: These fungi excel in outcompeting pathogens for nutrients in the soil.

- » **Secondary Metabolites**: The production of antifungal compounds further suppresses pathogen growth.

Applications:

Gliocladium spp. are commonly applied as soil amendments or seed treatments to protect crops like peppers, cucumbers, and melons from damping-off and root rot diseases.

3. Viral BCAs

Viral BCAs, though less commonly used than bacterial and fungal agents, offer a unique approach to managing plant pathogens. These agents primarily target fungal pathogens by infecting and reducing their virulence.

3.1 Key Characteristics of Viral BCAs

- » **Host-Specificity**: Viruses often target specific pathogens, ensuring minimal impact on non-target organisms.
- » **Hypovirulence**: Virally infected pathogens exhibit reduced virulence, enabling plants to tolerate their presence.
- » **Sustainability**: Viral BCAs persist in the environment, providing long-term suppression of pathogens.

3.2 Key Example

3.2.1 Hypoviruses

Hypoviruses are the most well-known viral BCAs, used extensively to manage chestnut blight caused by *Cryphonectria parasitica*.

- » **Mechanism of Action**: Hypoviruses infect *Cryphonectria parasitica* and disrupt its metabolic pathways, reducing its ability to produce toxins and cause disease.
- » **Ecological Benefits**: The use of hypoviruses restores the balance between chestnut trees and the fungal pathogen, allowing for coexistence and forest recovery.

Applications:

Hypovirulent strains of *Cryphonectria parasitica* have been introduced in North America and Europe to combat chestnut blight, resulting in significant reductions in disease severity and tree mortality.

4.2.2 Mechanisms of Action

Biological control agents (BCAs) employ a diverse array of mechanisms to suppress pathogens and promote plant health. These mechanisms are not only specific to the type of BCA but often work synergistically, targeting multiple aspects of pathogen biology to enhance overall efficacy. The key mechanisms of action include competition for resources, antibiosis, mycoparasitism, induced resistance, and quorum sensing disruption.

1. Competition for Resources

Competition for essential nutrients and ecological niches is one of the most fundamental ways BCAs suppress pathogens. Both antagonists and pathogens rely on limited resources such as carbon, nitrogen, and micronutrients like iron for growth and reproduction. BCAs with superior competitive abilities dominate these resources, effectively starving the pathogens and limiting their proliferation.

1.1 Siderophore Production

Iron is a critical micronutrient for microbial growth, but its availability in the soil is often limited. Many BCAs, particularly bacteria such as *Pseudomonas fluorescens*, produce siderophores—high-affinity iron-chelating compounds.

Mechanism of Suppression:

» Siderophores bind to iron in the rhizosphere, rendering it unavailable to pathogens. For instance, *Pseudomonas fluorescens* produces pyoverdines, which inhibit the growth of soil-borne pathogens like *Fusarium oxysporum*.

Applications:

» In wheat cultivation, siderophore-producing strains of *Pseudomonas fluorescens* have been used to control take-all disease caused by *Gaeumannomyces graminis*.

1.2 Carbon and Nitrogen Competition

BCAs like *Trichoderma spp.* excel in utilizing carbon and nitrogen resources, outcompeting pathogens in nutrient-limited environments. This metabolic dominance deprives pathogens of the energy required for infection and propagation.

Example:

» In strawberry fields, the application of *Trichoderma harzianum* reduced *Verticillium dahliae* populations by outcompeting them for carbon-rich substrates in crop residues.

2. Antibiosis

Antibiosis is the production of bioactive compounds by BCAs that directly inhibit or kill pathogens. These compounds include antibiotics, volatile organic compounds (VOCs), and toxins, which target critical cellular functions in pathogens.

2.1 Antibiotics

Antibiotics are secondary metabolites produced by BCAs such as *Bacillus subtilis* and *Streptomyces spp.* that disrupt pathogen metabolism and growth.

Examples of Antibiotics:

» Iturin: Produced by *Bacillus subtilis*, this lipopeptide disrupts fungal cell membranes, causing leakage and cell death.

- Phenazines: Produced by *Pseudomonas fluorescens*, these compounds inhibit fungal respiration and growth.
- Streptomycin: Produced by *Streptomyces spp.*, it is effective against bacterial pathogens.

Case Study:

- In cucumber cultivation, *Bacillus subtilis* application suppressed *Pythium ultimum* through the production of iturin and surfactin, reducing root rot incidence by 60%.

2.2 Volatile Organic Compounds (VOCs)

VOCs are gaseous metabolites that antagonistic microbes release to inhibit pathogens in the surrounding environment.

- **Hydrogen Cyanide (HCN)**: Produced by *Pseudomonas spp.*, HCN inhibits respiratory enzymes in pathogens, leading to their death.
- **2,3-Butanediol**: This VOC not only suppresses pathogen growth but also promotes plant growth by stimulating hormonal pathways.

3. Mycoparasitism

Mycoparasitism is a direct antagonistic interaction in which fungal BCAs parasitize pathogenic fungi, breaking down their cellular structures and extracting nutrients. This mechanism is highly effective against soil-borne fungal pathogens.

3.1 Enzymatic Degradation

Fungal BCAs like *Trichoderma spp.* produce hydrolytic enzymes that target the structural components of pathogen cell walls, leading to cell lysis.

- **Chitinases**: Degrade chitin, a key component of fungal cell walls.
- **Glucanases**: Break down glucan, another critical structural polymer in fungal pathogens.

3.2 Antibiosis in Mycoparasitism

In addition to enzymatic action, mycoparasitic fungi produce secondary metabolites that inhibit pathogen growth, enhancing their antagonistic activity.

Example:

- *Trichoderma harzianum* effectively parasitizes *Rhizoctonia solani*, suppressing damping-off disease in vegetables by producing chitinases and secondary metabolites like gliotoxin.

4. Induced Resistance

Some BCAs enhance the plant's innate immune system, preparing it to resist future pathogen attacks. This priming of plant defenses occurs through two major pathways: Induced Systemic Resistance (ISR) and Systemic Acquired Resistance (SAR).

4.1 Induced Systemic Resistance (ISR)

ISR is mediated by rhizosphere-associated microbes that activate jasmonic acid (JA) and ethylene (ET) signaling pathways in plants.

Mechanism:

» ISR primes the plant to produce defense-related enzymes and compounds when attacked by a pathogen. For example, *Pseudomonas fluorescens* triggers ISR, enhancing the production of peroxidases and phenylalanine ammonia-lyase in plants.

Applications:

» ISR induced by *Pseudomonas fluorescens* has been shown to reduce sheath blight severity in rice by 50%.

4.2 Systemic Acquired Resistance (SAR)

SAR is typically induced by pathogens or BCAs like *Trichoderma spp.* through the accumulation of salicylic acid (SA) and the activation of PR (pathogenesis-related) genes.

Example:

» SAR triggered by *Trichoderma spp.* in tomato plants provided long-term resistance against *Fusarium oxysporum*, reducing wilt incidence by 40%.

5. Quorum Sensing Disruption

Quorum sensing is a communication mechanism used by bacterial pathogens to coordinate activities such as biofilm formation and toxin production. Certain BCAs interfere with this process, effectively disarming the pathogens.

5.1 Quorum Quenching

BCAs like *Bacillus spp.* produce enzymes that degrade quorum-sensing molecules, such as N-acyl homoserine lactones (AHLs), preventing pathogens from coordinating their virulence.

5.2 Disruption of Biofilms

Biofilms are protective structures formed by pathogens to enhance their survival and virulence. By interfering with biofilm formation, BCAs expose pathogens to natural suppression mechanisms.

Example:

» In potato cultivation, *Bacillus subtilis* disrupted quorum sensing in *Pectobacterium carotovorum*, reducing soft rot incidence by 70%.

4.2.3 Resident vs. Introduced Biocontrol Agents

Biological control agents (BCAs) form the backbone of sustainable plant disease management strategies, and their deployment can follow two distinct approaches: utilizing resident antagonists or introducing external agents. Each approach offers unique benefits and presents its own set of challenges, making their comparative analysis essential for selecting optimal biocontrol strategies tailored to specific agroecosystems.

1. Resident Antagonists

Resident BCAs are naturally occurring microbes present within the soil or plant microbiome. These agents have evolved in the target environment, interacting with native pathogens and plant hosts over time, making them well-adapted to local conditions. Harnessing resident BCAs for biological control relies on managing their population dynamics and enhancing their activity through agricultural practices.

1.1 Advantages of Resident Antagonists

1.1.1 Adaptation to Local Conditions

Resident BCAs are inherently suited to the specific environmental conditions of their native habitats. Factors such as local soil pH, temperature, and cropping systems have already shaped their ecological fitness, ensuring their survival and activity under natural conditions.

Example:

» Native *Trichoderma spp.* in tropical climates exhibit superior competitive saprophytic ability (CSA) against soil-borne pathogens like *Fusarium oxysporum*, thriving in high-temperature environments where introduced species might falter.

1.1.2 Minimal Ecological Disruption

Utilizing resident BCAs maintains the natural balance of microbial communities. Since they are already integrated into the ecosystem, their deployment is unlikely to cause significant shifts in the native microbiome or unintended harm to non-target organisms.

Example:

» In organic farming systems, compost applications enhance the activity of native *Pseudomonas fluorescens* populations, boosting their ability to suppress root pathogens without disrupting soil biodiversity.

1.1.3 Cost-Effectiveness

The use of resident BCAs eliminates the need for costly production and transportation processes associated with introduced agents. Farmers can encourage their activity through cost-effective measures such as organic amendments and crop rotation.

Example:

» Incorporating crop residues into soil management practices in rice cultivation promotes the proliferation of native *Trichoderma harzianum*, reducing the inoculum of *Sclerotium rolfsii* at minimal expense.

1.2 Challenges of Resident Antagonists

1.2.1 Variable Efficacy

The effectiveness of resident BCAs can be inconsistent due to environmental variability, competition with other microbes, or fluctuations in soil health. Seasonal changes, nutrient availability, and climatic extremes can significantly impact their activity.

Example:

» While native *Streptomyces spp.* suppress *Verticillium dahliae* effectively under optimal conditions, their efficacy diminishes in drought-stressed soils.

1.2.2 Limited Commercial Exploitation

The site-specific nature of resident BCAs often limits their scalability for large-scale applications. Commercializing these agents requires extensive field trials to validate their effectiveness across diverse agroecosystems.

Example:

» Native *Pseudomonas spp.* populations that thrive in acidic soils may not perform as well in neutral or alkaline soils, restricting their broader application.

2. Introduced Antagonists

Introduced BCAs are external agents, often selected or engineered for their high efficacy against specific pathogens. These agents may originate from different ecological niches or undergo laboratory optimization to enhance their pathogen suppression capabilities.

2.1 Advantages of Introduced Antagonists

2.1.1 Targeted Action

Introduced BCAs are specifically selected for their ability to target particular pathogens. Their precision and potency ensure effective suppression of even aggressive pathogens, making them invaluable for high-risk scenarios.

Example:

» The commercial product Serenade®, based on *Bacillus subtilis*, is formulated for high antibiotic production, targeting both soil-borne and foliar pathogens in crops like tomatoes and strawberries.

2.1.2 Broad Application

Commercially produced introduced BCAs are often standardized for use in diverse cropping systems and agroecological zones. Their broad-spectrum activity and stability make them reliable across various conditions.

Example:

- *Trichoderma virens*, introduced into temperate regions, successfully suppresses *Pythium ultimum* and *Phytophthora spp.*, demonstrating adaptability to non-native environments.

2.1.3 Rapid Deployment

Introduced BCAs can be quickly mobilized to address emerging plant disease outbreaks, providing a flexible tool for disease management.

Example:

- During sudden outbreaks of sheath blight in rice, *Pseudomonas fluorescens* formulations have been rapidly deployed to mitigate losses, outperforming native microbial communities.

2.2 Challenges of Introduced Antagonists

2.2.1 Adaptation Issues

Introduced BCAs may struggle to establish themselves in unfamiliar environments. Factors such as soil type, climate, and microbial competition can impede their colonization and activity.

Example:

- Imported *Gliocladium spp.* formulations performed poorly in saline soils due to their limited tolerance to high salt concentrations.

2.2.2 Ecological Risks

The introduction of non-native BCAs can disrupt local microbial communities and pose risks to non-target organisms. Their potential invasiveness requires thorough ecological assessment before deployment.

Example:

- Overapplication of *Bacillus subtilis* in certain ecosystems has led to the suppression of beneficial native microbes, affecting overall soil health.

2.2.3 Higher Costs

The production, storage, and distribution of introduced BCAs involve significant investment. Additionally, their application may require specialized equipment or training, increasing operational costs for farmers.

Example:

- The use of *Bacillus subtilis* formulations in large-scale horticulture systems often necessitates cold storage and precision application technologies, raising costs for growers.

Comparative Analysis and Strategic Integration

A comparative evaluation of resident and introduced BCAs highlights their complementary strengths and challenges. Integrating both approaches can maximize biocontrol efficacy while minimizing risks.

1. Synergistic Use of Resident and Introduced BCAs

Combining native and introduced BCAs can provide synergistic effects, leveraging the strengths of each to suppress pathogens comprehensively.

Example:

- In maize cropping systems, co-inoculation with native *Trichoderma spp.* and introduced *Pseudomonas fluorescens* suppressed root rot caused by *Pythium spp.* more effectively than either agent alone.

2. Tailored Biocontrol Strategies

The selection of resident or introduced BCAs should be based on site-specific conditions, including soil health, climate, and the target pathogen. Resident BCAs may be more suitable for organic or smallholder farming systems, while introduced agents are better for large-scale, high-value crop production.

3. Enhancing Resident BCA Populations

Agronomic practices such as organic amendments, crop rotation, and reduced tillage can enhance the activity of resident BCAs, reducing the need for external inputs.

Example:

- Incorporating green manure in rice fields boosted the activity of resident *Pseudomonas fluorescens*, improving sheath blight suppression.

4.2.4 Comparative Strategies for Biocontrol Deployment

Biological control strategies have gained prominence as sustainable alternatives to chemical pesticides in plant disease management. By leveraging both resident (native) and introduced (non-native) biocontrol agents (BCAs), it is possible to create comprehensive, integrated strategies that maximize the efficacy of disease suppression while minimizing ecological and economic risks. Comparative deployment strategies emphasize the synergy between these two types of BCAs, tailoring their application to specific agricultural practices and environmental conditions.

1. Integration with Agricultural Practices

Integrating BCAs into existing agricultural practices provides a robust framework for enhancing their effectiveness and ensuring long-term sustainability. The synergy between biocontrol agents and agronomic techniques fosters microbial diversity, optimizes soil health, and creates conditions that favor beneficial microbes over pathogens.

1.1 Crop Rotation

Crop rotation disrupts the life cycles of host-specific pathogens, reducing their populations in the soil. This practice indirectly enhances the efficacy of both resident and introduced BCAs by decreasing pathogen pressure.

- **Resident BCAs**: Native *Trichoderma spp.* populations often thrive in rotation systems, as diverse root exudates from different crops provide varied nutrient sources.

- **Introduced BCAs**: Externally applied agents like *Pseudomonas fluorescens* benefit from reduced pathogen competition, enhancing their colonization and activity.

1.2 Organic Amendments

The addition of organic matter, such as compost or manure, enriches soil organic carbon and microbial activity, creating a favorable environment for BCAs.

Example:

- Compost application in tomato fields improved the population of native *Streptomyces spp.* while enhancing the effectiveness of introduced *Bacillus subtilis*, resulting in a 40% reduction in Fusarium wilt incidence.

1.3 Intercropping

Intercropping introduces multiple plant species into a single field, increasing biodiversity and altering the rhizosphere microbiome to favor beneficial microbes.

Example:

- In rice-legume intercropping systems, native *Pseudomonas spp.* populations were supplemented with *Trichoderma harzianum* formulations, leading to enhanced suppression of *Rhizoctonia solani*.

2. Co-Inoculation

- Co-inoculation refers to the simultaneous application of resident and introduced BCAs, leveraging their complementary mechanisms of action to achieve superior pathogen suppression. This approach targets multiple stages of pathogen development, creating a multi-pronged defense system.

2.1 Complementary Mechanisms

Combining BCAs that operate through different mechanisms enhances their overall efficacy. For instance:

- *Trichoderma spp.* suppress pathogens through mycoparasitism and enzymatic degradation, while *Pseudomonas fluorescens* produces antibiotics and VOCs to inhibit pathogen growth. Together, they provide comprehensive protection against soil-borne and foliar pathogens.

2.2 Enhanced Disease Suppression

- Co-inoculation provides synergistic effects, often achieving greater disease reduction than the sum of individual BCAs' impacts.

Example:

- In maize cultivation, co-inoculation with native *Gliocladium spp.* and introduced *Bacillus subtilis* reduced damping-off caused by *Pythium ultimum* by 70%, compared to 50% and 40% suppression when used individually.

2.3 Field Applications

- **Seed Coating**: Seeds can be coated with a combination of resident and introduced BCAs, ensuring early colonization of the rhizosphere and providing immediate protection against pathogens.
- **Soil Drench**: Applying co-inoculants as a soil drench targets soil-borne pathogens while promoting root health.

3. Tailored Formulations

Customizing formulations to combine resident and introduced BCAs ensures their stability and activity under diverse environmental conditions. These formulations incorporate protective carriers, growth stimulants, and delivery mechanisms to optimize the performance of BCAs.

3.1 Protective Carriers

Carriers like alginate beads, biochar, and vermiculite shield BCAs from environmental stressors such as UV radiation, desiccation, and pH fluctuations.

Example:

- Encapsulation of *Pseudomonas fluorescens* and native *Streptomyces spp.* in alginate beads increased their survival rates by 30% in saline soils, improving pathogen suppression.

3.2 Multi-Agent Formulations

Combining BCAs with different ecological niches and modes of action provides broader protection against pathogens. For instance:

- A formulation combining resident *Trichoderma spp.* (active in the rhizosphere) and introduced *Bacillus subtilis* (targeting foliar pathogens) can suppress multiple diseases in cropping systems.

3.3 Precision Delivery

Advancements in biocontrol delivery systems, such as drone-based spraying and drip irrigation integration, allow for precise application of tailored formulations, reducing wastage and improving field efficacy.

4. Monitoring and Adaptation

Regular monitoring of pathogen dynamics and environmental conditions is critical for the adaptive deployment of BCAs. This approach ensures long-term efficacy and allows for real-time adjustments to biocontrol strategies.

4.1 Pathogen Monitoring

Tracking pathogen populations through techniques such as PCR and ELISA enables the timely application of BCAs.

Example:

» In wheat fields, monitoring of *Fusarium graminearum* spore loads guided the application of *Pseudomonas fluorescens*, reducing Fusarium head blight by 50%.

4.2 Environmental Adaptation

Adjusting BCA application schedules and methods based on environmental conditions, such as soil moisture and temperature, enhances their effectiveness.

Example:

» Drip irrigation systems tailored to maintain optimal soil moisture levels improved the colonization of *Bacillus subtilis* in the rhizosphere, enhancing its ability to suppress *Ralstonia solanacearum*.

4.3 Field Trials and Feedback

Conducting field trials with combinations of resident and introduced BCAs provides valuable insights into their performance under local conditions. Feedback from these trials can inform future formulations and deployment strategies.

4.4 Biological Control of Soil-Borne and Foliar Diseases

The management of soil-borne and foliar diseases has long been a challenge in agriculture due to their devastating impact on crop yields and quality. Biological control provides an environmentally friendly and sustainable approach to managing these diseases, leveraging the capabilities of beneficial organisms to suppress pathogens through various mechanisms.

4.4.1 Biological Control of Soil-Borne Diseases

Soil-borne diseases are caused by pathogens that persist in the soil, often infecting plant roots and causing wilting, damping-off, and root rot. These pathogens include fungi, bacteria, and nematodes, many of which can survive as spores or saprophytes for extended periods. BCAs combat these pathogens by disrupting their survival strategies, colonizing the rhizosphere, and enhancing plant defenses.

1. Key Soil-Borne Pathogens

» *Fusarium spp.*: Cause Fusarium wilt in crops like tomatoes, bananas, and cucumbers.

» *Rhizoctonia solani*: Responsible for damping-off and root rot in crops such as beans and cotton.

» *Pythium spp.*: Induce seedling damping-off, especially in greenhouse and field crops.

» *Sclerotium rolfsii*: Affects a wide range of crops, leading to stem rot and damping-off.

2. Mechanisms of Biocontrol Against Soil-Borne Diseases

2.1 Rhizosphere Colonization

BCAs such as *Trichoderma spp.* and *Pseudomonas fluorescens* colonize the rhizosphere, competing with pathogens for nutrients and space.

Example:

- *Pseudomonas fluorescens* forms biofilms on root surfaces, creating a protective barrier that suppresses *Fusarium oxysporum*.

2.2 Antibiosis

Many soil-borne pathogens are suppressed by BCAs producing antimicrobial compounds, such as antibiotics and volatile organic compounds (VOCs).

Example:

- *Bacillus subtilis* produces iturin and surfactin, which disrupt fungal membranes, effectively reducing *Rhizoctonia solani* populations in bean fields.

2.3 Mycoparasitism

Fungal BCAs like *Trichoderma spp.* parasitize soil-borne fungal pathogens, breaking down their cell walls using enzymes such as chitinases and glucanases.

Example:

- *Trichoderma harzianum* suppresses *Sclerotium rolfsii* by degrading its sclerotia, preventing the pathogen's survival in the soil.

2.4 Induced Systemic Resistance (ISR)

Certain BCAs prime plant defense mechanisms, enabling crops to resist soil-borne infections.

Example:

- *Pseudomonas fluorescens* activates jasmonic acid and ethylene pathways in tomato plants, reducing the severity of Fusarium wilt.

3. Case Studies in Soil-Borne Disease Control

3.1 Suppression of Fusarium Wilt in Tomatoes

In tomato cultivation, applying *Trichoderma spp.* as a soil treatment reduced Fusarium wilt incidence by 60%. This was achieved through a combination of mycoparasitism, competition, and ISR activation.

3.2 Control of Pythium Damping-Off in Cucumbers

A study in cucumber greenhouses demonstrated that applying *Pseudomonas fluorescens* as a seed treatment reduced Pythium damping-off by 50%, highlighting the effectiveness of BCAs in protected cultivation systems.

4.4.2 Biological Control of Foliar Diseases

Foliar diseases affect the aerial parts of plants, including leaves, stems, and flowers. Common pathogens causing foliar diseases include fungi and bacteria, which thrive in humid environments and can rapidly spread across plant canopies. BCAs deployed as foliar sprays or systemic agents play a critical role in managing these diseases.

1. Key Foliar Pathogens

- *Botrytis cinerea*: Causes gray mold in crops like strawberries, grapes, and tomatoes.
- *Alternaria spp.*: Responsible for early blight in potatoes and tomatoes.
- *Xanthomonas spp.*: Induce bacterial blight in rice and other cereals.
- *Powdery Mildew Pathogens*: Affect a wide range of crops, including cucurbits and cereals.

2. Mechanisms of Biocontrol Against Foliar Diseases

2.1 Antibiosis and VOCs

BCAs such as *Bacillus subtilis* and *Pseudomonas fluorescens* produce antimicrobial compounds that suppress foliar pathogens.

Example:

- Foliar application of *Bacillus subtilis* reduced gray mold severity in strawberries by 50%, attributed to the production of lipopeptides and VOCs.

2.2 Competitive Exclusion

BCAs colonize leaf surfaces, occupying niches that would otherwise be exploited by pathogens.

Example:

- *Pseudomonas fluorescens* forms a protective biofilm on wheat leaves, suppressing infections caused by *Alternaria spp.*

2.3 Enzymatic Activity

Enzymes produced by BCAs degrade pathogen cell walls, reducing their infectivity.

Example:

- *Trichoderma viride* produces chitinases that break down the cell walls of powdery mildew fungi on cucumber leaves.

2.4 Induction of Systemic Acquired Resistance (SAR)

Certain BCAs activate plant defense pathways, leading to enhanced resistance against foliar pathogens.

Example:

- *Bacillus subtilis* triggers SAR in rice plants, reducing bacterial blight severity caused by *Xanthomonas oryzae*.

3. Case Studies in Foliar Disease Control

3.1 Management of Powdery Mildew in Grapes

Applying *Trichoderma harzianum* as a foliar spray in grapevines reduced powdery mildew incidence by 45%, demonstrating its efficacy in controlling fungal pathogens in perennial crops.

3.2 Control of Early Blight in Tomatoes

In a trial, *Bacillus subtilis* sprays reduced early blight severity in tomatoes by 40%, improving both yield and fruit quality.

4.4.3 Integrated Management Strategies

The effectiveness of biological control for soil-borne and foliar diseases is enhanced when integrated with other disease management practices, such as crop rotation, organic amendments, and resistant varieties.

1. Synergistic Use of BCAs

Combining soil-applied and foliar-applied BCAs provides comprehensive protection against both root and aerial pathogens.

Example:

- » Co-application of *Trichoderma harzianum* (soil treatment) and *Bacillus subtilis* (foliar spray) reduced the combined severity of root rot and gray mold in cucumber fields by 60%.

2. Integration with Cultural Practices

Practices such as proper irrigation management, canopy pruning, and crop residue incorporation create conditions that favor BCAs while suppressing pathogens.

3. Biopesticide Formulations

Commercial products like Serenade® (*Bacillus subtilis*) and RootShield® (*Trichoderma harzianum*) provide ready-to-use solutions for managing soil-borne and foliar diseases.

4.5 Compatibility of Bioagents with Agrochemicals and Other Microbial Antagonists

The integration of biological control agents (BCAs) with conventional agrochemicals and other microbial antagonists is essential for developing comprehensive and sustainable plant disease management strategies. Effective compatibility ensures that BCAs retain their efficacy when combined with fungicides, insecticides, or fertilizers, as well as when used alongside other BCAs.

4.5.1 Compatibility with Agrochemicals

Agrochemicals, including fungicides, herbicides, and insecticides, remain integral to many cropping systems. Their combined use with BCAs can be beneficial or detrimental depending on the interaction. While some agrochemicals may inhibit BCAs, others can act synergistically, enhancing the overall disease management efficacy.

1. Positive Interactions with Agrochemicals

Certain agrochemicals, especially low-toxicity formulations, can coexist with BCAs without compromising their activity.

Example:

- » Copper-based fungicides are compatible with *Pseudomonas fluorescens*, and their combined use has been shown to reduce bacterial blight in rice without affecting the bacterium's colonization or activity.

2. Negative Interactions with Agrochemicals

Some agrochemicals, particularly broad-spectrum fungicides, can inhibit the growth or metabolic activity of BCAs.

Example:

- The application of benzimidazole fungicides such as carbendazim can suppress the growth of *Trichoderma spp.*, limiting its ability to control soil-borne pathogens.

3. Strategies to Improve Compatibility

- **Selection of Low-Toxicity Agrochemicals**: Using selective agrochemicals minimizes their impact on BCAs. For instance, systemic fungicides like azoxystrobin are often less inhibitory to microbial agents compared to contact fungicides.
- **Timing of Application**: Staggering the application of BCAs and agrochemicals reduces direct interactions. For example, applying *Bacillus subtilis* 48 hours before a fungicide allows the bacterium to establish itself and reduce potential negative impacts.
- **Encapsulation Technologies**: Protective encapsulation of BCAs in carriers such as alginate beads can shield them from the harmful effects of agrochemicals.

4.5.2 Compatibility with Fertilizers and Soil Amendments

Fertilizers and soil amendments can influence the survival and activity of BCAs in the soil. Balanced nutrient management supports the coexistence of BCAs and fertilizers, enhancing disease suppression and crop productivity.

1. Synergistic Effects

Organic fertilizers, such as compost and manure, provide substrates for BCAs, promoting their proliferation and activity.

Example:

- The application of compost enriched with *Trichoderma harzianum* not only improved nutrient availability but also suppressed *Fusarium oxysporum* in tomato cultivation.

2. Potential Inhibitory Effects

Inorganic fertilizers, especially those with high salt concentrations, may negatively affect microbial populations by altering soil pH or osmotic potential.

Example:

- Excessive application of ammonium-based fertilizers can inhibit the activity of nitrogen-fixing bacteria such as *Rhizobium spp.*, reducing their effectiveness in legume-based cropping systems.

3. Strategies for Co-Use

- **Optimized Fertilizer Dosage**: Avoiding overapplication of chemical fertilizers ensures that BCAs are not suppressed by adverse soil conditions.
- **Integration with Organic Amendments**: Combining inorganic fertilizers with organic matter buffers their effects on microbial communities, supporting BCAs.

4.5.3 Compatibility Among Microbial Antagonists

Combining multiple BCAs that employ complementary mechanisms of action can provide synergistic benefits, improving disease suppression and promoting plant growth. However, not all microbial antagonists are compatible, and their interactions must be carefully assessed.

1. Synergistic Interactions

BCAs with different ecological niches or modes of action can complement each other, targeting pathogens at multiple stages of their life cycle.

Example:

- Co-inoculation of *Trichoderma harzianum* (a mycoparasitic fungus) with *Pseudomonas fluorescens* (a rhizosphere-colonizing bacterium) suppressed both root and foliar pathogens in maize, reducing disease incidence by 60%.

2. Antagonistic Interactions

Some BCAs may compete for the same resources or produce inhibitory compounds that suppress each other's activity.

Example:

- The application of *Bacillus subtilis* in combination with *Streptomyces spp.* sometimes results in reduced efficacy, as both agents compete for the same ecological niche in the rhizosphere.

3. Strategies to Improve Compatibility

- **Complementary Selection**: Choosing BCAs with distinct ecological roles or mechanisms minimizes competition. For instance, combining a root-colonizing bacterium with a phyllosphere-targeting fungus ensures effective disease suppression across different plant zones.
- **Sequential Application**: Applying BCAs in succession, rather than simultaneously, can reduce competition and enhance their individual efficacy.
- **Formulated Products**: Commercial formulations combining compatible BCAs ensure that antagonistic interactions are minimized while maintaining synergistic benefits.

4.5.4 Methods for Assessing Compatibility

Understanding the interactions between BCAs, agrochemicals, and other antagonists requires rigorous compatibility testing. These assessments help predict potential synergies or conflicts, guiding their effective co-use in the field.

1. In Vitro Testing

Laboratory assays are used to evaluate the effects of agrochemicals or other BCAs on the growth, metabolism, and activity of target BCAs.

Example:

» Disk diffusion assays assess the inhibitory effects of fungicides on *Trichoderma spp.* by measuring the growth inhibition zone.

2. Soil Microcosm Studies

Simulated soil environments are used to study interactions between BCAs, agrochemicals, and native microbial communities.

Example:

» A soil microcosm study demonstrated that *Bacillus subtilis* retained its biocontrol efficacy in the presence of copper-based fungicides when applied at reduced doses.

3. Field Trials

Field evaluations under real-world conditions provide insights into the long-term compatibility and efficacy of combined treatments.

Example:

» Field trials with co-applied *Trichoderma harzianum* and organic amendments in cucumber fields revealed enhanced suppression of *Rhizoctonia solani.*

4.5.5 Case Studies in Compatibility

1. Integrated Use of BCAs and Fungicides

In strawberry fields, the co-application of *Pseudomonas fluorescens* with a reduced dose of azoxystrobin fungicide suppressed gray mold by 70%, demonstrating that low-toxicity fungicides can enhance BCA activity.

2. Co-Inoculation of BCAs

In greenhouse trials, *Trichoderma harzianum* and *Bacillus subtilis* were successfully combined to manage root rot in tomatoes, reducing disease incidence by 50% compared to individual treatments.

3. Synergistic Effects of Organic Amendments and BCAs

In rice cultivation, combining compost enriched with *Streptomyces spp.* and organic manure enhanced nitrogen availability while suppressing sheath blight caused by *Rhizoctonia solani.*

Chapter - 5

Commercial Production and Delivery Systems of Bioagents

The commercial production and delivery of bioagents are critical components of the biological control industry, enabling the large-scale application of biocontrol strategies to manage pests and pathogens effectively. These systems involve the mass production, formulation, packaging, and distribution of bioagents to ensure their viability, stability, and ease of use across diverse agricultural settings. Bioagents, including beneficial microorganisms such as bacteria, fungi, and viruses, as well as macroscopic organisms like predatory insects and nematodes, require specialized production methods tailored to their biological characteristics and ecological functions. The production process begins with the isolation and selection of highly effective strains or species, which undergo extensive screening for traits such as pathogen specificity, environmental adaptability, and compatibility with existing farming practices. For instance, strains of *Bacillus subtilis* and *Trichoderma harzianum* are selected for their ability to suppress soil-borne pathogens and promote plant growth, while predatory mites like *Phytoseiulus persimilis* are chosen for their efficacy against spider mites in horticultural crops.

The production of microbial bioagents typically involves fermentation technology, wherein large quantities of microorganisms are cultivated in bioreactors under controlled conditions. Factors such as temperature, pH, oxygen levels, and nutrient availability are optimized to maximize biomass yield and metabolite production. Solid-state fermentation is often used for fungi like *Trichoderma spp.* and *Beauveria bassiana*, while submerged fermentation is preferred for bacterial agents such as *Pseudomonas fluorescens* and *Bacillus spp.*. Once the desired biomass is obtained, it is processed into various formulations, including powders, granules, liquid suspensions, and encapsulated products. These formulations are designed to enhance the shelf life, stability, and ease of application of the bioagents. For example, encapsulation techniques protect bioagents from environmental stressors such as ultraviolet radiation and desiccation, ensuring their viability during storage and transport. The use of carrier materials such as peat, vermiculite, or alginate further improves the delivery and establishment of bioagents in the field, providing a favorable environment for their growth and activity.

The production of macroscopic bioagents, such as predatory insects and parasitic nematodes, involves rearing the organisms under controlled conditions that mimic their natural habitats. For instance, predatory insects like *Cryptolaemus montrouzieri* and parasitoids like *Encarsia formosa* are mass-reared on host insects, which serve as their food source during production. Similarly, entomopathogenic nematodes like *Steinernema spp.* are cultivated in artificial media or live insect hosts. The production systems are designed to ensure high reproductive rates, genetic consistency, and behavioral effectiveness of the bioagents. After production, these bioagents are carefully packaged to maintain their viability during storage and transport. Cold storage or refrigeration is often used to extend the shelf life of macroscopic bioagents, allowing farmers to access them when needed.

The delivery systems for bioagents are equally important in ensuring their successful application in the field. Various methods are employed depending on the type of bioagent, the target pest or pathogen, and the crop management system. For microbial bioagents, delivery systems include seed coatings, soil drenches, foliar sprays, and drip irrigation. Seed coatings, for example, involve applying bioagents to seeds before planting, allowing them to colonize the rhizosphere as the plant grows. This method is commonly used for bacterial agents like *Rhizobium spp.* and *Azospirillum spp.*, which enhance nitrogen fixation and root health. Soil drenches are applied directly to the soil to target soil-borne pathogens, while foliar sprays are used to protect above-ground plant parts from diseases and pests. For macroscopic bioagents, release techniques such as inundative releases and inoculative releases are employed. Inundative releases involve the large-scale introduction of bioagents, such as predatory insects, to achieve immediate pest suppression, while inoculative releases introduce smaller numbers of bioagents to establish a long-term population that provides ongoing control.

The scalability and success of commercial production and delivery systems are influenced by various factors, including cost-effectiveness, ease of use, and farmer adoption. Developing cost-effective production methods is essential to make bioagents accessible to smallholder farmers, who often face financial constraints. For instance, low-cost fermentation systems and locally sourced carrier materials can significantly reduce production costs without compromising the quality of the bioagents. The ease of use of bioagent formulations and delivery methods is another critical factor. Ready-to-use formulations that are compatible with existing farming equipment and practices are more likely to be adopted by farmers. For example, bioagent formulations that can be applied using conventional sprayers or integrated into drip irrigation systems streamline the application process, making biological control more practical for large-scale farming.

In addition to technical considerations, regulatory frameworks and quality assurance play a vital role in the commercial production and delivery of bioagents. Governments and industry bodies establish guidelines for the registration, labeling, and distribution of bioagents to ensure their safety, efficacy, and environmental compatibility. Quality control measures, such as testing for microbial purity, viability, and absence of contaminants, are implemented to maintain high standards and build farmer confidence in the products. Furthermore, extension services and training programs are essential for educating farmers about the proper use of bioagents, including timing, dosage, and application methods. These initiatives help bridge the gap between commercial production and on-farm implementation, maximizing the impact of biological control.

5.1 Techniques and Processes for Commercial Production of Antagonists

The commercial production of microbial antagonists is a sophisticated process that integrates microbiological, biochemical, and industrial methodologies. These antagonists, including bacteria, fungi, and other beneficial microorganisms, are crucial for sustainable agricultural practices as they suppress plant pathogens and reduce reliance on chemical pesticides. The techniques and processes involved in their production must ensure high efficacy, stability, scalability, and compatibility with field applications.

5.1.1 Isolation and Screening of Antagonists

The initial step in the commercial production of microbial antagonists involves the meticulous process of isolating effective strains from natural environments. These microorganisms, which include bacteria, fungi, and actinomycetes, are sourced from ecological niches where natural antagonistic interactions with pathogens occur. The objective is to identify and cultivate microbial populations with the inherent ability to suppress plant pathogens, either by direct inhibition or through promoting plant health.

Sourcing Microbial Antagonists

Natural environments serve as a vast reservoir of microbial diversity. Common sources include:

- **Soil Samples:** Microorganisms in the soil are particularly well-adapted to compete with pathogens. Rhizospheric soil, in proximity to plant roots, often harbours antagonists due to its nutrient-rich and competitive environment.
- **Compost and Organic Matter:** Decaying plant material and compost piles provide a habitat for microbes that decompose organic matter and compete with pathogens.
- **Plant Tissues and Surfaces:** Endophytes and epiphytes associated with healthy plants often exhibit antagonistic traits, making them valuable sources for isolation.
- **Extreme Environments:** Microorganisms from harsh environments, such as saline soils or arid regions, may exhibit unique traits like stress tolerance and robust biocontrol capabilities.

Sampling strategies emphasize regions with high biodiversity and minimal anthropogenic disturbance, as such areas are likely to yield unique and potent microbial strains.

Isolation of Microorganisms

Microbial isolation begins with the preparation of environmental samples, which are serially diluted and plated on selective media to cultivate diverse microbial populations. The choice of media, such as nutrient agar, potato dextrose agar, or King's B medium, is tailored to favor the growth of specific microbial groups. To ensure a representative microbial population, techniques like soil suspension preparation, root washing, and direct plating of plant tissues are used. Colony morphology, pigmentation, and growth characteristics guide the preliminary selection of isolates.

Screening for Biocontrol Potential

Once isolates are obtained, they are subjected to rigorous screening tests to assess their ability to suppress plant pathogens. These tests include:

1. **Dual-Culture Assays:**
   - In this widely used method, the potential antagonist is co-cultured with the target pathogen on an agar plate. Zones of inhibition, where pathogen growth is restricted, indicate antagonistic activity.
   - This technique also helps evaluate the interaction dynamics between the antagonist and the pathogen, such as direct competition or antibiosis.

2. **Production of Volatile Organic Compounds (VOCs):**
   » Certain microorganisms release VOCs that inhibit pathogen growth. Isolates are tested for their ability to produce such compounds in closed systems where pathogen suppression occurs without physical contact.
3. **Enzymatic Activity Assays:**
   » Enzymes such as chitinase, cellulase, and protease play a crucial role in degrading the cell walls of fungal and bacterial pathogens. Microorganisms are tested for the production of these enzymes on substrate-specific media.
   » Isolates with high enzymatic activity are prioritized for further study.
4. **Siderophore Production:**
   » Siderophores are iron-chelating compounds that limit the availability of this essential nutrient to pathogens, effectively suppressing their growth. Tests like the Chrome Azurol S (CAS) assay are used to identify siderophore-producing strains.
5. **Competition for Nutrients and Space:**
   » Antagonists are evaluated for their ability to outcompete pathogens for essential nutrients and colonization sites. This involves co-cultivation experiments and studies on substrate utilization.

Characterization of Mechanisms

Understanding the modes of action employed by antagonists is critical for their successful application. Mechanisms such as antibiosis, competition, parasitism, and induction of systemic resistance in plants are characterized using:

» **Microscopy:** Scanning and transmission electron microscopy reveal physical interactions like mycoparasitism or the formation of biofilms.

» **Biochemical Assays:** The production of secondary metabolites, such as antibiotics or antifungal compounds, is quantified using chromatography and spectrometry techniques.

» **Molecular Tools:** Advanced genomic, transcriptomic, and proteomic analyses help identify genes and proteins involved in biocontrol. This provides insights into the metabolic pathways and regulatory networks underpinning antagonistic activity.

High-Throughput Screening

Modern advancements in technology enable high-throughput screening of large microbial libraries. Automated systems equipped with robotics and imaging tools allow researchers to quickly evaluate thousands of isolates for traits like pathogen inhibition, enzyme production, and stress tolerance.

» **Microplate Assays:** These miniaturized assays facilitate the rapid assessment of microbial activity, reducing time and labor requirements.

» **Omics Approaches:** Metagenomics, metatranscriptomics, and metabolomics uncover functional traits within complex microbial communities, helping identify promising candidates for isolation.

Strain Selection

After initial screening, the most effective strains are selected for further study. Selection criteria include:

- Broad-spectrum activity against multiple pathogens.
- Compatibility with agricultural practices, such as soil or foliar application.
- Stability and viability under storage and field conditions.
- Non-toxicity to plants, humans, and the environment.

Selected strains undergo additional testing under greenhouse and field conditions to validate their performance in real-world scenarios. Isolates that consistently demonstrate high efficacy and environmental adaptability progress to the next stages of strain optimization and mass production.

5.1.2 Strain Selection and Improvement

Once microbial antagonists are isolated and screened for their biocontrol potential, the most promising strains are selected for further refinement and enhancement. This process, termed strain selection and improvement, is critical for optimizing the performance of microbial antagonists in diverse agricultural settings. These strains are evaluated based on their antagonistic efficacy, resilience to environmental stresses, ability to colonize plant systems, and compatibility with commercial production requirements. Advanced techniques are employed to improve their characteristics, ensuring higher efficacy, stability, and adaptability.

Criteria for Strain Selection

The selection of microbial strains for biocontrol applications involves evaluating several key traits, including:

- **Antagonistic Activity:** The ability to suppress multiple pathogens through mechanisms like antibiosis, competition, or parasitism.
- **Environmental Resilience:** The strain's ability to survive under varying temperatures, pH levels, and moisture conditions.
- **Colonization Ability:** Efficient root, shoot, or rhizosphere colonization without adversely affecting plant health.
- **Shelf-Life Stability:** High viability during storage and after formulation.
- **Non-Toxicity:** Absence of harmful effects on plants, non-target organisms, and the environment.

Only strains that meet these stringent criteria advance to the improvement phase.

Techniques for Strain Improvement

Strain improvement is a critical step in enhancing the efficacy and field performance of microbial antagonists. Various methodologies are employed to enhance desirable traits such as antimicrobial compound production, stress tolerance, and colonization efficiency.

1. Mutagenesis and Selection

Mutagenesis is a classical technique used to induce genetic variation in microbial strains. It involves exposing microorganisms to mutagens, which can be physical (e.g., UV radiation, X-rays) or chemical (e.g., ethyl methane-sulfonate, nitrosoguanidine). These mutagens introduce random mutations in the microbial genome, some of which may enhance desirable traits.

Procedure:

» The microbial culture is exposed to mutagens under controlled conditions.

» Mutant strains are isolated and screened for improved traits such as increased production of secondary metabolites or enhanced pathogen suppression.

» Desirable mutants are subjected to stability testing to ensure the heritability of the improved traits.

Applications:

- Enhanced production of antibiotics, enzymes, and siderophores.
- Improved tolerance to environmental stresses like drought, salinity, and heat.

2. Protoplast Fusion

Protoplast fusion is a technique used to combine the genetic material of two different strains, creating hybrid strains with superior traits. This method is particularly useful for microorganisms where sexual reproduction is absent or limited.

Procedure:

» Cell walls of two microbial strains are enzymatically removed to create protoplasts.

» The protoplasts are fused using chemical agents (e.g., polyethylene glycol) or electrical pulses (electrofusion).

» The fused protoplasts are regenerated into hybrid strains on selective media.

Applications:

» Development of strains with combined biocontrol mechanisms, such as simultaneous antibiosis and nutrient competition.

» Introduction of beneficial traits like higher sporulation efficiency or enhanced root colonization.

3. Genetic Engineering

Genetic engineering offers precise and targeted methods to improve microbial strains by modifying their DNA. This technique involves inserting, deleting, or modifying specific genes responsible for traits like antimicrobial compound production, stress resistance, or plant colonization.

Applications:

» Enhancing the biosynthesis of antibiotics, such as pyrrolnitrin and phenazines, in bacterial strains like *Pseudomonas* and *Bacillus* species.

- Introducing genes for stress tolerance (e.g., heat-shock proteins) to improve strain viability under field conditions.
- Engineering fungi like *Trichoderma* to overproduce enzymes such as cellulases and chitinases, increasing their mycoparasitic activity.

Emerging Tools:

- CRISPR-Cas9 technology allows for precise genome editing to create strains with enhanced biocontrol efficacy and environmental resilience.
- Synthetic biology approaches enable the design of novel metabolic pathways for improved secondary metabolite production.

4. Adaptive Laboratory Evolution (ALE)

Adaptive laboratory evolution involves subjecting microbial strains to specific environmental conditions over an extended period to select for individuals that exhibit improved traits. This technique mimics natural selection in a controlled environment.

Procedure:

- The strain is continuously cultured under stress conditions, such as high salinity, drought-like conditions, or pathogen presence.
- Over time, microbial populations adapt to the imposed conditions, resulting in strains with enhanced resilience or biocontrol activity.
- Selected strains are sequenced to identify genetic changes associated with the improved traits.

Applications:

- Development of strains that can withstand harsh environmental conditions in agricultural fields.
- Selection of antagonists with enhanced colonization and persistence in the rhizosphere.

Rigorous Testing for Improved Strains

Once improved strains are developed through the aforementioned techniques, they undergo comprehensive testing under controlled and field conditions to ensure their efficacy, stability, and safety.

1. **Controlled Environment Testing:**
    - Laboratory assays, such as dual-culture tests and enzyme activity assays, validate the enhanced biocontrol properties.
    - Greenhouse experiments evaluate the strains' ability to suppress pathogens and promote plant growth under semi-natural conditions.
2. **Field Trials:**
    - Large-scale field trials assess the performance of the improved strains under real-world conditions, including varying soil types, climatic conditions, and crop varieties.

- » Parameters such as pathogen suppression, crop yield, and environmental impact are monitored over multiple growing seasons.

3. **Safety Assessments:**

- » Toxicological studies ensure that the improved strains do not produce harmful metabolites or pose risks to non-target organisms.
- » Environmental impact assessments evaluate potential effects on soil microbial communities and ecosystem balance.

5.1.3 Mass Production Techniques

The mass production of microbial antagonists is a critical step in the commercialization of biological control agents. This process requires the optimization of growth conditions, fermentation parameters, and bioreactor design to achieve high yields of viable, active microorganisms. The chosen production method must ensure the economic viability of the process while maintaining the effectiveness and stability of the antagonists. Various fermentation techniques are employed, each tailored to the specific needs of the microorganism and the intended application.

Submerged Fermentation (SmF)

Submerged fermentation is one of the most widely used methods for producing microbial antagonists, particularly bacteria and certain fungi. This technique involves growing microorganisms in liquid nutrient media within bioreactors. SmF is ideal for organisms that thrive in aqueous environments and require precise control over growth conditions.

Process Overview:

- » Microbial cultures are inoculated into sterilized liquid media containing essential nutrients such as carbon, nitrogen, and trace elements.
- » Bioreactors are equipped with aeration systems, agitators, and sensors to regulate pH, temperature, dissolved oxygen, and agitation speed.
- » Growth is monitored to ensure optimal biomass production, with harvest occurring when the microorganism reaches its peak activity.

Advantages:

- » High level of control over growth conditions, ensuring consistent quality and yield.
- » Suitability for large-scale production due to automated systems and ease of monitoring.
- » Effective for producing secondary metabolites, such as antibiotics and enzymes.

Applications:

- » Production of bacterial antagonists like *Bacillus subtilis*, known for its broad-spectrum antifungal activity and ability to produce lipopeptides.

» Cultivation of fungal antagonists like *Trichoderma spp.*, which produce cellulases and chitinases for pathogen suppression.

Solid-State Fermentation (SSF)

Solid-state fermentation involves the growth of microorganisms on a moist solid substrate without free-flowing liquid. This method is particularly suited for filamentous fungi and actinomycetes, which naturally grow in such environments. SSF replicates the natural habitat of these microorganisms, making it an efficient technique for their mass production.

Process Overview:

- Substrates such as wheat bran, rice husk, or soybean meal are moistened and sterilized before inoculation with microbial cultures.
- The inoculated substrate is incubated under controlled conditions of temperature and humidity.
- Periodic mixing may be performed to ensure uniform growth and prevent substrate drying.

Advantages:

- Low water and energy requirements compared to SmF.
- Reduced risk of contamination due to the absence of free water.
- High yields of spores, enzymes, and secondary metabolites.

Applications:

- Large-scale production of *Trichoderma spp.* for soil and seed treatments.
- Cultivation of actinomycetes like *Streptomyces*, known for their ability to produce antibiotics and siderophores.

Fed-Batch Fermentation

Fed-batch fermentation combines aspects of batch and continuous fermentation, offering flexibility and efficiency in microbial production. In this method, nutrients are periodically added to the bioreactor to sustain microbial growth and prolong activity.

Process Overview:

- Initial inoculation occurs in a batch mode, with microorganisms growing in a defined volume of liquid media.
- As nutrients are depleted, additional feed is introduced at intervals to maintain microbial growth.
- The process continues until biomass production reaches the desired level, after which the culture is harvested.

Advantages:

- Prolonged microbial activity and higher yields compared to batch fermentation.
- Better control over substrate concentration, reducing the risk of byproduct accumulation.
- Suitable for microorganisms requiring specific growth phases for metabolite production.

Applications:

- Production of antagonists like *Pseudomonas fluorescens*, which require controlled conditions to produce bioactive metabolites.
- Enhanced production of fungal biomass and spores for agricultural formulations.

Continuous Fermentation

Continuous fermentation ensures the steady production of microbial antagonists by maintaining a constant supply of fresh media and simultaneous harvesting of microbial biomass. This method is particularly useful for high-demand production where consistency is critical.

Process Overview:

- Fresh media are continuously supplied to the bioreactor, maintaining optimal nutrient levels.
- Microbial biomass is harvested at regular intervals while the remaining culture is retained in the system.
- Advanced monitoring systems regulate growth conditions to ensure consistent output.

Advantages:

- Steady and predictable production rates, making it ideal for large-scale operations.
- Efficient use of resources, as the process minimizes downtime between batches.
- Uniform quality of the final product due to constant operating conditions.

Applications:

- Industrial production of microbial antagonists for use in integrated pest management (IPM) systems.
- Large-scale production of bioactive metabolites for commercial formulations.

Key Factors in Mass Production

The success of microbial antagonist production depends on optimizing several critical factors:

1. **Growth Media:**

   » The composition of the media affects microbial growth, metabolite production, and yield.

   » Cost-effective substrates, such as agricultural byproducts, are often used to reduce production costs.

2. **Bioreactor Design:**
   - Bioreactors must provide uniform mixing, efficient aeration, and precise control of environmental conditions.
   - Specialized designs, such as air-lift or packed-bed bioreactors, may be employed for specific microorganisms.
3. **Fermentation Parameters:**
   - Parameters such as pH, temperature, oxygen concentration, and agitation speed are optimized for each microorganism.
   - Scale-up studies ensure that conditions optimized at the laboratory scale can be replicated in industrial settings.
4. **Downstream Processing:**
   - After fermentation, microbial biomass or metabolites are harvested, concentrated, and processed into stable formulations.
   - Techniques such as centrifugation, filtration, and spray-drying are commonly used.

Choice of Fermentation Technique

The selection of an appropriate fermentation technique depends on several factors, including:

- **Type of Microorganism:** Bacteria and yeast typically thrive in SmF, while filamentous fungi and actinomycetes are better suited for SSF.
- **Cost Considerations:** SSF is often more economical for small-scale operations, whereas SmF and fed-batch fermentation are preferred for large-scale production.
- **Intended Formulation:** Techniques that produce high spore counts are favored for dry formulations, while liquid formulations require high cell viability.

5.1.4 Formulation Development

Formulation development is a crucial step in the commercial production of microbial antagonists. After mass production, these beneficial microorganisms must be converted into stable, effective, and easy-to-use products suitable for agricultural applications. The formulation process ensures that the microbial antagonists remain viable during storage, transportation, and application, while also protecting them from environmental stresses. Additionally, the formulation type is tailored to the target crop, pathogen, and application method, enhancing the efficacy and usability of the product.

Powder Formulations

Powder formulations are among the most widely used types of microbial products due to their ease of handling, storage, and application. In this method, microbial cells or spores are blended with inert carriers such as talc, clay, or peat to create a fine, dry powder.

Process Overview:

- Microbial biomass, harvested from fermentation, is washed and concentrated.

- » The concentrated biomass is mixed with a sterile carrier material to create a uniform powder.
- » Additives such as desiccants and dispersing agents may be included to enhance the formulation's performance.

Advantages:

- » Long shelf life under proper storage conditions.
- » Ease of application, particularly for seed treatment and soil incorporation.
- » Compatibility with existing agricultural equipment, such as seed drills and spreaders.

Applications:

- » Powder formulations of *Trichoderma spp.* for seed coating to protect against soilborne pathogens.
- » Talc-based *Bacillus subtilis* products for foliar sprays and soil treatments.

Granular Formulations

Granular formulations are larger particles designed for direct soil application. These formulations provide a slow release of microbial antagonists, making them ideal for sustained action against soilborne pathogens.

Process Overview:

- » Microbial cells or spores are mixed with a granular carrier material, such as clay, corn cob grits, or vermiculite.
- » The mixture is shaped into granules using extrusion or compaction processes and dried to achieve the desired moisture content.

Advantages:

- Easy to apply using conventional methods like broadcasting or incorporation into the soil.
- Gradual release of microbial cells, ensuring prolonged activity in the soil.
- Reduced risk of environmental contamination due to localized application.

Applications:

- Granular formulations of *Pseudomonas fluorescens* for controlling root rot and damping-off diseases.
- Soil-applied *Trichoderma spp.* granules for protecting root systems against nematodes and fungi.

Liquid Formulations

Liquid formulations consist of microbial cells or spores suspended in a liquid medium. These formulations are highly versatile and can be applied as foliar sprays, seed treatments, or through irrigation systems.

Process Overview:

- Microbial biomass is suspended in a liquid medium containing protective agents, such as glycerol, to enhance stability.
- Emulsifiers and surfactants are added to ensure uniform distribution during application.

Advantages:

- High viability of microbial cells due to the absence of desiccation stress.
- Easy to mix with other agricultural inputs, such as fertilizers and pesticides.
- Ideal for precision applications, such as drip irrigation and foliar spraying.

Applications:

- Liquid formulations of *Bacillus subtilis* for foliar application against leaf spot and blight diseases.
- Suspension concentrates of *Trichoderma spp.* for seed treatments to enhance germination and disease resistance.

Encapsulation

Encapsulation techniques involve enclosing microbial cells or spores in protective materials, such as alginate or synthetic polymers. This method safeguards the microorganisms from environmental stresses, such as desiccation, UV radiation, and temperature fluctuations, while also enhancing their shelf life.

Process Overview:

- Microbial cells are mixed with a gelling agent, such as sodium alginate.
- The mixture is dropped into a cross-linking solution (e.g., calcium chloride), forming beads or capsules.
- The encapsulated product is dried or stored in a liquid medium, depending on the intended use.

Advantages:

- Enhanced stability and viability under harsh environmental conditions.
- Controlled release of microbial cells, ensuring prolonged activity.
- Improved compatibility with mechanical application systems.

Applications:

- Encapsulated *Bacillus thuringiensis* for targeting insect pests in soil and foliage.
- Alginate beads of *Trichoderma spp.* for slow-release soil application.

Pellets and Tablets

Pellets and tablets are compact, solid formulations designed for specific applications, such as root dipping or direct soil placement. These formulations are easy to handle and apply, offering a precise dosage of microbial antagonists.

Process Overview:

- Microbial cells or spores are mixed with binding agents and compressed into pellets or tablets.
- The products are dried to reduce moisture content and enhance shelf life.

Advantages:

- Convenient for localized applications, such as root dipping or placing at planting sites.
- Reduced wastage due to precise application.
- Prolonged activity in the field due to slow release of microbial cells.

Applications:

- Tablets of *Azospirillum* for enhancing nitrogen fixation in legume crops.
- Pellets of *Trichoderma spp.* for treating transplant roots to prevent damping-off diseases.

Critical Parameters in Formulation Development

Several factors influence the stability, efficacy, and usability of microbial formulations. Optimizing these parameters ensures the success of the final product:

1. **Water Activity:**
   » Low water activity reduces microbial metabolism during storage, prolonging shelf life.
   » Desiccants or drying techniques are employed to achieve optimal moisture levels.
2. **pH:**
   » The pH of the formulation medium is adjusted to maintain microbial viability and prevent spoilage.
   » Buffering agents are added to stabilize pH during storage.
3. **Protective Additives:**
   » Cryoprotectants, such as glycerol and trehalose, protect microbial cells during freeze-drying or storage.
   » UV protectants, such as titanium dioxide, shield microorganisms from harmful radiation during field application.
4. **Storage and Transportation:**
   » Packaging materials and conditions, such as vacuum sealing or refrigeration, are optimized to maintain product integrity.
   » Formulations are tested under simulated storage and transport conditions to ensure robustness.

5.1.5 Quality Control and Assurance

Quality control and assurance are fundamental to the commercial production of microbial antagonists, ensuring that products consistently meet efficacy, safety, and regulatory standards. The complex nature of microbial products necessitates rigorous quality control protocols at every stage of production, from fermentation to formulation, packaging, and distribution. A well-designed quality assurance program not only safeguards the product's effectiveness but also ensures its compatibility with agricultural practices and environmental safety.

Viability and Count

One of the most important aspects of quality control is monitoring the viability and concentration of microbial cells throughout the production process. This ensures that the final product contains an adequate population of active microorganisms to exert biocontrol effects.

» **Assessment Methods:**
- **Colony-Forming Units (CFU):** Serial dilution and plating methods are used to quantify the number of viable cells in a sample. CFU counts provide an estimate of the microbial population.
- **Flow Cytometry:** Advanced flow cytometry techniques enable rapid and precise assessment of cell viability and population density.
- **Molecular Methods:** Quantitative PCR (qPCR) and other molecular tools are used to detect and quantify specific microbial strains, especially in mixed cultures.

» **Monitoring Stages:**
- During fermentation, to ensure optimal growth and timely harvest.
- Post-formulation, to assess the impact of drying, encapsulation, or other processes on cell viability.

» **Importance:**
- High viability ensures that the microbial antagonist is effective upon application.
- Regular monitoring helps identify and address issues in the production process, such as contamination or suboptimal growth conditions.

Purity Testing

Purity testing is critical to ensure that microbial products are free from contaminants that could compromise their safety, efficacy, or regulatory compliance. Contaminants may include unwanted microorganisms, chemical residues, or impurities introduced during production.

» **Testing for Contaminants:**
- **Microbial Contaminants:** Culture-based methods and molecular tools like 16S rRNA sequencing are used to detect the presence of non-target microorganisms.
- **Chemical Residues:** Analytical techniques such as gas chromatography-mass spectrometry (GC-MS) or high-performance liquid chromatography (HPLC) detect residual solvents, antibiotics, or other chemicals.

- **Particulate Impurities:** Filtration and visual inspection ensure the absence of unwanted particulates in liquid formulations.

» **Preventative Measures:**

- Sterile production environments and aseptic techniques during fermentation and formulation.
- Regular equipment cleaning and sterilization protocols.

» **Importance:**

- Purity testing ensures the safety of the product for plants, humans, and the environment.
- Regulatory compliance often hinges on the absence of harmful contaminants.

Efficacy Testing

Efficacy testing validates the ability of microbial antagonists to suppress target pathogens and promote plant health under controlled conditions. This step is critical for ensuring that the product delivers the intended benefits in agricultural settings.

» **Laboratory Bioassays:**

- Dual-culture assays evaluate the antagonist's ability to inhibit pathogen growth on nutrient media.
- Enzyme activity assays test for the production of hydrolytic enzymes, such as chitinase and cellulase, which degrade pathogen cell walls.
- Volatile compound production tests assess the antagonist's ability to suppress pathogens through the release of antimicrobial volatiles.

» **Greenhouse Trials:**

- Microbial formulations are tested on crops grown in greenhouse conditions to evaluate their efficacy against specific pathogens.
- Parameters such as disease incidence, severity, and plant growth promotion are measured.

» **Field Validation:**

- Large-scale field trials assess the product's performance under real-world conditions, including variable climates, soil types, and agricultural practices.

» **Importance:**

- Efficacy testing ensures that the product meets the expectations of farmers and agricultural professionals.
- Continuous validation helps maintain market trust and regulatory approval.

Shelf-Life Studies

Shelf-life studies assess the stability and viability of microbial formulations over time and under different storage conditions. This is essential for determining expiration dates and ensuring consistent product performance.

- **Testing Conditions:**
  - Formulations are stored under varying conditions of temperature, humidity, and light exposure to simulate real-world scenarios.
  - Parameters such as microbial viability, physical integrity, and efficacy are monitored at regular intervals.
- **Key Considerations:**
  - **Water Activity:** Low water activity is critical for dry formulations to prevent microbial metabolism and spoilage.
  - **Protective Additives:** Cryoprotectants (e.g., glycerol, trehalose) and stabilizers (e.g., polysaccharides) are added to enhance stability.
  - **Packaging:** The choice of packaging materials, such as laminated foils or UV-resistant bottles, impacts shelf life.
- **Importance:**
  - Shelf-life studies help producers provide accurate usage guidelines to farmers.
  - Stable formulations reduce waste and ensure consistent field performance.

Adherence to International Quality Standards

Compliance with international quality standards is a cornerstone of commercial microbial production. Regulatory bodies like the Food and Agriculture Organization (FAO), the Organisation for Economic Co-operation and Development (OECD), and national agencies oversee the safety and efficacy of biocontrol products.

- **Documentation Requirements:**
  - Detailed reports on microbial strain identification, production methods, and quality control protocols.
  - Data from efficacy and safety studies, including toxicological and environmental impact assessments.
- **Certifications:**
  - Good Manufacturing Practices (GMP) certification ensures that production processes meet quality and safety standards.
  - ISO certifications, such as ISO 9001 (quality management) and ISO 14001 (environmental management), enhance market credibility.
- **Importance:**
  - Adhering to regulatory standards facilitates market acceptance and international trade.
  - Compliance minimizes the risk of product recalls or legal disputes.

Integrated Quality Control Systems

Modern production facilities employ integrated quality control systems that monitor all aspects of the production process. Advanced technologies, such as automation, sensors, and data analytics, ensure real-time monitoring and decision-making.

» **Process Monitoring:**
  - Sensors in bioreactors track parameters like temperature, pH, and dissolved oxygen, ensuring optimal growth conditions.
  - Automated sampling systems provide real-time data on microbial viability and purity.

» **Data Management:**
  - Quality control data are stored in centralized databases for analysis and reporting.
  - Predictive analytics identify potential issues before they impact product quality.

5.1.6 Packaging and Storage

Packaging and storage are critical components in the commercial production and delivery of microbial antagonists. Effective packaging ensures the stability, viability, and ease of use of the product, while proper storage conditions extend shelf life and maintain the antagonists' efficacy. Microbial antagonists, being living organisms, are sensitive to environmental factors such as moisture, temperature, oxygen, and light. Hence, packaging and storage systems must be meticulously designed to protect the product from these adverse conditions.

Packaging Materials and Techniques

Packaging materials for microbial antagonists are selected based on the formulation type (e.g., powder, liquid, granule) and the specific needs of the microorganism. These materials must provide a barrier against environmental stresses while maintaining product integrity during transportation and storage.

1. **Laminated Foil Pouches:**
   » Commonly used for dry formulations such as powders and granules, laminated foil pouches offer excellent moisture and light resistance.
   » The multi-layer construction includes materials such as aluminum foil, polyethylene, and polyester, which protect the product from humidity and UV exposure.
   » These pouches are lightweight, durable, and cost-effective, making them ideal for large-scale packaging.

2. **Airtight Containers:**
   » Airtight containers, made of high-density polyethylene (HDPE) or polypropylene (PP), are widely used for both dry and liquid formulations.
   » These containers prevent moisture ingress and oxygen exposure, maintaining the viability of the microbial cells.

3. UV-Resistant Bottles:

- For liquid formulations, UV-resistant bottles protect the microorganisms from harmful UV radiation, which can damage cellular structures and reduce viability.
- These bottles are typically made of amber-colored or opaque plastics that block UV light while being chemically inert.

4. Vacuum Packaging:

- Vacuum packaging removes air from the package, reducing oxygen levels and microbial respiration.
- This technique is especially beneficial for anaerobic or facultative anaerobic microorganisms and helps minimize contamination risks.

5. Modified Atmosphere Packaging (MAP):

- MAP involves altering the composition of gases within the packaging to optimize storage conditions. For instance, reducing oxygen levels and increasing carbon dioxide can slow microbial respiration and prolong shelf life.
- This method is particularly useful for high-value microbial formulations that require extended storage periods.

6. Encapsulation-Specific Packaging:

- For encapsulated formulations, specialized packaging materials ensure the structural integrity of the capsules while protecting them from environmental factors.
- Materials such as laminated foils or vacuum-sealed pouches are often used for encapsulated beads or granules.

Packaging Design Considerations

Packaging design must balance protection, functionality, and cost-effectiveness. Key considerations include:

- **Barrier Properties:** Packaging materials should provide barriers against moisture, oxygen, and light to maintain microbial viability.
- **Compatibility:** Materials must be chemically inert and non-toxic to avoid interactions with the microbial formulation.
- **Ease of Use:** Packaging should allow for easy handling and application by farmers, with clear instructions and user-friendly designs.
- **Durability:** Packages should withstand mechanical stresses during transportation and handling to prevent leaks or damage.
- **Environmental Impact:** Eco-friendly and biodegradable packaging options are increasingly being adopted to reduce environmental footprints.

Storage Conditions

Proper storage is essential to maintain the viability and efficacy of microbial antagonists over time. Storage conditions are optimized based on the formulation type and the specific requirements of the microorganism.

1. **Dry Formulations (Powders and Granules):**
   - » Dry formulations are relatively stable and can be stored at room temperature.
   - » Low humidity conditions are crucial to prevent microbial metabolism and contamination.
   - » Desiccants, such as silica gel, are often included in packaging to absorb residual moisture and extend shelf life.
2. **Liquid Formulations:**
   - » Liquid formulations are more sensitive to environmental conditions and typically require refrigeration (4–10°C) to maintain viability.
   - » Protective additives, such as glycerol or trehalose, are added to stabilize microbial cells during storage.
   - » Proper sealing of containers prevents evaporation and contamination.
3. **Encapsulated Formulations:**
   - » Encapsulated products, such as alginate beads, are stored under conditions that maintain the integrity of the encapsulation material.
   - » Low temperatures and humidity levels are essential to prevent capsule degradation and microbial leakage.
4. **Refrigeration:**
   - » Refrigeration slows down microbial metabolism, extending shelf life without compromising viability.
   - » Cold storage is essential for liquid formulations and high-value microbial products requiring prolonged storage.
5. **Controlled Atmosphere Storage:**
   - » Controlled storage environments with regulated temperature, humidity, and gas composition ensure optimal conditions for microbial survival.
   - » For instance, reducing oxygen levels in storage facilities can limit oxidative damage to microbial cells.

Advanced Techniques for Stability Enhancement

To further enhance the stability of microbial antagonists during packaging and storage, advanced techniques are employed:

1. **Cryoprotectants:**
   - » Additives like glycerol, trehalose, and sucrose protect microbial cells from desiccation and temperature fluctuations.

- These substances stabilize cell membranes and proteins, reducing damage during storage.

2. **Freeze-Drying (Lyophilization):**
    - Freeze-drying removes moisture from microbial cells while maintaining their structural integrity.
    - The resulting dry product is highly stable and can be stored at room temperature for extended periods.
3. **Spray-Drying:**
    - Spray-drying involves atomizing a microbial suspension into a stream of hot air, producing dry particles.
    - This technique is cost-effective and suitable for large-scale production, though it may reduce viability for heat-sensitive microorganisms.
4. **Vacuum Sealing:**
    - Vacuum-sealing techniques remove air from packaging, reducing oxidative stress on microbial cells.
    - This method is especially beneficial for anaerobic and facultative anaerobic microorganisms.

Monitoring During Storage

Regular monitoring of packaged microbial products during storage ensures their quality and viability. Key parameters include:

- **Viability Tests:** Periodic CFU counts or molecular assays measure microbial viability over time.
- **Moisture Content:** Monitoring water activity prevents spoilage and contamination.
- **Physical Integrity:** Packaging materials are inspected for leaks, tears, or degradation.

Sustainability in Packaging

The growing emphasis on sustainability has led to the development of eco-friendly packaging options for microbial antagonists. Biodegradable materials, such as plant-based polymers and recycled paper, are increasingly being used to reduce environmental impact. Additionally, light weight and compact packaging designs minimize transportation costs and carbon footprints.

5.1.7 Delivery and Application Systems

The effectiveness of microbial antagonists in agricultural systems depends heavily on the efficiency of their delivery and application. Delivery systems are designed to ensure that the antagonists reach the target site in sufficient quantities, remain viable, and interact effectively with the target pathogens. Application methods vary depending on the type of formulation, crop, and disease being targeted. Recent advancements in technology, such as precision agriculture and automated systems, have revolutionized the delivery of microbial antagonists, improving their efficiency and adoption by farmers.

Seed Treatment

Seed treatment is a widely used method for delivering microbial antagonists directly to the crop at its earliest stage. In this approach, seeds are coated with a microbial formulation, providing protection against soilborne pathogens and enhancing early plant growth.

» **Process:**
- Seeds are mixed with a microbial suspension (liquid or powder) in seed treatment equipment.
- Adhesives, such as gum arabic or polymer coatings, are used to ensure uniform coating and adherence of the antagonist to the seed surface.
- Treated seeds are dried to stabilize the formulation before storage or planting.

» **Advantages:**
- Direct application of antagonists to the rhizosphere ensures early protection against pathogens like *Rhizoctonia* and *Fusarium*.
- Cost-effective and requires lower quantities of microbial formulations compared to other methods.
- Enhances germination and seedling vigor by promoting plant growth and suppressing pathogens.

» **Applications:**
- *Trichoderma spp.* formulations for controlling damping-off diseases in vegetable crops.
- *Pseudomonas fluorescens* for protecting cereals from root rot pathogens.

Soil Application

Soil application involves introducing microbial antagonists into the soil where they colonize plant roots and suppress soilborne pathogens. This method is particularly effective for crops affected by diseases that thrive in the root zone or soil environment.

» **Process**:
- Microbial formulations (granular or liquid) are incorporated into the soil using conventional farming equipment, such as seed drills or furrow applicators.
- For granules, broadcasting or banding methods are employed, while liquid formulations are applied through drenching or spraying.

» **Advantages:**
- Provides sustained protection by establishing a population of beneficial microbes in the root zone.
- Enhances nutrient availability through mechanisms such as phosphate solubilization and nitrogen fixation.
- Reduces the buildup of soilborne pathogens over time, promoting long-term soil health.

» **Applications:**

- Granular *Trichoderma spp.* for controlling nematodes and fungal pathogens in horticultural crops.
- Liquid *Bacillus subtilis* for improving root colonization and suppressing wilt pathogens in legumes.

Foliar Application

Foliar application delivers microbial antagonists directly to the aerial parts of the plant, targeting pathogens that affect leaves, stems, and flowers. This method is particularly effective for controlling foliar diseases caused by fungi, bacteria, and viruses.

» **Process:**

- Microbial formulations are prepared as a spray solution, often mixed with surfactants to enhance adhesion and spread on leaf surfaces.
- The solution is applied using sprayers, such as backpack sprayers for small-scale use or tractor-mounted sprayers for larger fields.

» **Advantages:**

- Direct targeting of foliar pathogens such as *Alternaria*, *Botrytis*, and *Powdery Mildew*.
- Compatible with integrated pest management (IPM) systems, allowing for combination with other control measures.
- Prevents secondary infections by creating a protective microbial barrier on plant surfaces.

» **Applications:**

- Liquid formulations of *Bacillus amyloliquefaciens* for controlling leaf spots in cereals.
- *Pseudomonas spp.* sprays to suppress bacterial blight in rice.

Drip Irrigation

Drip irrigation systems provide an efficient means of delivering liquid formulations of microbial antagonists to the root zone. This method integrates irrigation and biocontrol, ensuring uniform distribution of the antagonists and water.

» **Process:**

- Microbial suspensions are mixed into the irrigation water using fertigation systems or injectors.
- The solution is delivered through drip emitters, ensuring targeted application to the root zone.

» **Advantages:**

- Reduces water usage and minimizes waste compared to conventional application methods.
- Ensures consistent delivery of microbial antagonists throughout the growing season.

- Enhances root colonization by maintaining a favorable environment for microbial growth.

» **Applications:**

- Liquid *Trichoderma spp.* for protecting vegetable crops from root rot pathogens.
- *Azospirillum spp.* for promoting growth and yield in sugarcane.

Advanced Delivery Systems

Modern technology has introduced innovative delivery systems that enhance the precision and efficiency of microbial antagonist application:

1. **Drones:**
   - Drones equipped with spray systems deliver microbial formulations over large areas with minimal labor and time.
   - Precision mapping ensures targeted application, reducing wastage and environmental impact.
2. **Precision Agriculture Tools:**
   - GPS-enabled equipment and variable rate technology (VRT) optimize the application of microbial antagonists based on field conditions.
   - Soil and crop sensors provide real-time data, ensuring accurate delivery to areas with the highest need.
3. **Seed-Embedded Delivery:**
   - Advanced techniques embed microbial formulations within seed coatings, eliminating the need for separate application processes.
   - Slow-release systems ensure prolonged activity throughout early plant growth stages.

5.1.8 Regulatory Compliance

The commercialization of microbial antagonists requires adherence to stringent regulatory frameworks to ensure their safety, efficacy, and environmental compatibility. Regulatory compliance involves evaluating the microbial strain, production process, quality control measures, and field performance data. This is critical for gaining market approval and building trust among stakeholders.

Regulatory Evaluation

Regulatory bodies assess the following aspects of microbial antagonists before granting approval:

1. **Microbial Identification**:
   » Accurate identification of the microbial strain, including its taxonomic classification and genetic profile.
   » Molecular tools such as 16S rRNA sequencing and whole-genome sequencing (WGS) provide detailed strain characterization.

2. **Safety Assessments:**
   - Toxicological studies ensure that the product is safe for humans, animals, and non-target organisms.
   - Environmental impact assessments evaluate the potential risks to soil microbial communities and ecosystems.
3. **Efficacy Validation:**
   - Field trials and laboratory bioassays demonstrate the product's effectiveness against target pathogens.
   - Long-term studies assess the stability and consistency of the antagonist under different conditions.

Documentation Requirements

Comprehensive documentation is essential for regulatory approval and includes the following:

- **Strain Information:** Detailed profiles of the microbial strain, including its origin, genetic characteristics, and mechanisms of action.
- **Production Details:** Description of the production process, quality control protocols, and formulation development.
- **Field Performance Data:** Results from greenhouse and field trials demonstrating the product's efficacy and safety.

Compliance with International Standards

Compliance with international quality standards facilitates market acceptance and trade. Key standards include:

- **FAO Guidelines:** Ensure safe and effective use of microbial antagonists in agriculture.
- **OECD Standards:** Provide protocols for testing biocontrol agents' safety and efficacy.
- **National Regulations:** Each country has specific frameworks, such as the Environmental Protection Agency (EPA) in the U.S. or the Central Insecticides Board (CIB) in India.

5.2 Delivery Systems for Biocontrol Agents

The success of biocontrol agents in agricultural systems depends significantly on the delivery mechanisms used to apply them effectively to target areas. Delivery systems are designed to ensure that the agents reach the intended site in sufficient quantities, remain viable, and interact successfully with plant pathogens or pests. The efficiency of these systems impacts the agents' effectiveness, ease of use, and overall adoption by farmers. Modern advancements in biotechnology, precision agriculture, and application technologies have revolutionized the delivery of biocontrol agents, making them more reliable and efficient.

5.2.1 Seed Treatment Delivery Systems

Seed treatment is a foundational method for delivering biocontrol agents efficiently to the soil environment and plant roots during the earliest stages of plant development. This method involves coating seeds with formulations of microbial antagonists or other biocontrol agents to create a protective shield against pathogens. By placing beneficial microorganisms in close proximity to the seed and the developing root system, seed treatments establish an early presence of biocontrol agents in the rhizosphere, providing both disease protection and growth enhancement.

Process Overview

The process of seed treatment involves several critical steps designed to ensure effective delivery and long-term stability of the biocontrol agents:

1. **Preparation of Seeds:**
   - » Seeds are cleaned to remove debris, dust, and microbial contaminants. This step ensures the coating adheres uniformly and prevents interference from unwanted microorganisms.
   - » Drying the seeds to an optimal moisture level enhances their compatibility with the formulation.
2. **Preparation of Formulations:**
   - » The biocontrol agent is formulated as a powder, liquid suspension, or encapsulated product. Additives such as binders, dispersants, and stabilizers may be included to enhance performance and durability.
   - » Adhesive agents, such as polymers, gums, or starches, are prepared to ensure that the microbial formulation adheres uniformly to the seed surface.
3. **Application of Formulation:**
   - » The seed treatment process typically involves specialized equipment that tumbles the seeds while spraying or dusting them with the microbial formulation. This ensures even coating across the seed surface.
   - » Depending on the type of formulation, additional processes such as drying or curing may follow to stabilize the treated seeds.
4. **Post-Treatment Handling:**
   - » The treated seeds are dried in controlled conditions to maintain the viability of the microbial agents while preventing clumping or moisture-related spoilage.
   - » Treated seeds are stored in moisture-proof, breathable packaging to preserve their quality during transportation and storage.

Advantages of Seed Treatment

Seed treatment delivery systems offer numerous benefits, making them a preferred method for applying biocontrol agents in agricultural systems:

1. **Early Protection Against Soilborne Pathogens:**
   - » Coated seeds establish a zone of microbial activity around the seed and emerging roots. This shields the young plant from soilborne pathogens like *Fusarium*, *Pythium*, and *Rhizoctonia* during its most vulnerable growth stages.

2. **Promotion of Early Root Colonization:**
   » Beneficial microbes introduced through seed treatment rapidly colonize the rhizosphere. This early establishment improves root health, enhances nutrient uptake, and stimulates overall plant growth.
3. **Reduction in Field Applications:**
   » By applying biocontrol agents directly to the seed, the need for additional field applications is minimized. This saves time, labor, and resources while reducing the risk of uneven application in the field.
4. **Economic and Environmental Benefits:**
   » Seed treatments use smaller quantities of biocontrol agents compared to foliar or soil applications, making them cost-effective.
   » The localized application of biocontrol agents reduces environmental exposure, minimizing the risk of off-target effects.
5. **Compatibility with Existing Systems:**
   » Treated seeds are compatible with conventional planting equipment, ensuring ease of integration into existing farming practices.

Applications in Agriculture

Seed treatment delivery systems are versatile and widely used across various crops and agricultural settings. Examples of their application include:

1. **Cereal and Legume Crops:**
   » **Microbial Agents:** *Trichoderma spp.* and *Pseudomonas fluorescens* are commonly used for cereals and legumes to combat damping-off and root rot diseases.
   » **Pathogen Targets:** These agents suppress pathogens like *Fusarium* and *Rhizoctonia* while promoting root health and early growth.
2. **Vegetable Crops:**
   » **Microbial Agents:** *Bacillus subtilis* is applied to vegetable seeds for controlling seedborne fungal pathogens, such as *Alternaria* and *Aspergillus*.
   » **Additional Benefits:** Enhanced germination rates and improved seedling vigor are often observed with treated seeds.
3. **Specialty and High-Value Crops:**
   » Seed treatments are also employed in high-value crops like horticultural plants and ornamentals, where disease prevention and uniform growth are critical.
4. **Organic Farming:**
   » Seed treatment with microbial biocontrol agents is especially suitable for organic farming systems, where chemical pesticides and fungicides are restricted. These treatments provide a natural and sustainable alternative for disease management.

Technological Advancements in Seed Treatment

Advances in seed treatment technology are further enhancing the precision and efficacy of this delivery system:

1. **Encapsulation Techniques:**
    » Encapsulated formulations protect microbial agents from environmental stressors, such as temperature and desiccation, while providing a slow-release effect around the seed.
2. **Polymer Coatings:**
    » Advanced polymer coatings improve the adherence of formulations to seeds, ensuring even distribution and long-term stability during storage and planting.
3. **Bio-Priming:**
    » This innovative technique involves hydrating seeds with a suspension of biocontrol agents before planting. Bio-priming accelerates microbial colonization and promotes early germination.
4. **Seed Coating Machines:**
    » Modern seed treatment equipment integrates tumbling, spraying, and drying processes to achieve uniform and efficient coating at a commercial scale.

Challenges in Seed Treatment Delivery Systems

Despite its many advantages, seed treatment systems face certain challenges that require attention:

1. **Viability of Microbial Agents:**
    » Ensuring that microbial cells remain viable during storage and handling is critical. Advances in protective formulations, such as freeze-dried powders and cryoprotectants, are addressing this challenge.
2. **Compatibility with Chemical Treatments:**
    » Seeds often receive additional chemical treatments, such as insecticides or nutrients. Compatibility studies are necessary to ensure that these chemicals do not interfere with the biocontrol agents' efficacy.
3. **Environmental Sensitivity:**
    » Environmental factors such as soil pH, moisture, and temperature can influence the performance of biocontrol agents delivered via seed treatments. Research into strain improvement and formulation optimization is helping to overcome these limitations.

5.2.2 Soil Application Systems

Soil application systems are among the most effective methods for delivering biocontrol agents to target pathogens, particularly soilborne ones. By introducing microbial antagonists directly into the soil, these systems ensure close interaction with the plant root zone (rhizosphere), where a dynamic exchange of nutrients and microbial activity takes place. This approach not only suppresses pathogens but also enhances soil health, contributing to the sustainability of agricultural systems. Tailored techniques such as broadcasting, band application, and drenching provide flexibility in application, depending on the crop, pathogen, and formulation type.

Techniques for Soil Application

Soil application systems employ several techniques to introduce biocontrol agents into the soil effectively. These techniques are selected based on the type of formulation (granular, liquid, or encapsulated) and the specific requirements of the crop and target pathogens.

1. **Broadcasting:**
   - **Process:** Granular formulations of biocontrol agents are spread uniformly across the field using manual or mechanical spreaders.
   - **Advantages:**
     - Provides broad coverage, making it suitable for field crops and areas with widespread pathogen presence.
     - Ensures an even distribution of biocontrol agents throughout the soil.
   - **Challenges**:
     - Requires higher volumes of formulation compared to other methods.
     - Less targeted, which may reduce efficiency against localized pathogen hotspots.
2. **Band Application:**
   - **Process:** Granules or liquid formulations are applied in narrow bands along planting rows, either before planting or alongside seeds during sowing.
   - **Advantages:**
     - Targets the root zone directly, minimizing wastage and enhancing the interaction of biocontrol agents with roots.
     - Requires lower quantities of formulations, reducing costs.
   - **Applications:**
     - Commonly used for row crops such as maize, soybean, and sugarcane.
3. **Drenching:**
   - **Process:** Aqueous suspensions of biocontrol agents are poured or sprayed onto the soil near the plant roots. This technique is often employed in nurseries, greenhouses, and high-value crops.
   - **Advantages:**
     - Provides direct and immediate interaction with the root zone, ensuring rapid colonization.
     - Effective against localized pathogen infestations.
   - **Challenges:**
     - Requires precise application to avoid runoff or dilution in the soil.

Advantages of Soil Application Systems

Soil application systems offer several key benefits that make them an indispensable tool for sustainable agricultural practices:

1. **Direct Interaction with the Root Zone:**
   - Biocontrol agents applied through soil systems directly interact with the rhizosphere, a critical zone for plant-microbe-pathogen interactions.
   - This enhances the colonization of beneficial microbes on root surfaces, improving their effectiveness against pathogens.
2. **Long-Term Pathogen Suppression:**
   - Once established in the soil, biocontrol agents can persist and suppress pathogens over extended periods.
   - This long-lasting effect reduces the need for repeated applications and ensures continuous protection for the crop.
3. **Improvement of Soil Health:**
   - Biocontrol agents contribute to the balance of soil microbiota, promoting beneficial microbial communities while suppressing harmful ones.
   - They enhance nutrient cycling by producing enzymes, organic acids, and growth-promoting compounds.
4. **Eco-Friendly and Sustainable:**
   - Soil application systems minimize the need for chemical pesticides, reducing environmental contamination and promoting biodiversity.
   - They are compatible with organic farming systems and integrated pest management (IPM) strategies.

Applications in Agriculture

Soil application systems are widely used across various agricultural systems to manage soilborne diseases and improve crop productivity. Common examples include:

1. **Vegetable Crops:**
   - **Granular *Trichoderma spp.*:**
     - Applied to suppress nematodes and fungal pathogens in crops like tomatoes, peppers, and cucumbers.
     - Enhances plant vigor by promoting root growth and nutrient uptake.
2. **Banana Cultivation:**
   - **Liquid *Bacillus amyloliquefaciens*:**
     - Used for managing wilt pathogens such as *Fusarium oxysporum*.
     - Applied through drenching or drip irrigation to ensure effective delivery to the root zone.

3. **Cereal Crops:**
   » Biocontrol agents are applied through broadcasting to combat root rot and damping-off diseases in crops like wheat and rice.
4. **Nursery and Greenhouse Systems:**
   » Soil drenching is a preferred method for applying microbial formulations to seedlings and young plants.
   » Ensures early protection against pathogens like *Pythium* and *Phytophthora* in controlled environments.

Integration with Modern Technologies

Advancements in agricultural technology are enhancing the efficiency and precision of soil application systems:

1. **Automated Spreaders and Sprayers:**
   » GPS-enabled spreaders ensure uniform application of granular formulations across fields.
   » Automated sprayers deliver liquid formulations with precision, minimizing waste and maximizing coverage.
2. **Drip Irrigation Integration:**
   » Liquid biocontrol agents are increasingly applied through drip irrigation systems, combining irrigation and pest control in one operation.
   » This method ensures uniform distribution, reduces water usage, and minimizes labor costs.
3. **Encapsulated Formulations:**
   » Encapsulated biocontrol agents, such as alginate beads, provide a slow-release mechanism in the soil.
   » This enhances the longevity and efficacy of the agents, particularly in harsh environmental conditions.

Challenges in Soil Application Systems

While soil application systems are highly effective, they also present certain challenges that require attention:

1. **Environmental Variability:**
   » Factors such as soil pH, temperature, and moisture content can influence the performance of biocontrol agents.
   » Formulations need to be tailored to withstand diverse environmental conditions.
2. **Application Costs:**
   » Granular formulations can be costlier to produce and apply, particularly for large-scale operations.
   » Innovations in formulation technology and delivery systems are helping to address this issue.

3. **Pathogen-Specific Effectiveness:**
   » Not all biocontrol agents are equally effective against all pathogens. Identifying the right agent-pathogen pairing is critical for success.

5.2.3 Foliar Application Systems

Foliar application systems are a vital method for delivering biocontrol agents to the aerial parts of plants, including leaves, stems, and flowers. These systems target pathogens and pests that attack above-ground plant parts, such as powdery mildew, blights, and leaf spots. By directly applying biocontrol agents to the plant surface, foliar application ensures effective pathogen suppression and pest control while promoting environmental sustainability. This method is increasingly integrated into integrated pest management (IPM) strategies, reducing reliance on chemical pesticides and addressing concerns about chemical residues on crops.

Techniques for Foliar Application

The effectiveness of foliar application depends on the method used to deliver the biocontrol agents. Key techniques include:

1. **Spraying:**
   » **Process:**
   - Biocontrol agents are formulated into aqueous suspensions and applied to plant surfaces using sprayers.
   - Equipment ranges from manual backpack sprayers for small-scale farms to tractor-mounted boom sprayers for large agricultural fields.

   » **Advantages:**
   - Provides uniform coverage of plant surfaces, ensuring thorough contact with pathogens and pests.
   - Suitable for a wide range of crops and application scales.

   » **Considerations:**
   - Requires proper calibration of sprayers to avoid over- or under-application.
   - Drift control is essential to minimize off-target effects.

2. **Misting:**
   » **Process:**
   - Misting involves dispersing fine droplets of microbial formulations onto plant surfaces using specialized misting equipment.
   - This technique is commonly used in greenhouses and controlled environments.

   » **Advantages:**
   - Achieves better coverage of hard-to-reach areas, such as the undersides of leaves.
   - Reduces water usage compared to conventional spraying.

» **Considerations:**
  - Requires precise equipment to control droplet size and distribution.
  - May not be practical for large-scale field applications.

Advantages of Foliar Application Systems

Foliar application systems offer several key benefits, making them a preferred choice for managing foliar pathogens and pests in diverse agricultural settings:

1. **Effective Targeting of Foliar Pathogens and Pests:**
   » Direct application to the plant surface ensures immediate interaction with pathogens and pests, leading to rapid suppression.
   » Pathogens such as powdery mildew and leaf spots, which thrive on leaf surfaces, are effectively controlled.
2. **Reduced Chemical Residues:**
   » Biocontrol agents provide an eco-friendly alternative to chemical pesticides, reducing concerns about harmful residues on crops.
   » This benefit is especially important for export-oriented agriculture and organic farming systems.
3. **Integration with IPM Strategies:**
   » Foliar applications of biocontrol agents are easily integrated into IPM programs, complementing other control measures such as cultural practices and biological predators.
   » Biocontrol agents can be tank-mixed with other inputs, such as fertilizers and compatible pesticides, to optimize resource use.
4. **Promotion of Plant Health:**
   » Many biocontrol agents used in foliar applications produce secondary metabolites and enzymes that enhance plant defenses.
   » Certain formulations improve plant tolerance to environmental stresses, such as drought and salinity.

Applications of Foliar Application Systems

Foliar application systems are widely used across different crops and agro-ecosystems to manage specific diseases and pests. Common applications include:

1. **Pest Control:**
   » **Liquid *Bacillus thuringiensis* Sprays:**
     - Applied to control caterpillar pests such as *Helicoverpa* and *Spodoptera* in crops like cotton, maize, and vegetables.
     - Acts by producing insecticidal toxins that disrupt the pests' digestive systems.

2. **Disease Management:**
   » ***Pseudomonas fluorescens* Formulations:**
      - Used to manage bacterial leaf blight in rice, a major disease affecting crop yields.
      - Suppresses pathogens through competition, antibiosis, and induced systemic resistance.
3. **Horticultural Crops:**
   - Biocontrol agents such as *Trichoderma spp.* and *Bacillus subtilis* are applied to fruits and vegetables to control powdery mildew and botrytis blight.
   - Enhances post-harvest quality by reducing the incidence of storage diseases.
4. **Controlled Environments:**
   - Foliar application systems are particularly effective in greenhouses and polyhouses, where misting ensures even distribution of agents on crops like tomatoes, cucumbers, and ornamentals.

Integration with Modern Technologies

The adoption of advanced technologies is revolutionizing foliar application systems, enhancing their efficiency and precision:

1. **Electrostatic Spraying:**
   - Electrostatic sprayers charge droplets of biocontrol formulations, improving their adhesion to plant surfaces.
   - This method reduces drift and ensures even coverage, particularly for dense canopies.
2. **Drone-Based Application:**
   - Drones equipped with spray systems deliver biocontrol agents over large areas with high precision.
   - GPS mapping and real-time data integration enable site-specific applications, reducing wastage and environmental impact.
3. **Nano-formulations:**
   - Nanoencapsulation of biocontrol agents protects them from environmental stress and enhances their stability on leaf surfaces.
   - Controlled release mechanisms ensure prolonged activity against pathogens and pests.

5.2.4 Drip Irrigation Systems

Drip irrigation systems are an innovative and highly efficient method for delivering biocontrol agents directly to the root zone of plants through irrigation water. This approach combines irrigation with pest and disease management, ensuring that biocontrol agents are applied in a targeted manner with minimal waste. Drip irrigation systems are particularly suited for high-value crops, greenhouse cultivation, and regions with limited water availability. By integrating biocontrol delivery with irrigation, this system maximizes resource use efficiency, promotes plant health, and reduces operational costs.

The integration of biocontrol agents with drip irrigation involves precise steps to ensure even distribution and sustained efficacy:

1. **Preparation of Biocontrol Solutions:**
    - Liquid formulations of biocontrol agents, such as *Trichoderma spp.* or *Azospirillum spp.*, are prepared by diluting the concentrated product with water.
    - Compatibility testing is conducted to ensure that the biocontrol agents remain viable and effective when mixed with irrigation water or fertilizers.
2. **Injection into Irrigation System:**
    - The biocontrol solution is introduced into the drip irrigation system using fertigation equipment or injectors.
    - Specialized devices, such as Venturi injectors or diaphragm pumps, are used to regulate the flow and concentration of the biocontrol agents.
3. **Delivery Through Drip Lines:**
    - The treated water is delivered through drip lines fitted with emitters that release the solution directly to the root zone.
    - Uniform distribution is achieved by ensuring proper system calibration and maintenance of emitters.
4. **Monitoring and Optimization:**
    - Flow meters and sensors monitor the delivery rate and distribution of the biocontrol solution.
    - Regular checks ensure that the system is free from clogs and that the biocontrol agents reach the target area effectively.

Advantages of Drip Irrigation Systems

Drip irrigation systems offer several significant benefits for the delivery of biocontrol agents:

1. **Precise Delivery to the Target Area:**
    - By delivering biocontrol agents directly to the root zone, this method ensures optimal interaction between the agents and soilborne pathogens or pests.
    - Targeted application reduces the volume of biocontrol agents required, minimizing waste and lowering costs.
2. **Compatibility with Organic and Sustainable Farming**:
    - Drip irrigation systems align well with organic farming practices, as they promote eco-friendly pest and disease management.
    - This method also supports sustainable agriculture by conserving water and reducing the use of chemical inputs.

3. **Reduction in Labor and Operational Costs:**
   - Automated drip systems eliminate the need for manual application of biocontrol agents, saving time and labor.
   - The integration of irrigation and biocontrol delivery streamlines farm operations, improving overall efficiency.
4. **Uniform Distribution:**
   - Drip irrigation ensures consistent and uniform distribution of biocontrol agents across the field, reducing the risk of uneven treatment.
5. **Reduced Environmental Impact:**
   - The localized application minimizes runoff and leaching, protecting non-target organisms and surrounding ecosystems.
   - By reducing reliance on chemical pesticides, drip irrigation systems contribute to lower environmental contamination.

Applications in Agriculture

Drip irrigation systems are versatile and have been successfully used in various agricultural contexts for delivering biocontrol agents:

1. **Greenhouse Vegetables:**
   - **Biocontrol Agent:** *Trichoderma spp.* formulations.
   - **Target:** Soilborne pathogens such as *Pythium* and *Fusarium.*
   - **Outcome:** Improved root health, reduced disease incidence, and enhanced yield.
2. **Sugarcane Cultivation:**
   - **Biocontrol Agent:** *Azospirillum* inoculants.
   - **Target:** Growth promotion and enhanced nitrogen fixation.
   - **Outcome:** Increased biomass production and higher sugar content in the crop.
3. **Orchards and Vineyards**:
   - **Biocontrol Agent:** *Bacillus subtilis* formulations.
   - **Target:** Suppression of root rot and crown gall diseases.
   - **Outcome:** Improved plant vigor and fruit quality.
4. **Floriculture:**
   - **Biocontrol Agent:** *Pseudomonas fluorescens.*
   - **Target:** Soilborne pathogens in ornamental plants.
   - **Outcome:** Disease-free plants with better aesthetic value.

Integration with Advanced Technologies

Advancements in irrigation and agricultural technologies are enhancing the efficiency of drip irrigation systems for biocontrol delivery:

1. **Automated Fertigation Systems:**
   - Automated systems regulate the flow of biocontrol solutions based on real-time crop and soil needs, ensuring precise application.
2. **Sensor-Based Monitoring:**
   - Soil moisture sensors and flow meters provide real-time data on irrigation and biocontrol delivery, enabling farmers to make informed decisions.
3. **Nano-formulations:**
   - Nano-encapsulated biocontrol agents are compatible with drip irrigation systems, providing controlled and sustained release at the root zone.
4. **Remote Control and IoT Integration:**
   - Internet of Things (IoT) devices enable remote monitoring and control of irrigation and biocontrol applications, improving efficiency and reducing labor.

5.2.5 Trunk Injection Systems

Trunk injection systems represent a targeted and highly efficient method for delivering biocontrol agents directly into the vascular system of trees and perennial crops. This delivery method has gained popularity due to its ability to control systemic diseases and pests while minimizing environmental exposure and off-target effects. By bypassing external application methods and delivering the agents internally, trunk injection ensures that the biocontrol formulations reach all parts of the plant, providing comprehensive and sustained protection. This technique is particularly suitable for high-value crops, such as fruit trees, ornamentals, and palms, where maintaining plant health is critical for economic returns.

Process

The trunk injection process is straightforward yet requires precision and care to avoid damage to the tree and ensure effective delivery:

1. **Preparation of the Tree:**
   - The tree is assessed for health and suitability for injection. Specific attention is given to the type of disease or pest being targeted, as well as the tree's size and vascular structure.
   - Small holes, typically 1-2 cm in diameter, are drilled into the tree trunk at an angle to ensure proper uptake of the injected solution. The number of holes depends on the size of the tree and the dosage required.
2. **Preparation of the Biocontrol Formulation:**
   - The biocontrol agent is formulated into a liquid solution that is compatible with the tree's vascular system. Additives may be included to enhance solubility, stability, and uptake.

3. **Injection:**
   - The biocontrol solution is injected into the prepared holes using pressurized systems or manual syringes. Pressurized systems, such as infusion pumps or hydraulic injectors, are preferred for large-scale applications as they ensure rapid and uniform distribution of the formulation.
   - The injected solution is absorbed into the tree's xylem and phloem, spreading throughout the plant's vascular system.
4. **Post-Injection Care:**
   - The injection sites are sealed with plugs or natural resins to prevent infection or excessive sap loss. The tree is monitored to ensure recovery from the procedure and assess the effectiveness of the treatment.

Advantages of Trunk Injection Systems

Trunk injection systems offer several distinct advantages over conventional delivery methods, particularly for high-value and perennial crops:

1. **Systemic Protection:**
   - By directly delivering biocontrol agents into the vascular system, this method ensures that the agents are distributed to all parts of the plant, including roots, stems, and leaves.
   - This systemic movement makes trunk injection highly effective for controlling pests and diseases that are difficult to reach through external applications.
2. **Reduced Environmental Impact:**
   - Trunk injection minimizes environmental exposure and off-target effects, as the biocontrol agents are contained within the tree. This is particularly important in sensitive ecosystems or near water sources.
3. **Suitability for High-Value Crops:**
   - The precision and effectiveness of trunk injection make it ideal for high-value crops, such as fruit trees and ornamental plants, where maximizing plant health and yield is essential.
4. **Compatibility with Organic Farming:**
   - Trunk injection is an eco-friendly alternative to chemical pesticides and aligns well with organic farming practices, ensuring sustainable disease and pest management.
5. **Prolonged Efficacy:**
   - The systemic distribution of biocontrol agents provides long-lasting protection, reducing the frequency of applications and associated labor costs.

Applications of Trunk Injection Systems

1. **Control of Bacterial Wilt in Palms:**
    - **Biocontrol Agent:** *Pseudomonas fluorescens* formulations.
    - **Target Pathogen:** *Ralstonia solanacearum*, a major cause of bacterial wilt in palms.
    - **Outcome:** Effective suppression of the pathogen and enhanced palm health.
2. **Management of Systemic Fungal Diseases in Citrus:**
    - **Biocontrol Agent:** Endophytic microorganisms capable of colonizing the tree's vascular system.
    - **Target Pathogen:** *Phytophthora spp.*, which causes citrus gummosis.
    - **Outcome:** Reduced disease severity and improved yield quality.
3. **Suppression of Pest Infestations in Fruit Trees:**
    - Trunk injection systems are also used to deliver biocontrol agents targeting insect pests, such as borers and aphids, that affect the internal structures of trees.

5.2.6 Advanced Delivery Technologies

The development of advanced delivery technologies has revolutionized the application of biocontrol agents, enhancing their precision, efficacy, and scalability. These technologies leverage innovations in engineering, material science, and precision agriculture to overcome challenges associated with traditional methods. Advanced delivery systems not only optimize the performance of biocontrol agents but also address key environmental and economic concerns, making them increasingly popular in modern agriculture.

1. Drones

Drones have emerged as a game-changing technology in the agricultural sector, offering unparalleled efficiency and precision for the delivery of biocontrol agents.

» **Capabilities:**
- Equipped with spray systems, drones can apply biocontrol formulations over large areas quickly and uniformly.
- GPS mapping and data integration enable precise targeting of pest and disease hotspots, minimizing waste and environmental impact.

» **Advantages:**
- Ideal for inaccessible or uneven terrains where traditional equipment cannot operate.
- Reduces labor requirements and speeds up application processes.
- Enables targeted application, reducing costs and maximizing the efficacy of biocontrol agents.

» **Applications:**

- Spraying *Bacillus thuringiensis* formulations to control caterpillar infestations in field crops.
- Applying *Trichoderma spp.* to large orchard systems for disease prevention.

2. Seed-Embedded Delivery

Seed-embedded delivery involves incorporating microbial formulations into seed coatings, ensuring early establishment of biocontrol agents in the rhizosphere.

» **Capabilities:**

- Coatings are designed to release biocontrol agents gradually, providing sustained protection during germination and early plant growth stages.
- Compatible with a wide range of seed treatment techniques and planting equipment.

» **Advantages:**

- Eliminates the need for separate field applications, reducing labor and operational costs.
- Ensures uniform distribution of biocontrol agents across the field.

» **Applications:**

- Seed coatings with *Azospirillum* for nitrogen fixation and growth promotion in cereals.
- Embedded *Pseudomonas fluorescens* formulations for early protection against root rot pathogens.

3. Electrostatic Spraying

Electrostatic spraying enhances the adhesion of biocontrol formulations to plant surfaces, improving coverage and reducing wastage.

» **Capabilities:**

- Droplets of biocontrol agents are electrically charged, enabling them to adhere to plant surfaces, including hard-to-reach areas.
- Reduces drift and ensures uniform application.

» **Advantages:**

- Ideal for foliar applications in orchards and vineyards.
- Reduces formulation usage, lowering costs and environmental impact.

» **Applications:**

- Spraying *Trichoderma spp.* formulations to prevent powdery mildew in grapes.
- Applying *Bacillus subtilis* to control bacterial spots in tomatoes.

4. Gel and Capsule Delivery

Gel and capsule delivery systems provide controlled release of biocontrol agents, enhancing their longevity and efficacy in the field.

» **Capabilities:**
  - Gels and capsules protect microbial agents from environmental stresses, such as UV radiation and desiccation.
  - Gradual release mechanisms ensure sustained activity over time.

» **Advantages:**
  - Extends the shelf life and field efficacy of biocontrol formulations.
  - Reduces the frequency of applications, lowering labor costs.

» **Applications:**
  - Alginate gel formulations of *Trichoderma spp.* for soilborne pathogen control.
  - Capsule-based delivery of *Pseudomonas fluorescens* for systemic protection in ornamental plants.

5.2.7 Integration with Precision Agriculture

The integration of biocontrol delivery systems with precision agriculture tools has transformed their application, enabling site-specific management of pests and diseases. By leveraging data-driven technologies, precision agriculture ensures that biocontrol agents are applied efficiently, maximizing their impact while minimizing costs and environmental risks.

Capabilities of Precision Agriculture Integration

1. **GPS-Enabled Equipment:**
   » Enables site-specific delivery of biocontrol agents based on detailed field maps.
   » Reduces wastage by applying agents only where they are needed.
2. **Variable Rate Technology (VRT):**
   » Adjusts the application rate of biocontrol agents based on real-time soil and crop health data.
   » Ensures optimal use of resources, improving cost efficiency.
3. **Real-Time Monitoring:**
   » Sensors and drones provide real-time data on pest and disease outbreaks, guiding timely application of biocontrol agents.

Applications of Precision Agriculture Integration

1. **Site-Specific Biocontrol Delivery:**
   » Targeting root-knot nematodes in vegetables using variable rate soil applications.
   » Applying *Bacillus subtilis* formulations to localized disease hotspots in orchards.

2. **Automated Irrigation Systems:**
   - » Integrating drip irrigation with biocontrol delivery for managing soilborne pathogens in high-value crops.
   - » Ensuring consistent application of Trichoderma spp. formulations in greenhouse vegetables.

5.3 Application Strategies and Monitoring of Biocontrol Efficacy

The success of biocontrol agents in managing plant diseases and pests hinges on the adoption of effective application strategies and robust monitoring systems to evaluate their efficacy. These components are critical in ensuring that biocontrol agents reach their target pathogens or pests in sufficient quantities and remain viable under field conditions. Moreover, monitoring biocontrol efficacy provides essential feedback, allowing researchers and practitioners to optimize application methods and improve outcomes.

Application Strategies

The application of biocontrol agents must be tailored to the specific crop, pathogen or pest, environmental conditions, and the type of biocontrol formulation. Various strategies are employed to maximize the agents' efficacy while ensuring minimal environmental impact and cost-effectiveness.

One of the most common strategies involves **prophylactic applications**, where biocontrol agents are applied preventatively before the onset of disease or pest infestation. This approach is particularly effective in systems where the timing of pathogen activity or pest outbreaks is predictable, such as with seedborne or soilborne pathogens. By establishing the biocontrol agents in the rhizosphere or phyllosphere early in the cropping season, this strategy creates a protective barrier that suppresses pathogen establishment or pest colonization.

Curative applications, on the other hand, target active infections or infestations. In these scenarios, biocontrol agents are applied in response to observed symptoms or pest presence. This strategy is commonly used for foliar diseases and pests, where the application of biocontrol agents such as *Bacillus subtilis* or *Beauveria bassiana* can directly suppress pathogen growth or insect activity on plant surfaces. However, the success of curative applications often depends on the timing and frequency of applications, as delayed treatments may allow pathogens or pests to establish and spread.

Integrated application approaches combine biocontrol agents with other cultural, mechanical, or chemical control methods as part of an Integrated Pest Management (IPM) strategy. For instance, biocontrol agents may be applied in conjunction with crop rotation, resistant cultivars, or selective chemical treatments. This approach enhances the overall efficacy of pest and disease management while reducing the reliance on chemical inputs. Compatibility studies are essential in these systems to ensure that biocontrol agents are not adversely affected by other management practices, such as the use of fertilizers or fungicides.

Advances in delivery systems have also improved application strategies. Technologies like drip irrigation systems, drone-based spraying, and seed-embedded biocontrol formulations have enabled more precise and efficient delivery of biocontrol agents. These innovations not only enhance the efficacy of applications but also reduce labor and operational costs, making biocontrol agents more accessible to a broader range of agricultural systems.

Monitoring Biocontrol Efficacy

Monitoring the efficacy of biocontrol agents is an integral part of their application, providing critical data on their performance under field conditions. This process involves evaluating the impact of biocontrol agents on target pathogens or pests, as well as assessing their influence on crop health and yield. Field trials are one of the primary methods for monitoring biocontrol efficacy. These trials are designed to mimic real-world agricultural conditions while maintaining controlled variables for accurate assessment. Researchers evaluate parameters such as disease incidence, severity, pest populations, and crop yield in treated versus untreated plots. Standardized rating scales, such as the area under the disease progress curve (AUDPC), are often used to quantify disease suppression and compare the performance of different treatments.

Molecular techniques have revolutionized the monitoring of biocontrol efficacy, enabling precise tracking of microbial agents and their interactions with pathogens or pests. Techniques such as quantitative PCR (qPCR) and DNA fingerprinting allow researchers to quantify the population dynamics of biocontrol agents in the rhizosphere or phyllosphere over time. These methods also help identify whether biocontrol agents are effectively colonizing the target sites and producing the expected antimicrobial compounds.

In addition to pathogen or pest suppression, the impact of biocontrol agents on crop health and yield is a critical measure of their efficacy. Crop growth parameters, such as root biomass, leaf area, and nutrient uptake, are monitored to assess the indirect benefits of biocontrol applications. Enhanced crop performance often indicates successful suppression of pathogens or pests and the promotion of plant growth by biocontrol agents. Environmental monitoring is another essential component of efficacy evaluation. Biocontrol agents are living organisms that interact with the existing microbiota, and their application can influence soil and ecosystem health. Studies on soil microbial diversity, nutrient cycling, and non-target organism impacts provide insights into the broader ecological effects of biocontrol applications. This information is crucial for ensuring that biocontrol agents contribute to sustainable agricultural systems without causing unintended harm.

The integration of precision agriculture tools has further improved the monitoring of biocontrol efficacy. Sensors, drones, and remote sensing technologies provide real-time data on pest and disease outbreaks, allowing for timely adjustments in biocontrol applications. For instance, infrared imaging can detect early signs of plant stress caused by pathogens or pests, guiding targeted applications of biocontrol agents. Geographic Information Systems (GIS) and predictive modeling are also used to map disease and pest spread, enabling more strategic deployment of biocontrol agents.

Challenges and Future Directions

Despite the advancements in application strategies and monitoring techniques, several challenges remain in optimizing the use of biocontrol agents. Variability in field conditions, such as temperature, humidity, and soil properties, can influence the performance of biocontrol agents. Addressing this variability requires the development of resilient strains and formulations that remain effective under diverse conditions.

Another challenge lies in the compatibility of biocontrol agents with existing agricultural practices. Ensuring that biocontrol agents function synergistically with fertilizers, irrigation systems, and other inputs is critical for their widespread adoption. Additionally, the cost of monitoring systems, such as molecular assays and remote sensing technologies, may be prohibitive for small-scale farmers, necessitating the development of more affordable and accessible solutions. Looking ahead, advancements in synthetic biology, bioinformatics, and machine learning hold promise for further enhancing application strategies and monitoring systems. Synthetic biology can be used to engineer biocontrol agents with enhanced traits, such as increased stress tolerance or the ability to target specific pathogens or pests. Machine learning algorithms can analyze large datasets from field trials and precision agriculture tools, providing insights into the optimal conditions for biocontrol efficacy. The combination of effective application strategies and robust monitoring systems is essential for the success of biocontrol agents in modern agriculture. By continuously refining these components and integrating emerging technologies, biocontrol agents can become a cornerstone of sustainable pest and disease management, reducing reliance on chemical inputs and supporting global food security.

5.4 Challenges in Scaling Production and Deployment

The large-scale production and deployment of biocontrol agents are pivotal to their success as an integral component of sustainable agriculture. However, transitioning from laboratory-scale development to widespread agricultural use presents numerous challenges. These challenges encompass technical, economic, regulatory, and practical aspects that must be addressed to ensure the effective and reliable adoption of biocontrol technologies. This section delves into the key obstacles associated with scaling the production and deployment of biocontrol agents and explores potential solutions to overcome them.

Technical Challenges in Scaling Production

Scaling up the production of biocontrol agents involves replicating laboratory or pilot-scale processes in industrial settings, which often introduces unforeseen complexities. Microbial growth, formulation stability, and consistency are among the critical factors that pose significant technical challenges.

One of the primary technical issues is **strain optimization for mass production**. Many biocontrol agents, particularly microbial antagonists, are selected based on their efficacy under controlled laboratory conditions. However, these strains may not perform optimally under industrial fermentation conditions, where factors such as nutrient availability, pH, oxygen levels, and agitation differ significantly. For instance, fungal strains like *Trichoderma spp.* may exhibit reduced sporulation or altered metabolite production during large-scale fermentation, impacting their effectiveness as biocontrol agents. Another challenge is **formulation development and stability**. Biocontrol agents must be formulated to ensure their viability, stability, and ease of application under field conditions. Formulations such as powders, granules, and liquids require careful optimization of additives, carriers, and protective agents to enhance shelf life and field performance. However, maintaining microbial viability during processes like drying (e.g., spray-drying or freeze-drying) and storage remains a significant hurdle. Furthermore, formulations must be designed to withstand environmental stresses, such as temperature fluctuations and UV radiation, without compromising the agents' efficacy.

3. Contamination during large-scale production is another persistent issue. The industrial fermentation of biocontrol agents involves the use of large bioreactors and complex nutrient media, which can be prone to contamination by undesirable microorganisms. Contaminants not only compete with the target biocontrol agents for resources but may also produce toxic metabolites that affect the

quality of the final product. Ensuring sterile production environments and robust quality control protocols is essential to mitigate these risks.

Economic Constraints in Scaling Production

The economic feasibility of producing biocontrol agents at scale is a critical determinant of their adoption. High production costs can make biocontrol products less competitive than conventional chemical pesticides, limiting their uptake by farmers. One significant economic challenge is the **cost of raw materials** used in fermentation media. Many biocontrol agents require nutrient-rich substrates for growth, which can be expensive. Efforts to reduce costs by utilizing agricultural byproducts or waste materials as substrates are underway, but these alternatives often require additional processing or optimization to ensure compatibility with microbial growth.

The **cost of equipment and infrastructure** for large-scale production is another barrier. Biocontrol agent production facilities must include specialized equipment for fermentation, formulation, and packaging, which requires significant capital investment. Small and medium-sized enterprises (SMEs) that dominate the biocontrol industry may struggle to secure the funding needed to establish or expand production capacity.

Regulatory Challenges

The commercialization of biocontrol agents involves navigating complex and often fragmented regulatory frameworks that vary between countries and regions. These regulatory processes are designed to ensure the safety, efficacy, and environmental compatibility of biocontrol products, but they can also pose significant challenges for producers.

One of the key issues is the **lengthy and expensive registration process**. Regulatory bodies require extensive data on the microbial strain, production process, and field performance of biocontrol agents. This includes safety assessments to evaluate potential risks to human health, non-target organisms, and ecosystems. Generating this data requires time, resources, and expertise, which can delay the commercialization of new biocontrol products. Another challenge is the **lack of harmonized international standards** for biocontrol agents. While chemical pesticides often adhere to globally recognized standards, biocontrol agents are subject to varying requirements depending on the jurisdiction. This lack of consistency increases the regulatory burden for companies seeking to enter international markets.

Practical Challenges in Deployment

The deployment of biocontrol agents in agricultural systems presents a distinct set of challenges related to their application, integration into farming practices, and performance under field conditions. One of the primary challenges is **ensuring consistent field performance**. Biocontrol agents are living organisms, and their efficacy can be influenced by environmental factors such as temperature, humidity, soil type, and crop variety. For instance, microbial antagonists like *Bacillus subtilis* may exhibit reduced activity under extreme heat or drought conditions, limiting their effectiveness in arid regions. Developing robust strains and formulations that perform consistently across diverse environments is essential to address this issue.

Compatibility with existing farming practices is another challenge. Farmers often use chemical fertilizers, pesticides, and other inputs, which may interact with biocontrol agents and reduce their efficacy. For example, certain fungicides can inhibit the growth of beneficial fungi like *Trichoderma spp.*. Educating farmers on the compatibility of biocontrol agents with other inputs and promoting integrated pest management (IPM) practices are critical for successful deployment. The **logistics of storage and transportation** also pose practical challenges. Biocontrol agents, particularly microbial formulations, require specific storage conditions to maintain viability. For example, liquid formulations often need refrigeration, which may not be feasible in regions with limited cold storage infrastructure. Ensuring the stability of biocontrol products during transportation and storage is crucial for maintaining their quality and effectiveness. Farmer awareness and acceptance are equally important factors in the deployment of biocontrol agents. Many farmers are unfamiliar with the use of biocontrol products and may be hesitant to adopt them due to concerns about efficacy, cost, or complexity. Building trust through demonstration trials, training programs, and extension services is vital for encouraging the adoption of biocontrol technologies.

Potential Solutions and Future Directions

Addressing the challenges associated with scaling the production and deployment of biocontrol agents requires a multifaceted approach that combines scientific innovation, regulatory reform, and stakeholder engagement. On the technical front, advances in **synthetic biology** and **metabolic engineering** hold promise for improving the scalability of biocontrol agents. These tools can be used to develop microbial strains with enhanced growth rates, stress tolerance, and metabolite production. For example, genetic modifications can enable biocontrol agents to thrive under suboptimal conditions or resist contamination during fermentation.

The development of **cost-effective substrates** and alternative fermentation technologies, such as solid-state fermentation, can help reduce production costs. Using agricultural byproducts, such as molasses or rice husks, as nutrient sources for microbial growth is a promising strategy that aligns with the principles of circular economy and sustainability. Regulatory reform is essential to streamline the approval process for biocontrol agents. Harmonizing international standards and adopting risk-based assessment frameworks can reduce the regulatory burden on producers while maintaining safety and efficacy standards. Collaboration between regulatory bodies, industry stakeholders, and researchers is critical to achieving these goals.

To address deployment challenges, **extension services, and capacity-building programs** should be prioritized. These initiatives can educate farmers about the benefits of biocontrol agents, demonstrate their application methods, and provide technical support. Public-private partnerships can play a pivotal role in funding and implementing these programs, particularly in resource-constrained regions. Finally, the integration of **precision agriculture tools** can enhance the deployment of biocontrol agents by enabling site-specific applications. Technologies such as drones, GPS-enabled equipment, and soil sensors can improve the efficiency and accuracy of biocontrol applications, reducing waste and maximizing impact.

Chapter - 6

Integration of Biological Control in Modern Farming Systems

The integration of biological control into modern farming systems represents a paradigm shift toward sustainable agriculture, emphasizing ecological balance, resource efficiency, and long-term productivity. Biological control, which employs natural enemies such as predators, parasitoids, and microbial agents to regulate pest populations, aligns seamlessly with the goals of modern farming by reducing reliance on chemical pesticides and promoting environmentally friendly practices. This integration is particularly critical in the face of global challenges such as climate change, biodiversity loss, and the intensification of agricultural practices, which demand innovative solutions to ensure food security without compromising ecosystem health. By embedding biological control within a holistic framework that includes cultural, physical, and chemical methods, modern farming systems can achieve comprehensive pest and disease management while enhancing resilience and sustainability.

A key aspect of integrating biological control is its compatibility with Integrated Pest Management (IPM) strategies, which combine multiple pest management tactics to minimize economic, environmental, and health risks. In an IPM framework, biological control serves as a cornerstone, complementing cultural practices like crop rotation, intercropping, and habitat manipulation. For example, planting flowering plants alongside crops provides food and refuge for natural enemies, such as parasitoid wasps and predatory beetles, enhancing their effectiveness in controlling pests. Similarly, reduced tillage and the use of cover crops improve soil health and microbial diversity, creating favorable conditions for microbial biocontrol agents such as *Trichoderma spp.* and *Bacillus subtilis* to suppress soil-borne pathogens. These synergies between biological control and cultural practices not only improve pest management outcomes but also contribute to the overall health and resilience of agricultural ecosystems.

The integration of biological control into modern farming systems also involves the judicious use of chemical pesticides to minimize their impact on beneficial organisms. Selective or reduced-risk pesticides, which target specific pests while sparing natural enemies, are increasingly incorporated into IPM programs to preserve the efficacy of biocontrol agents. For instance, the use of insect growth regulators and botanical pesticides can suppress pest populations without harming predators like ladybird beetles or parasitoids like *Encarsia formosa*. This selective approach reduces the risks of pesticide resistance and environmental contamination, ensuring that biological control agents remain viable and effective over the long term. Furthermore, the adoption of precision agriculture technologies, such as remote sensing, drone-based spraying, and IoT-enabled monitoring systems, enhances the integration of biological control by enabling targeted applications of biocontrol agents and pesticides. These technologies provide real-time data on pest populations, crop health, and environmental conditions, allowing farmers to make informed decisions that optimize resource use and minimize ecological disruption.

Another critical element of integrating biological control into modern farming systems is its role in organic and regenerative agriculture. These farming approaches prioritize the use of natural processes and inputs to improve soil health, biodiversity, and ecosystem services. Biological control aligns perfectly with these principles, offering effective pest and disease management solutions that comply with organic certification standards. For example, microbial agents such as *Beauveria bassiana* and *Metarhizium anisopliae* are widely used in organic farming to control insect pests, while mycorrhizal fungi enhance nutrient uptake and disease resistance in a variety of crops. The use of natural enemies like predatory mites and parasitic nematodes further reduces the need for synthetic inputs, making biological control an indispensable tool for organic and regenerative farmers. By fostering healthy soils, diverse cropping systems, and functional ecosystems, the integration of biological control in these farming systems contributes to long-term agricultural sustainability.

The economic and market benefits of integrating biological control into modern farming systems are also significant. As consumer demand for pesticide-free and sustainably produced food continues to grow, farmers who adopt biological control gain access to premium markets and export opportunities. Crops grown with biocontrol practices often meet stringent food safety standards, such as those set by the European Union, which limit chemical residues in agricultural products. This compliance not only enhances marketability but also builds consumer trust in the safety and quality of the produce. Moreover, the cost-effectiveness of biological control, particularly in reducing expenditure on chemical pesticides and mitigating crop losses, further supports its integration into modern farming systems. For smallholder farmers, in particular, biological control offers an affordable and accessible solution that aligns with limited resources and promotes economic resilience.

Education and capacity building are crucial for the successful integration of biological control into modern farming systems. Farmers need access to training programs that provide knowledge on the identification, application, and management of biocontrol agents. Extension services and demonstration farms play a vital role in showcasing the benefits of biological control and encouraging its adoption at scale. Collaborative efforts between researchers, policymakers, and industry stakeholders are also essential for addressing challenges such as the availability and affordability of biocontrol products, regulatory hurdles, and market incentives. Public-private partnerships can drive innovation in the production and delivery of biocontrol agents, ensuring that farmers have access to high-quality products and support services.

6.1 Role of Biological Control in Integrated Disease Management (IDM)

The integration of biological control into Integrated Disease Management (IDM) represents a transformative shift in modern agricultural practices. By combining biocontrol agents with other disease management strategies, IDM promotes sustainable, eco-friendly approaches to crop protection while reducing dependence on chemical pesticides. This holistic strategy leverages the natural mechanisms of biocontrol agents to suppress pathogens, improve plant resilience, and enhance the long-term health of agricultural ecosystems. The role of biological control in IDM is multifaceted, encompassing pathogen suppression, plant health promotion, environmental preservation, and compatibility with broader farming systems.

6.1.1. Biological Control as a Core Component of IDM

Biological control serves as a cornerstone of Integrated Disease Management (IDM), a holistic approach to plant disease management that prioritizes sustainability, reduced reliance on chemical inputs, and long-term agricultural productivity. By leveraging beneficial microorganisms, macro-organisms, and their metabolic products, biological control effectively suppresses pathogens, enhances plant resilience, and contributes to a balanced agricultural ecosystem. These agents—including fungi, bacteria, nematodes, and natural predators—operate through diverse mechanisms such as antibiosis, competition for nutrients and space, parasitism, and the induction of systemic resistance. Their integration into IDM systems addresses the limitations of chemical and cultural controls, offering a synergistic approach to disease management that ensures crop protection while maintaining environmental health.

Pathogen Suppression Through Natural Antagonism

One of the primary contributions of biological control in IDM is its ability to suppress pathogen populations using natural antagonistic mechanisms. This suppression is achieved through direct and indirect interactions between biocontrol agents and plant pathogens. Direct mechanisms include antibiosis, parasitism, and enzymatic degradation, while indirect mechanisms involve competition for resources and space. These interactions disrupt the growth, reproduction, and virulence of pathogens, reducing their ability to infect crops.

Bacterial antagonists such as *Pseudomonas fluorescens* are exemplary biocontrol agents that employ multiple mechanisms to suppress soilborne pathogens like *Fusarium spp.*, *Pythium spp.*, and *Rhizoctonia solani*. One key mode of action is the production of antibiotics such as phenazines, pyoluteorin, and pyrrolnitrin. These compounds inhibit pathogen growth by disrupting cell membranes, interfering with essential enzymatic pathways, or inducing oxidative stress. For example, phenazines produced by *Pseudomonas fluorescens* disrupt fungal membranes, leading to leakage of cellular contents and eventual cell death. In addition to antibiosis, *Pseudomonas fluorescens* produces siderophores—iron-chelating compounds that sequester iron from the soil. By reducing the availability of this essential nutrient, siderophores create an environment that is hostile to pathogenic microbes while favoring the growth of the biocontrol agent.

Fungal biocontrol agents, such as *Trichoderma spp.*, contribute to pathogen suppression through parasitism and enzymatic degradation. *Trichoderma spp.* are known for their ability to colonize the rhizosphere and outcompete pathogens for resources and space. When encountering pathogenic fungi, *Trichoderma spp.* deploy lytic enzymes such as chitinases, glucanases, and proteases, which degrade the cell walls of the target pathogen. This enzymatic action not only halts the growth of the pathogen but also limits its ability to infect plant roots. Moreover, *Trichoderma spp.* form physical attachments to the mycelium of pathogenic fungi, effectively parasitizing them and reducing their viability. This dual mode of action—parasitism and competition—makes *Trichoderma spp.* highly effective against a wide range of soilborne pathogens.

Another notable example is the use of endophytic fungi such as *Piriformospora indica*, which colonize plant roots and provide protection against soilborne pathogens. These fungi establish mutualistic relationships with plants, where they enhance nutrient uptake and induce systemic resistance while suppressing pathogens through competition and antibiosis. Endophytes are particularly effective in environments with high pathogen pressure, as they create a protective microbial barrier that limits pathogen access to plant tissues.

Enhancement of Plant Resilience Through Induced Systemic Resistance (ISR)

Beyond direct pathogen suppression, biological control agents play a crucial role in enhancing plant resilience to diseases through the induction of systemic resistance. Induced systemic resistance (ISR) is a physiological state in which plants are primed to activate their defense mechanisms more rapidly and robustly upon pathogen attack. This phenomenon is triggered by specific biocontrol agents that interact with plant roots, releasing signaling molecules that stimulate defense-related pathways.

For instance, *Bacillus subtilis*, a well-studied biocontrol bacterium, is known to trigger ISR in various crops. Upon colonizing the root surface, *Bacillus subtilis* produces elicitors such as lipopeptides, volatile organic compounds, and bacterial exopolysaccharides. These elicitors activate signaling pathways in the plant, leading to the synthesis of defense compounds such as phytoalexins, phenolic acids, and pathogenesis-related (PR) proteins. Phytoalexins are antimicrobial compounds that inhibit pathogen growth, while PR proteins degrade pathogen cell walls and enhance the structural integrity of plant tissues. The activation of these defenses reduces the susceptibility of plants to a broad spectrum of pathogens, including bacteria, fungi, and viruses.

ISR is not limited to bacterial agents; fungal biocontrol agents such as *Trichoderma harzianum* also induce similar responses. *Trichoderma spp.* colonize plant roots and release metabolites that trigger signaling cascades associated with systemic resistance. These cascades involve the activation of hormonal pathways, including jasmonic acid (JA) and ethylene (ET) signaling, which play a central role in plant defense against necrotrophic pathogens and insect pests. By enhancing the plant's innate defense capabilities, ISR reduces the need for chemical interventions, aligning with the goals of IDM to promote sustainable disease management.

Integration with Soil Health and Ecosystem Balance

Biological control agents not only suppress pathogens but also contribute to the broader health of agricultural ecosystems. Their activities improve soil microbiota diversity, enhance nutrient cycling, and promote plant-microbe symbiosis, all of which support long-term crop productivity and resilience. The application of biocontrol agents such as *Trichoderma spp.*, mycorrhizal fungi, and nitrogen-fixing bacteria enriches the soil microbiome, creating a favorable environment for plant growth.

For example, mycorrhizal fungi, including species of *Glomus* and *Rhizophagus*, form symbiotic associations with plant roots, extending the root system's functional area through fungal hyphae. This association improves the plant's access to nutrients such as phosphorus and micronutrients, while also providing protection against root pathogens. Similarly, nitrogen-fixing bacteria such as *Rhizobium* and *Azospirillum* contribute to soil fertility by converting atmospheric nitrogen into forms that plants can assimilate. When integrated into IDM systems, these microbial agents not only control diseases but also enhance soil health, reducing the need for synthetic fertilizers.

Biological control agents also play a role in mitigating abiotic stresses, such as drought and salinity, which exacerbate disease susceptibility in plants. For instance, endophytic fungi like *Piriformospora indica* enhance the water uptake capacity of plants, reducing drought-induced stress and improving overall plant vigor. By promoting plant health under challenging environmental conditions, biocontrol agents extend the scope of IDM beyond disease suppression to holistic crop management.

Compatibility with Integrated Disease Management Practices

The success of biological control in IDM lies in its compatibility with other disease management practices, creating a cohesive and sustainable framework for crop protection. Biocontrol agents complement cultural, mechanical, and chemical methods, enhancing the overall efficacy of IDM systems. Cultural practices such as crop rotation, intercropping, and residue management create conditions that favor the establishment of biocontrol agents while suppressing pathogens. For instance, rotating crops with legumes that harbor nitrogen-fixing bacteria enriches the soil microbiome, creating a supportive environment for biocontrol agents. Similarly, incorporating organic amendments like compost or green manure provides nutrients and habitats for beneficial microbes, enhancing their activity in the rhizosphere.

Mechanical practices, including pruning, soil tillage, and irrigation management, can be integrated with biological control to create unfavorable conditions for pathogens. For example, pruning infected plant parts followed by the application of biocontrol agents such as *Bacillus amyloliquefaciens* prevents the spread of secondary infections. Drip irrigation systems can be used to deliver liquid formulations of biocontrol agents directly to the root zone, ensuring precise application and minimizing resource wastage. Chemical inputs, when used judiciously, can be integrated with biological control to enhance IDM outcomes. Reduced-risk fungicides and biopesticides that are compatible with biocontrol agents allow for targeted pathogen suppression without disrupting beneficial microbial populations. For example, combining *Trichoderma spp.* with specific fungicides can improve disease control while reducing chemical usage, aligning with the goals of IDM to minimize environmental impact.

Examples of Biological Control in IDM

The integration of biological control in IDM has demonstrated success across diverse cropping systems and geographic regions. In rice cultivation, foliar sprays of *Pseudomonas fluorescens* combined with resistant varieties and water management practices have significantly reduced bacterial leaf blight incidence in Asia. In greenhouse tomato production, the use of *Trichoderma harzianum* as a soil amendment, paired with precision irrigation systems, has minimized gray mold infections and improved yields. Similarly, in citrus orchards, endophytic biocontrol agents like *Burkholderia spp.* have been used alongside nutrient management practices to combat citrus greening disease, enhancing tree health and fruit quality.

6.1.2. Compatibility with Other IDM Practices

Biological control harmoniously integrates with various components of Integrated Disease Management (IDM), forming a robust and multifaceted strategy for disease suppression and sustainable crop health. By complementing cultural, mechanical, and chemical methods, biological control agents enhance the effectiveness of IDM systems while minimizing environmental impacts. This compatibility allows farmers to create a synergistic framework where different management practices work together to achieve comprehensive disease control.

Integration with Cultural Practices

Cultural practices form the foundation of many IDM strategies, and biological control agents amplify the benefits of these methods by improving soil microbiota and nutrient cycling. Practices such as crop rotation, intercropping, and soil health management create conditions that favor the establishment and activity of biocontrol agents, further suppressing pathogens and promoting plant health. For example, crop rotation is a widely used practice to break the life cycle of pathogens and reduce their buildup in the soil. When combined with biological control, this method becomes even more effective. The application of *Trichoderma spp.* to rotational crops enhances soil health by increasing microbial diversity and producing enzymes that degrade pathogen cell walls. Additionally, crop residues enriched with biocontrol agents create a hostile environment for soilborne pathogens, reducing their persistence.

Intercropping, another essential cultural practice, benefits significantly from the addition of biological control agents. Leguminous crops, often used in intercropping systems, harbor nitrogen-fixing bacteria such as *Rhizobium* and *Azospirillum*, which not only improve soil fertility but also create favorable conditions for biocontrol agents. For instance, *Pseudomonas fluorescens* can thrive in the rhizosphere of legumes, suppressing root pathogens and promoting plant growth. The mutual reinforcement of intercropping and biological control results in healthier crops and reduced disease pressure. Soil health management practices, such as the incorporation of organic amendments like compost, green manure, and biochar, provide an excellent medium for biological control agents. Organic amendments enrich the soil with nutrients and organic matter, supporting the growth of beneficial microbes. When applied alongside biocontrol agents like *Trichoderma spp.*, these amendments enhance soil suppressiveness, reducing the incidence of soilborne diseases such as damping-off and root rot. The combined approach improves soil structure, fertility, and disease resistance, creating a resilient agricultural ecosystem.

Synergy with Mechanical Control Measures

Mechanical control measures, including pruning, rogueing, and tillage, create environments that are less conducive to pathogen proliferation. These practices disrupt the physical habitat of pathogens and pests, limiting their ability to survive and spread. The integration of biological control agents into these systems further amplifies their effectiveness, ensuring long-lasting disease suppression.

Pruning, a common practice in orchard and vineyard management, removes infected plant parts and opens the canopy to improve air circulation and light penetration. After pruning, the application of biocontrol agents like *Bacillus amyloliquefaciens* as foliar sprays prevents secondary infections by airborne or wound-invading pathogens. These agents colonize the pruned surfaces, producing antimicrobial compounds and creating a protective barrier against pathogens such as *Botrytis cinerea* in grapes and *Alternaria* in citrus. Rogueing, the removal of diseased plants from the field, is particularly effective in managing systemic infections like bacterial wilt and viral diseases. The integration of biocontrol agents into rogueing practices enhances the overall strategy. For example, after removing infected plants, the soil can be treated with *Trichoderma spp.* or *Pseudomonas fluorescens* to suppress residual pathogen populations and prevent reinfection. This combination ensures that the field remains disease-free and productive.

Tillage, a practice that disrupts soil structure to bury crop residues and pathogens, can be paired with biological control agents to improve disease suppression. Incorporating *Trichoderma spp.* or other biocontrol formulations into the soil during tillage accelerates the decomposition of organic matter and reduces pathogen survival. These agents colonize the soil environment, competing with pathogens for nutrients and producing enzymes that degrade pathogen propagules.

Complementation with Chemical Management

The integration of biological control with chemical management requires careful consideration to avoid negative interactions. While chemical inputs such as fungicides and insecticides play an important role in IDM, their indiscriminate use can harm beneficial microorganisms and reduce the efficacy of biocontrol agents. Therefore, compatibility studies are essential to identify chemical formulations that allow for the coexistence of biocontrol agents and chemical products. Reduced-risk fungicides and biopesticides have emerged as effective tools for integrating chemical and biological control. These products are designed to target pathogens without adversely affecting non-target organisms, including biocontrol agents. For example, reduced-risk fungicides can be applied alongside *Trichoderma spp.* or *Bacillus subtilis* to manage foliar diseases while preserving the activity of these beneficial microbes. Such combinations maintain disease control while reducing the overall chemical load on the environment.

Biopesticides, derived from natural sources such as plant extracts, microbial metabolites, and mineral oils, are inherently compatible with biocontrol agents. When used in tandem, biopesticides and biological control agents create a multi-faceted defense against pathogens. For instance, combining neem oil with *Pseudomonas fluorescens* enhances the suppression of root-knot nematodes, as the oil disrupts nematode activity while the bacterium colonizes the rhizosphere and outcompetes the nematodes for resources. The timing and sequence of applications are critical for ensuring compatibility between chemical and biological inputs. For example, applying a fungicide before introducing a biocontrol agent can reduce the agent's viability, whereas sequential application with adequate intervals allows both products to function effectively. Adopting integrated schedules that account for the persistence and activity of both chemical and biological agents ensures that disease control is maximized without compromising the sustainability of the IDM program.

Advantages of Compatibility in IDM Systems

The seamless integration of biological control with cultural, mechanical, and chemical practices offers several advantages that enhance the overall effectiveness and sustainability of IDM systems.

1. **Enhanced Disease Suppression:** The complementary mechanisms of action provided by biocontrol agents and other IDM components create a multi-layered defense system that is more effective than any single method. This synergy ensures robust protection against a wide range of pathogens and pests.
2. **Reduced Environmental Impact:** By reducing the reliance on chemical inputs, biological control minimizes the risks of pesticide residues, groundwater contamination, and non-target effects. This aligns with the goals of sustainable agriculture and supports ecological balance.

3. **Increased Crop Resilience:** The integration of biocontrol agents with soil health management practices improves soil microbiota diversity and plant-microbe interactions, resulting in healthier crops that are more resistant to diseases and environmental stresses.
4. **Cost Efficiency:** The use of biological control agents alongside cultural and mechanical practices reduces the frequency and volume of chemical applications, lowering input costs for farmers. Additionally, improved soil health and plant resilience translate into higher yields and better economic returns.
5. **Adaptability to Diverse Systems:** The compatibility of biological control with various IDM practices makes it adaptable to a wide range of cropping systems, from smallholder farms to large-scale commercial agriculture. This versatility ensures that IDM systems can be tailored to specific crops, pathogens, and environmental conditions.

6.1.3 Environmental and Economic Benefits

The integration of biological control into Integrated Disease Management (IDM) provides profound environmental and economic advantages, making it a cornerstone of sustainable agricultural practices. These benefits are critical for addressing the pressing challenges of modern agriculture, including the need to reduce chemical pesticide reliance, safeguard natural resources, and ensure economic viability for farmers. Biological control not only mitigates the environmental impact of traditional disease management practices but also contributes to cost savings and market competitiveness, particularly in systems transitioning to organic and sustainable production.

Environmental Benefits

One of the most significant environmental contributions of biological control in IDM is its ability to reduce the dependency on chemical pesticides. Chemical pesticides, though effective, pose several risks to the environment, including contamination of water bodies, degradation of soil quality, and harm to non-target organisms, such as beneficial insects and soil microbes. The widespread and prolonged use of chemical pesticides also accelerates the development of resistant pathogen and pest strains, undermining their long-term effectiveness. Biological control agents, by contrast, rely on natural mechanisms to suppress pathogens, offering a safer and more sustainable alternative.

Biological control agents, including bacteria, fungi, and natural predators, are inherently selective in their actions. For example, microbial antagonists such as *Trichoderma spp.* and *Bacillus subtilis* specifically target plant pathogens without adversely affecting beneficial soil microbiota. This selectivity helps preserve the ecological balance of agricultural ecosystems, maintaining a diverse and functional microbial community that supports plant growth and soil health. By promoting biodiversity, biological control enhances the resilience of agroecosystems to environmental stressors and pathogen outbreaks. The application of biological control also reduces the risk of pesticide residues in crops, an increasingly important consideration in both domestic and export markets. Regulatory standards for pesticide residues in agricultural products are becoming more stringent, particularly in high-value export-oriented systems. By replacing or supplementing chemical pesticides with biocontrol agents, farmers can meet these regulatory requirements, ensuring the safety and marketability of their produce. Additionally, residue-free production aligns with consumer preferences for organic and sustainably grown food, bolstering market opportunities for farmers adopting IDM practices.

Groundwater contamination is another critical environmental concern associated with chemical pesticide use. Many synthetic pesticides leach into the soil and contaminate groundwater, posing risks to both ecosystems and human health. Biological control agents, being living organisms or their byproducts, do not persist in the environment in the same way as chemical pesticides. Their natural degradation reduces the risk of long-term contamination, making biological control a more environmentally friendly option. For instance, the use of *Pseudomonas fluorescens* in soilborne disease management has been shown to suppress pathogens effectively without contributing to groundwater pollution. In addition to reducing chemical inputs, biological control agents contribute to sustainable soil management. Many biocontrol agents, such as *Trichoderma spp.* and mycorrhizal fungi, enhance soil structure, nutrient availability, and organic matter content. These improvements not only suppress pathogens but also create a healthier soil environment that supports long-term agricultural productivity. This dual benefit of disease suppression and soil health promotion underscores the role of biological control in achieving sustainable farming systems.

Economic Benefits

From an economic perspective, the inclusion of biological control in IDM offers tangible cost savings and revenue opportunities for farmers. While the initial cost of biocontrol agents may be higher than chemical pesticides, their long-term benefits far outweigh these upfront expenses. Biological control agents reduce the frequency and volume of chemical pesticide applications, leading to significant savings in input costs over time.

One of the primary ways biological control lowers costs is by improving soil health and crop productivity. Healthy soils supported by biocontrol agents require fewer inputs, such as fertilizers and soil amendments, reducing overall production costs. For example, *Azospirillum* and *Rhizobium* not only suppress soilborne pathogens but also enhance nitrogen fixation, reducing the need for synthetic nitrogen fertilizers. Similarly, *Trichoderma spp.* and mycorrhizal fungi improve nutrient uptake and water retention, leading to better crop growth and higher yields. These benefits translate into improved economic returns for farmers, especially in resource-limited agricultural systems. Another economic advantage of biological control is its alignment with the growing consumer demand for organic and sustainably produced food. As consumers become increasingly aware of the environmental and health impacts of chemical pesticide use, there is a rising preference for residue-free and organically certified produce. This trend creates lucrative market opportunities for farmers who adopt IDM practices incorporating biological control. Certified organic produce often commands premium prices, providing farmers with an incentive to transition to sustainable disease management strategies.

Biological control also enhances market access for farmers in export-oriented agricultural systems. Many countries, particularly in the European Union, impose strict limits on pesticide residues in imported agricultural products. Compliance with these regulations is essential for accessing high-value export markets. By incorporating biological control into IDM, farmers can meet these stringent requirements, ensuring the competitiveness of their produce in international markets. For example, Indian mango exporters have successfully used biological control agents to manage post-harvest diseases while complying with European residue standards, allowing them to expand their market presence. In addition to direct economic benefits, biological control contributes to the long-term resilience and profitability of agricultural systems. By reducing the reliance on chemical pesticides, farmers mitigate the risks associated with pesticide resistance, ensuring the continued effectiveness of disease management practices. This proactive approach reduces the likelihood of yield losses due to resistant pathogens, protecting farmers' incomes and food security. The economic benefits of biological control are also evident in the context of government and industry support for sustainable agriculture. Many governments and organizations offer subsidies, grants, and technical assistance to promote the adoption of biocontrol agents and IDM practices. These incentives reduce the financial burden on farmers and encourage widespread adoption of sustainable disease management strategies. For instance, programs in countries like Brazil and the Netherlands provide financial support to farmers transitioning to biological control, accelerating the adoption of IDM at a national scale.

6.1.4 Case Studies of Biological Control in IDM

The integration of biological control into Integrated Disease Management (IDM) has demonstrated remarkable success across diverse cropping systems and regions. These case studies illustrate how the strategic use of biocontrol agents in combination with other IDM practices can suppress diseases, enhance crop health, and improve yields while reducing reliance on chemical inputs. Each example highlights the adaptability and effectiveness of biological control in addressing specific disease challenges, showcasing its potential as a cornerstone of sustainable agriculture.

Rice Systems in Asia

Bacterial leaf blight (BLB), caused by *Xanthomonas oryzae pv. oryzae*, is one of the most devastating diseases affecting rice crops in Asia. It leads to significant yield losses and compromises food security in rice-dependent regions. IDM programs for managing bacterial leaf blight have successfully integrated biological control agents with complementary practices, demonstrating the power of a multifaceted approach.

In these systems, the bacterium *Pseudomonas fluorescens* plays a pivotal role as a biocontrol agent. Foliar sprays of *Pseudomonas fluorescens* are applied to rice crops during critical growth stages to suppress the pathogen. This bacterium produces a range of antimicrobial metabolites, including siderophores and phenazines, which inhibit the growth of *Xanthomonas* on rice leaves. Additionally, *Pseudomonas fluorescens* induces systemic resistance in rice plants, priming them to activate their innate defense mechanisms upon pathogen attack.

To enhance the effectiveness of *Pseudomonas fluorescens*, IDM programs integrate its use with resistant rice cultivars and water management practices. Resistant cultivars provide a genetic barrier against the pathogen, reducing its ability to infect plants. Water management practices, such as maintaining proper drainage and avoiding prolonged leaf wetness, create unfavorable conditions for the proliferation of *Xanthomonas*. The combined use of these practices has led to a significant reduction in bacterial leaf blight incidence and severity across multiple regions in Asia. Field studies in countries like India and Vietnam have documented yield increases of up to 30% in rice fields managed with this IDM approach compared to conventional systems. Furthermore, the reduction in chemical pesticide use has lowered production costs and minimized environmental impacts, making this strategy highly sustainable for smallholder farmers.

Vegetable Crops in Europe

Gray mold, caused by the fungus *Botrytis cinerea*, is a major disease affecting greenhouse vegetable crops, particularly tomatoes. This pathogen thrives in the humid and densely planted environments of greenhouses, leading to extensive crop losses if not effectively managed. IDM strategies for gray mold management in Europe have successfully incorporated biological control agents alongside precision farming techniques, creating an eco-friendly and highly effective disease management system.

One of the cornerstone biocontrol agents in these systems is *Trichoderma harzianum*, a soilborne fungus with exceptional antagonistic properties. *Trichoderma harzianum* is applied as a soil amendment to suppress *Botrytis cinerea* and other fungal pathogens. It acts through multiple mechanisms, including mycoparasitism, competition for nutrients, and the production of antifungal compounds. By colonizing the rhizosphere, *Trichoderma harzianum* establishes a protective barrier around plant roots, preventing the pathogen from accessing nutrients and infecting the host plant.

The integration of *Trichoderma harzianum* with precision irrigation systems further enhances its effectiveness. Precision irrigation ensures that water is delivered directly to the root zone, reducing humidity levels in the greenhouse environment. This creates conditions that are less favorable for *Botrytis cinerea*, limiting its sporulation and spread. Additionally, biological ventilation systems are employed to improve air circulation, further reducing humidity and the likelihood of fungal infections. This IDM approach has been widely adopted across European greenhouse systems, significantly reducing gray mold-related losses. By minimizing the need for chemical fungicides, this strategy has also lowered production costs and supported the transition to residue-free vegetable production. Growers have reported improved fruit quality and extended shelf life, enhancing the market value of their produce.

Citrus Orchards in the United States

Citrus greening disease, or Huanglongbing (HLB), is one of the most serious threats to citrus production worldwide. Caused by the bacterium *Candidatus Liberibacter asiaticus* and transmitted by the Asian citrus psyllid (*Diaphorina citri*), this disease causes fruit deformities, yield reductions, and tree decline. In the United States, particularly in Florida and California, IDM programs have been developed to combat citrus greening using an integrated approach that includes biological control.

Endophytic biocontrol agents, such as *Burkholderia spp.*, have emerged as key components of IDM strategies for citrus greening. These bacteria colonize the internal tissues of citrus trees, providing systemic protection against *Candidatus Liberibacter asiaticus*. *Burkholderia spp.* produce antimicrobial compounds that inhibit the growth of the pathogen within the plant's vascular system, reducing disease progression and symptom severity.

Cultural practices, such as optimized tree spacing and nutrient management, are integrated with the application of *Burkholderia spp.* to enhance the overall effectiveness of the IDM program. Proper tree spacing improves air circulation and light penetration, creating an environment that is less conducive to psyllid activity and disease spread. Nutrient management strategies, including the application of micronutrients like zinc and manganese, improve tree vigor and resilience, helping trees withstand the effects of citrus greening. The adoption of these IDM practices has yielded significant benefits for citrus growers in the United States. Field trials have shown that trees treated with *Burkholderia spp.* and managed with cultural practices exhibit better growth, reduced disease symptoms, and higher fruit quality compared to untreated trees. Additionally, the reduction in pesticide use has supported the sustainability of citrus production, preserving beneficial insect populations and reducing environmental impacts.

6.2 Integration of Biocontrol Agents in Integrated Pest Management (IPM)

The integration of biocontrol agents into Integrated Pest Management (IPM) represents a pivotal step in transforming agricultural practices toward sustainability. IPM is a holistic approach to pest control that combines biological, cultural, mechanical, and chemical methods to manage pest populations at economically acceptable levels while minimizing environmental harm. Within this framework, biocontrol agents play a critical role as eco-friendly alternatives or supplements to chemical pesticides. Their integration into IPM systems not only reduces reliance on synthetic inputs but also enhances the resilience of agroecosystems through the promotion of biodiversity and natural pest regulation mechanisms.

6.2.1 Role of Biocontrol Agents in IPM

Biocontrol agents play a pivotal role in Integrated Pest Management (IPM), serving as natural regulators of pest populations through various ecological and biological mechanisms. These agents include predatory insects, parasitic wasps, microbial antagonists, and entomopathogenic fungi, each with specific roles in suppressing pests effectively. Their inclusion in IPM reduces reliance on chemical pesticides, minimizes environmental impacts, and enhances the ecological balance of agroecosystems. By targeting specific pests and preserving beneficial organisms, biocontrol agents contribute to a sustainable and resilient approach to pest management.

A critical strength of biocontrol agents in IPM is their ability to target pests with precision while sparing non-target organisms, including pollinators and natural predators. Predatory insects like ladybird beetles (*Coccinella septempunctata*) and lacewing larvae (*Chrysoperla carnea*) are widely recognized for their efficiency in feeding on soft-bodied pests such as aphids, whiteflies, and mealybugs. These predators are particularly effective in diverse cropping systems, where they can exploit a variety of pest species as food sources while coexisting with other beneficial insects. Parasitic wasps, such as those from the *Trichogramma* genus, provide another layer of specificity in pest suppression. Female *Trichogramma* wasps lay their eggs inside the eggs of lepidopteran pests like armyworms and corn borers. The developing parasitoid larvae consume the host egg contents, effectively neutralizing the pest before it emerges. This natural control mechanism is highly targeted and scalable, making it a staple in many IPM programs. It is particularly effective in high-value crops such as cotton and maize, where pest populations can cause significant economic damage.

Microbial biocontrol agents further enhance the IPM framework by providing biological alternatives to chemical pesticides. Bacterial agents such as *Bacillus thuringiensis* (Bt) are among the most widely used microbial biocontrol tools. Bt produces crystalline toxins that are ingested by specific insect pests, such as caterpillars, disrupting their digestive systems and causing mortality. Bt-based biopesticides have become indispensable in managing pests in crops like maize, cotton, and vegetables, offering effective control with minimal environmental impact.

Fungal biocontrol agents, including *Beauveria bassiana* and *Metarhizium anisopliae*, target a broad range of insect pests. These entomopathogenic fungi infect pests through spore attachment and germination on the insect's cuticle, followed by colonization of internal tissues. This process leads to the pest's death within days, making these fungi effective tools for managing pests such as whiteflies, thrips, and locusts. The advantage of these fungal agents lies in their ability to persist in the environment, providing long-term pest suppression in both field and greenhouse systems. Viruses, such as nucleopolyhedroviruses (NPVs) and granuloviruses (GVs), also contribute to pest management in IPM. These viral agents are highly specific to their target pests, such as the cotton bollworm (*Helicoverpa armigera*) and diamondback moth (*Plutella xylostella*). Upon ingestion, the viruses replicate within the pest, disrupting physiological functions and causing death. Their specificity ensures that non-target organisms remain unaffected, preserving ecological balance. In addition to their direct role in pest suppression, biocontrol agents promote the long-term stability of pest management systems by fostering natural pest-predator relationships. These relationships maintain pest populations at sub-economic levels, reducing the need for repeated interventions. For example, the use of predatory mites in orchards can keep spider mite populations in check, preventing outbreaks that would otherwise require chemical treatments. Similarly, the introduction of biocontrol agents into a cropping system enhances biodiversity, creating a more resilient agroecosystem.

Biocontrol agents also mitigate the risk of pesticide resistance, a growing concern in conventional pest management. Continuous reliance on chemical pesticides exerts selective pressure on pest populations, leading to the evolution of resistant strains. By incorporating biocontrol agents into IPM, farmers disrupt this selection pressure, prolonging the effectiveness of chemical controls when they are used judiciously. For instance, rotating Bt-based biopesticides with other biocontrol agents helps prevent resistance development in caterpillar pests, ensuring sustainable pest management over time. Biocontrol agents form the backbone of IPM by providing eco-friendly, effective, and sustainable pest management solutions. Their ability to work synergistically with other IPM components, such as cultural practices, mechanical controls, and selective chemical treatments, makes them indispensable in modern agriculture.

6.2.4. Case Studies of Biocontrol Integration in IPM

The successful integration of biocontrol agents in IPM has been demonstrated in various crops and regions, providing evidence of their effectiveness and adaptability.

Cotton Systems in India: The pink bollworm (*Pectinophora gossypiella*) is a major pest in cotton production, causing significant yield losses. IPM programs in India have integrated *Trichogramma chilonis* (a parasitic wasp) with Bt cotton varieties and pheromone traps. This combination has reduced bollworm infestations by over 70%, minimizing pesticide use and increasing farmer profits.

Greenhouse Vegetables in Europe: Whiteflies are a persistent pest in greenhouse tomato and cucumber production. IPM strategies have incorporated *Encarsia formosa* (a parasitoid wasp) and *Beauveria bassiana* alongside sticky traps and biological ventilation. These practices have successfully controlled whiteflies, reduced losses, and supported the transition to organic greenhouse production.

Maize Systems in Africa: Fall armyworm (*Spodoptera frugiperda*) has emerged as a devastating pest in maize fields across Africa. IPM programs have deployed *Telenomus remus* (a parasitic wasp) along with cultural practices such as intercropping maize with legumes. This integrated approach has improved pest control and stabilized maize yields in affected regions.

6.2.2 Integration with Other IPM Components

The integration of biocontrol agents into Integrated Pest Management (IPM) enhances pest management systems by combining biological, cultural, mechanical, and chemical strategies. This synergistic approach leverages the strengths of each component while addressing their limitations, ensuring a comprehensive and sustainable pest control framework. Biocontrol agents, when integrated effectively, amplify the benefits of other IPM components, reducing reliance on chemical inputs, promoting ecological balance, and improving pest suppression outcomes.

Integration with Cultural Practices

Cultural practices are foundational in IPM systems, altering the agricultural environment to suppress pest populations and enhance the efficacy of natural enemies. The integration of biocontrol agents with cultural practices fosters a dynamic synergy that optimizes pest management. Crop rotation, a well-established cultural practice, disrupts the life cycles of pests by alternating host and non-host crops across growing seasons. When combined with biocontrol agents, crop rotation becomes even more effective. For instance, introducing *Beauveria bassiana* or *Metarhizium anisopliae* into rotated fields not only targets residual soilborne pests but also establishes a beneficial microbial population that suppresses pathogens and pests in subsequent crops. This integration is particularly effective in controlling root-knot nematodes and other soilborne pests.

Intercropping systems, which involve growing two or more crops together, enhance habitat complexity and resource availability for natural enemies. For example, intercropping maize with cowpeas provides shelter and alternative food sources for predatory beetles, parasitic wasps, and other beneficial organisms. These natural enemies are crucial for controlling stem borers, aphids, and leafhoppers. The presence of biocontrol agents like *Trichogramma spp.* in these systems further amplifies pest suppression, as the wasps parasitize pest eggs, breaking their reproductive cycle. Cover crops also play a pivotal role in supporting biocontrol agents. By providing habitat and food for natural enemies, cover crops create a favorable environment for predators and parasitoids. For instance, planting legumes as cover crops enriches the soil with nitrogen, supporting microbial biocontrol agents like *Pseudomonas fluorescens*. These bacteria, in turn, suppress soilborne pests while enhancing plant health.

Integration with Mechanical and Physical Controls

Mechanical and physical controls in IPM systems include practices such as handpicking pests, installing traps, and using physical barriers to limit pest movement. The integration of biocontrol agents into these practices enhances their effectiveness by targeting residual pest populations and preventing reinfestations.

Yellow sticky traps, commonly used to monitor and control whiteflies, thrips, and aphids, can be paired with biocontrol agents for improved pest management. These traps capture flying pests, reducing their populations while allowing biocontrol agents such as *Beauveria bassiana* to establish and target surviving pests. The spores of *Beauveria bassiana* infect and kill pests that evade mechanical traps, ensuring comprehensive pest suppression. In greenhouse systems, physical barriers such as insect-proof nets and controlled ventilation prevent pest entry while creating favorable conditions for biocontrol agents. The controlled environment enhances the efficacy of entomopathogenic fungi like *Metarhizium anisopliae*, which thrive in the consistent humidity and temperature levels of greenhouses.

This combination is particularly effective for managing pests such as whiteflies and spider mites, which are common in protected cultivation systems. Handpicking, a labor-intensive but effective method for removing visible pests, can be complemented by releasing biocontrol agents into the field. For instance, after handpicking caterpillars or large pests, the introduction of *Trichogramma spp.* or predatory beetles can target remaining pests at different life stages, such as eggs or smaller larvae, ensuring complete pest management.

Integration with Chemical Controls

Chemical controls, while effective, often pose risks to non-target organisms and the environment. Integrating biocontrol agents with selective chemical inputs mitigates these risks while preserving the efficacy of both components. Careful selection and timing of chemical applications are critical to maintaining compatibility with biocontrol agents. Selective insecticides, such as neem-based products or insect growth regulators (IGRs), are particularly compatible with biocontrol agents. Neem extracts disrupt pest feeding and reproduction without harming natural enemies, making them an ideal choice for integration with biocontrol agents. For instance, combining neem-based insecticides with *Bacillus thuringiensis* (Bt) formulations enhances the control of caterpillar pests in vegetables and field crops. The Bt toxins target the digestive systems of caterpillars, while neem products reduce their reproductive capacity, creating a multifaceted approach to pest suppression.

The judicious use of reduced-risk insecticides in rotation with biocontrol agents also minimizes the risk of pest resurgence and resistance development. For example, *Trichogramma spp.*, a parasitic wasp used extensively in IPM systems, can be integrated with reduced-risk insecticides to control lepidopteran pests in rice and cotton. The insecticides target larval stages, while *Trichogramma spp.* parasitize pest eggs, breaking the reproductive cycle and preventing future infestations. Proper timing and sequencing of chemical and biological inputs are essential for successful integration. Applying insecticides with short residual activity before releasing biocontrol agents ensures that the chemical does not harm beneficial organisms. Alternatively, biocontrol agents can be introduced first to establish a population that suppresses pests naturally, followed by targeted chemical treatments to address specific pest outbreaks.

Maximizing Synergy in IPM Systems

The integration of biocontrol agents with cultural, mechanical, and chemical practices creates a synergistic pest management system that maximizes the strengths of each component. This synergy enhances pest suppression, reduces the need for repeated interventions, and promotes ecological balance. The use of biocontrol agents as a cornerstone of IPM ensures that pest management systems remain sustainable, effective, and adaptable to diverse agricultural conditions.

6.2.3 Benefits of Biocontrol Integration in IPM

The integration of biocontrol agents into Integrated Pest Management (IPM) offers a wide array of benefits that enhance the sustainability, efficiency, and economic viability of agricultural systems. These advantages not only address the limitations of traditional pest management approaches but also support the broader goals of environmental conservation and food security. By providing eco-friendly alternatives to synthetic pesticides and promoting long-term pest regulation, biocontrol agents contribute to the resilience and productivity of modern farming practices.

Reduced Dependence on Chemical Pesticides

One of the most significant advantages of incorporating biocontrol agents into IPM is the reduced reliance on synthetic chemical pesticides. Overreliance on chemical inputs has long been associated with adverse environmental and health consequences, including pesticide residues in food, contamination of water sources, and the decline of non-target organisms. Biocontrol agents, as natural regulators of pest populations, mitigate these risks by offering effective pest suppression without introducing harmful residues. Biocontrol agents such as *Bacillus thuringiensis* (Bt), *Beauveria bassiana*, and parasitic wasps like *Trichogramma spp.* target specific pests while leaving beneficial organisms unharmed. This specificity ensures that pest populations are controlled without disrupting the delicate balance of agroecosystems. Furthermore, the reduced need for chemical pesticides aligns with the increasing global demand for residue-free produce, particularly in markets where stringent regulations and consumer preferences prioritize food safety and environmental sustainability.

Environmental Conservation

The use of biocontrol agents in IPM supports environmental conservation by maintaining ecological balance and preserving biodiversity. Chemical pesticides often have unintended impacts on non-target organisms, including beneficial insects, pollinators, and soil microbiota. These non-target effects disrupt ecosystem services, such as pollination, natural pest regulation, and nutrient cycling, which are essential for sustainable agriculture.

Biocontrol agents, by contrast, enhance these ecosystem services. Predatory insects like ladybird beetles (*Coccinella septempunctata*) and lacewing larvae (*Chrysoperla carnea*) feed on pest populations while coexisting with other beneficial organisms. Similarly, microbial agents such as *Metarhizium anisopliae* and *Pseudomonas fluorescens* contribute to soil health by suppressing pathogens and promoting beneficial microbial communities. The integration of biocontrol agents into IPM also reduces the risk of non-target impacts, such as the decline of pollinator populations. Pollinators like bees and butterflies are critical for the reproduction of many crops, and their decline poses a significant threat to global food security. By minimizing the use of broad-spectrum chemical pesticides, biocontrol agents help protect these vital organisms, ensuring the continued provision of pollination services.

Economic Savings

Although the initial cost of biocontrol agents may be higher than chemical pesticides, their integration into IPM results in significant long-term economic savings. Biocontrol agents reduce the need for frequent pesticide applications, lowering input costs for farmers. Additionally, enhanced crop health and yield, driven by the synergistic effects of biocontrol and other IPM practices, improve economic returns.

The use of biocontrol agents also helps mitigate the economic losses associated with pesticide resistance. Resistant pest populations often require higher doses or more frequent applications of chemical pesticides, increasing production costs. By incorporating biocontrol agents into pest management strategies, farmers reduce the selection pressure for resistance, prolonging the effectiveness of chemical inputs and ensuring more stable pest control outcomes. Moreover, the economic benefits of biocontrol integration extend beyond cost savings. Crops produced with biocontrol agents are often eligible for premium markets and certification programs, such as organic and sustainable agriculture labels. These certifications not only enhance market access but also allow farmers to command higher prices for their produce, further boosting profitability.

Sustainability

Biocontrol agents are inherently aligned with the principles of sustainability, as they leverage natural pest regulation mechanisms to achieve long-term control. Unlike chemical pesticides, which often require repeated applications to manage pest outbreaks, biocontrol agents establish self-sustaining interactions with pest populations. Predators, parasitoids, and microbial agents contribute to the natural regulation of pest dynamics, maintaining populations at levels that do not cause economic damage.

The sustainability of biocontrol integration is particularly evident in its ability to prevent resistance development. Continuous use of chemical pesticides exerts selective pressure on pest populations, leading to the evolution of resistant strains. This resistance undermines the effectiveness of chemical inputs, necessitating the development of new pesticides and increasing costs for farmers. Biocontrol agents, by disrupting pest life cycles through natural mechanisms, reduce this selective pressure, ensuring the continued efficacy of IPM strategies. In addition to pest suppression, biocontrol agents promote the sustainability of agricultural systems by improving soil health and ecosystem resilience. For example, fungal biocontrol agents like *Trichoderma spp.* enhance soil structure and nutrient availability, supporting healthy plant growth. These benefits contribute to the overall productivity and sustainability of farming systems, reducing the need for external inputs such as fertilizers and soil amendments.

Consumer and Market Benefits

The integration of biocontrol agents into IPM aligns with evolving consumer preferences for environmentally friendly and residue-free agricultural practices. As awareness of the environmental and health impacts of chemical pesticides grows, consumers are increasingly demanding produce that is cultivated using sustainable methods. Farmers who adopt IPM with biocontrol agents can meet these demands, positioning themselves as providers of safe and eco-friendly food.

Market trends also reflect the value of biocontrol integration. Certification programs for organic and sustainably produced crops often require the use of non-chemical pest management methods, creating opportunities for farmers to access premium markets. These certifications not only enhance marketability but also build consumer trust, fostering long-term demand for sustainable products. Additionally, the adoption of biocontrol agents in IPM supports compliance with regulatory standards for pesticide residues. Many countries, particularly in Europe and Asia, have established strict limits on pesticide residues in food products. By reducing or eliminating the use of chemical pesticides, biocontrol agents help farmers meet these standards, ensuring access to export markets and avoiding penalties associated with non-compliance.

Impact on Global Agriculture

The benefits of biocontrol integration in IPM extend beyond individual farms, contributing to the sustainability and resilience of global agricultural systems. By reducing chemical pesticide use, preserving biodiversity, and enhancing the long-term productivity of farming systems, biocontrol agents address key challenges in food production. Their integration into IPM represents a critical step toward achieving sustainable agriculture that balances productivity, environmental health, and economic viability.

6.3 Biological Control in Organic Farming Systems

Biological control serves as a cornerstone in organic farming systems, aligning perfectly with the principles of sustainability, environmental stewardship, and the avoidance of synthetic chemicals. Organic farming systems rely on holistic management practices that prioritize ecological balance and the health of soils, plants, animals, and people. In this context, biological control agents, including beneficial microorganisms, insects, and other natural enemies of pests, play a critical role in maintaining crop health and productivity. These agents provide effective pest and disease management solutions while adhering to the stringent standards of organic certification, which prohibit the use of most synthetic pesticides and fertilizers.

6.3.1 Role of Biological Control in Organic Farming

Biological control serves as a cornerstone of pest and disease management in organic farming systems, aligning seamlessly with the principles of ecological balance, environmental sustainability, and chemical-free agriculture. The absence of synthetic pesticides in organic farming elevates the significance of biocontrol agents, which rely on natural mechanisms to manage pests and pathogens effectively. By providing eco-friendly alternatives to chemical inputs, biological control agents support the goals of organic agriculture, ensuring sustainable crop production, biodiversity preservation, and long-term soil and ecosystem health.

The Importance of Biocontrol Agents in Organic Farming Systems

In organic farming, pest and disease management must comply with stringent standards that prohibit most synthetic chemicals. As a result, biocontrol agents play a pivotal role in addressing pest and disease pressures. These agents include predatory insects, parasitoids, microbial antagonists, and entomopathogenic fungi, all of which contribute to natural pest suppression without compromising the organic integrity of the farming system.

Predatory insects, such as ladybird beetles (*Coccinella septempunctata*) and predatory mites (*Phytoseiulus persimilis*), are widely used in organic farming to target soft-bodied pests like aphids, whiteflies, and spider mites. These predators are highly efficient in reducing pest populations and preventing infestations from reaching economic thresholds. For instance, *Phytoseiulus persimilis* is particularly effective in controlling spider mite outbreaks in crops such as strawberries and cucumbers, where chemical options are not viable under organic certification. Parasitoids, such as *Trichogramma spp.*, play an equally important role in organic pest management. These tiny wasps parasitize the eggs of lepidopteran pests, such as caterpillars and moths, effectively neutralizing them before they hatch. The use of *Trichogramma spp.* in organic systems is particularly common in crops like maize and sugarcane, where lepidopteran pests cause significant damage. By preventing pest populations from establishing, these parasitoids reduce the need for intervention during later stages of the crop cycle.

Microbial biocontrol agents, including bacteria, fungi, and viruses, are integral to pest suppression in organic systems. *Bacillus thuringiensis* (Bt), for example, is one of the most widely used microbial agents for controlling caterpillar pests. Bt produces specific toxins that disrupt the digestive systems of targeted insects, causing mortality while leaving beneficial organisms unharmed. Its specificity makes Bt an ideal tool for organic farming, where preserving biodiversity and non-target organisms is a priority. Similarly, entomopathogenic fungi like *Beauveria bassiana* infect and kill pests through spore germination and colonization, providing long-term control of pests such as thrips, whiteflies, and locusts.

Role in Disease Suppression

In addition to pest control, biological control agents are instrumental in managing plant diseases in organic farming. These agents suppress soilborne and foliar pathogens through mechanisms such as competition, antibiosis, and mycoparasitism. By targeting pathogens naturally, biocontrol agents not only reduce disease incidence but also promote soil health and microbial diversity.

Fungal antagonists, such as *Trichoderma spp.*, are among the most effective biocontrol agents for disease suppression. These fungi colonize plant roots and outcompete soilborne pathogens like *Fusarium spp.*, *Pythium spp.*, and *Rhizoctonia solani*. Through the production of lytic enzymes and secondary metabolites, *Trichoderma spp.* degrade the cell walls of pathogenic fungi, preventing their proliferation. Additionally, these fungi stimulate plant growth by enhancing nutrient uptake and inducing systemic resistance, which improves the plant's ability to defend against future infections. Bacterial biocontrol agents, such as *Pseudomonas fluorescens* and *Bacillus subtilis*, also play a critical role in disease management. These bacteria suppress pathogens through the production of antibiotics, such as phenazines and pyoluteorin, which inhibit the growth of fungi and bacteria. For example, *Pseudomonas fluorescens* has been shown to control damping-off disease in vegetables caused by *Pythium spp.* by competing for resources and producing antimicrobial compounds. Furthermore, these bacteria improve soil fertility by promoting nutrient cycling and nitrogen fixation, enhancing overall plant health and resilience.

Entomopathogenic nematodes, such as *Steinernema spp.* and *Heterorhabditis spp.*, provide a unique form of biological control in organic systems. These nematodes infect and kill insect pests that dwell in the soil, such as rootworms and weevils, reducing damage to crops without the need for synthetic insecticides. Their ability to persist in the soil and adapt to different environmental conditions makes them a reliable tool for pest management in organic farming.

Enhancing Soil Health Through Biological Control

Biological control agents not only suppress pests and diseases but also contribute significantly to soil health, a core principle of organic farming. By fostering a diverse and active soil microbiome, biocontrol agents improve soil structure, nutrient availability, and water retention, creating a more supportive environment for plant growth.

Microbial biocontrol agents like *Trichoderma spp.* and *Bacillus subtilis* enhance soil health by promoting beneficial microbial communities. These microbes decompose organic matter, release nutrients, and produce growth-promoting substances that improve plant vigor. For instance, *Trichoderma spp.* secrete enzymes that break down organic materials into simpler compounds, making nutrients more accessible to plants. Additionally, the presence of these beneficial microbes suppresses harmful pathogens, reducing the risk of disease outbreaks. Mycorrhizal fungi, although not strictly biocontrol agents, play a complementary role in organic systems by improving nutrient uptake and protecting plants from root diseases. These fungi form symbiotic associations with plant roots, extending their reach through fungal hyphae and accessing nutrients like phosphorus and micronutrients. The improved nutrient availability enhances plant growth, making crops more resistant to pests and diseases.

Promoting Biodiversity and Ecosystem Services

One of the key benefits of biological control in organic farming is its role in promoting biodiversity and supporting ecosystem services. By relying on natural enemies and microbial agents, organic systems create habitats that sustain diverse populations of beneficial organisms. These organisms, in turn, provide essential services such as pollination, pest regulation, and nutrient cycling. Predatory insects, parasitoids, and microbial antagonists coexist with other beneficial species in organic systems, creating a balanced ecosystem that reduces the risk of pest outbreaks. For example, planting cover crops or maintaining hedgerows around organic fields provides habitats for predators like ladybird beetles and lacewings, which control pest populations naturally. Similarly, soil amendments that promote microbial activity enhance the abundance of biocontrol agents, creating a self-sustaining system that requires minimal external inputs.

Compatibility with Organic Certification Standards

The use of biological control agents is fully compatible with organic certification standards, which prohibit most synthetic inputs and prioritize natural and sustainable practices. These agents meet the requirements of organic farming by being derived from natural sources and posing minimal risks to the environment, human health, and non-target organisms. In organic farming systems, biocontrol agents are often incorporated into integrated management plans that include cultural practices, crop rotation, and mechanical controls. For instance, combining *Bacillus thuringiensis* with crop rotation and intercropping can significantly reduce pest populations while maintaining compliance with organic standards. Similarly, releasing *Trichogramma spp.* in conjunction with cover cropping enhances pest suppression while preserving the ecological integrity of the farm.

6.3.2 Advantages of Biological Control in Organic Farming

The integration of biological control into organic farming systems offers a multitude of advantages that are deeply aligned with the core principles of organic agriculture. These benefits extend across ecological, environmental, economic, and soil health dimensions, reinforcing the role of biological control as a cornerstone of sustainable farming practices. By leveraging natural mechanisms to suppress pests and diseases, biocontrol agents contribute to the productivity, resilience, and marketability of organic farming systems while preserving biodiversity and environmental health.

Ecological Balance

One of the primary advantages of biological control in organic farming is its role in maintaining ecological balance. By relying on predators, parasitoids, and microbial antagonists to suppress pests and pathogens, biological control fosters the coexistence of beneficial organisms within the agroecosystem. This natural balance is crucial for the stability and sustainability of organic farming systems.

Predatory insects such as ladybird beetles (*Coccinella septempunctata*) and lacewing larvae (*Chrysoperla carnea*) feed on soft-bodied pests like aphids, whiteflies, and mealybugs, effectively regulating their populations without the need for chemical interventions. Similarly, parasitoids such as *Trichogramma spp.* target the eggs of lepidopteran pests, preventing their emergence and reducing the likelihood of outbreaks. These biocontrol agents not only control pests but also coexist with other beneficial organisms, such as pollinators and decomposers, creating a resilient agroecosystem capable of self-regulation. Microbial antagonists, including *Trichoderma spp.* and *Pseudomonas fluorescens*, contribute to ecological balance by suppressing soilborne pathogens and promoting beneficial microbial communities. These agents enhance the diversity and functionality of the soil microbiome, creating an environment that supports healthy plant growth while naturally mitigating disease pressures. This balance reduces the risk of pest and disease outbreaks, stabilizing the productivity of organic farming systems.

Environmental Sustainability

The use of biological control agents in organic farming minimizes the environmental impact of pest and disease management, making it a sustainable alternative to synthetic pesticides. Synthetic chemicals, while effective, often leave harmful residues in the soil, water, and surrounding ecosystems, posing risks to non-target organisms and long-term agricultural viability. In contrast, biocontrol agents are biodegradable and naturally decompose in the environment, leaving no toxic residues behind. For example, entomopathogenic fungi such as *Beauveria bassiana* and *Metarhizium anisopliae* infect and kill insect pests through natural processes without affecting beneficial insects, pollinators, or soil organisms. Similarly, microbial agents like *Bacillus thuringiensis* (Bt) produce specific toxins that target pest species while sparing non-target organisms, ensuring minimal disruption to the agroecosystem.

Biological control also reduces the risk of environmental contamination. By eliminating or significantly reducing the use of synthetic pesticides, biocontrol agents prevent pesticide runoff into water bodies, protecting aquatic ecosystems and ensuring the safety of drinking water sources. This aligns with the broader goals of organic agriculture to preserve natural resources and maintain ecosystem services.

Compliance with Organic Standards

Biological control agents are inherently compatible with organic certification standards, which prohibit most synthetic inputs and emphasize the use of natural and sustainable practices. These standards, established by organizations such as the USDA National Organic Program (NOP) and the European Union Organic Certification, require farmers to adopt eco-friendly pest and disease management strategies that align with the principles of organic farming. Biological control agents meet these requirements by being derived from natural sources and demonstrating minimal risks to human health, non-target organisms, and the environment. Their integration into organic farming systems ensures compliance with certification standards, enabling farmers to maintain their organic status and access premium markets for certified organic produce.

For example, the use of *Trichogramma spp.* to control caterpillar pests in organic vegetable production or the application of *Trichoderma spp.* to suppress soilborne pathogens in organic fruit orchards adheres to organic standards while effectively managing pest and disease pressures. This compatibility eliminates the need for synthetic alternatives, ensuring that farmers can achieve sustainable pest control while maintaining their certification.

Consumer and Market Benefits

The integration of biological control into organic farming systems addresses the growing consumer demand for safe, residue-free, and environmentally friendly food products. As awareness of the health and environmental impacts of synthetic pesticides increases, consumers are seeking alternatives that align with their values. Organic produce, grown using biocontrol agents and other sustainable practices, meets these expectations, offering a competitive edge in domestic and international markets. Organic produce often commands premium prices, providing farmers with economic incentives to adopt biological control as a central component of their production systems. For instance, organic fruits and vegetables treated with *Pseudomonas fluorescens* or *Beauveria bassiana* to manage pests and diseases are highly valued in markets where consumers prioritize food safety and sustainability. These premium prices offset the initial costs associated with biocontrol agents, making them a cost-effective choice for organic farmers. In addition to price premiums, the use of biological control enhances the marketability of organic produce by aligning with certification requirements and consumer preferences. Farmers who adopt biocontrol agents gain access to certification programs such as USDA Organic, EU Organic, and Fair Trade, which provide additional opportunities for market differentiation and branding.

Enhanced Soil Health

Microbial biocontrol agents contribute significantly to improved soil health, a foundational principle of organic agriculture. By promoting microbial diversity, nutrient cycling, and organic matter decomposition, these agents create a fertile and resilient soil environment that supports long-term productivity. For example, *Trichoderma spp.* not only suppress soilborne pathogens like *Fusarium spp.* and *Pythium spp.* but also enhance root growth and nutrient uptake in crops. These fungi produce enzymes and secondary metabolites that decompose organic matter, releasing essential nutrients into the soil and making them available to plants. This dual role in disease suppression and soil health improvement makes *Trichoderma spp.* a valuable tool in organic farming systems.

Bacterial biocontrol agents such as *Bacillus subtilis* and *Pseudomonas fluorescens* further enhance soil health by fixing nitrogen, solubilizing phosphorus, and producing growth-promoting substances. These activities improve soil fertility and plant vigor, reducing the need for external inputs such as synthetic fertilizers. The presence of these beneficial microbes also suppresses harmful pathogens, creating a soil environment that is naturally resistant to diseases. The integration of biocontrol agents into organic farming systems not only addresses immediate pest and disease pressures but also builds long-term soil resilience. Healthy soils enriched with biocontrol agents are better equipped to withstand environmental stresses such as drought, salinity, and temperature fluctuations, ensuring the sustainability of organic farming systems.

Economic and Social Impacts

The use of biological control in organic farming also has positive economic and social implications. By reducing dependency on chemical inputs, farmers lower production costs associated with purchasing, applying, and managing synthetic pesticides. The premium prices associated with organic produce further enhance profitability, providing a sustainable income for farmers. Socially, biological control contributes to safer working conditions for farm laborers by eliminating exposure to toxic pesticides. It also supports community health by reducing pesticide residues in the environment, improving the quality of air, water, and soil in agricultural regions. These benefits align with the broader goals of organic agriculture to promote the well-being of farmers, consumers, and communities.

6.3.3 Applications of Biological Control in Organic Systems

Biological control is integral to organic farming practices, offering versatile and eco-friendly solutions to manage pests and diseases effectively. These agents are deployed in various ways across organic systems, ensuring compliance with organic standards while addressing specific challenges associated with pests, pathogens, and soil health. Their role extends beyond pest suppression to enhancing soil fertility, promoting plant vigor, and sustaining biodiversity. Key applications of biological control in organic systems include pest management in horticultural crops, soilborne disease control, the management of lepidopteran pests, and the control of nematodes.

Pest Management in Horticultural Crops

Horticultural crops, including vegetables, fruits, and ornamentals, are highly susceptible to pest infestations due to their intensive cultivation and high economic value. In organic farming systems, biological control agents provide an effective and sustainable solution to these challenges, ensuring pest suppression without compromising environmental health or organic certification standards.

In controlled environments such as greenhouses, biological control agents like *Encarsia formosa* and *Phytoseiulus persimilis* are extensively used to manage whiteflies and spider mites, respectively. *Encarsia formosa*, a parasitoid wasp, specifically targets whitefly nymphs by laying its eggs inside the host, leading to the development of parasitoid larvae that consume the nymphs from within. This method is particularly effective in managing whiteflies in crops like tomatoes, cucumbers, and peppers, where chemical interventions are restricted in organic systems.

Similarly, *Phytoseiulus persimilis*, a predatory mite, preys on spider mites, one of the most destructive pests in greenhouse horticulture. *P. persimilis* thrives in the warm, humid conditions of greenhouses and rapidly establishes populations that suppress spider mites effectively. The introduction of *P. persimilis* in crops such as strawberries, roses, and eggplants has significantly reduced the need for chemical inputs, aligning with the principles of organic farming.

These biocontrol agents are often integrated with cultural practices, such as maintaining optimal humidity and temperature levels, to enhance their effectiveness. For example, improving air circulation in greenhouses not only reduces pest pressure but also creates a favorable environment for biocontrol agents to thrive. This synergy between biological control and cultural practices ensures sustainable pest management in organic horticulture.

Soilborne Disease Control

Soilborne pathogens, such as *Rhizoctonia spp.*, *Fusarium spp.*, and *Pythium spp.*, pose significant challenges in organic farming systems, where chemical fungicides are not an option. Biological control agents, particularly fungal antagonists like *Trichoderma spp.* and bacterial species such as *Pseudomonas fluorescens*, play a pivotal role in suppressing these pathogens and improving soil health.

Trichoderma spp. are widely recognized for their ability to combat soilborne pathogens through mechanisms such as mycoparasitism, competition, and the production of lytic enzymes. These fungi colonize plant roots, forming a protective barrier that prevents pathogens from establishing infections. Additionally, *Trichoderma spp.* produce secondary metabolites that inhibit the growth of pathogens and enhance plant resilience. Their dual role in disease suppression and growth promotion makes them invaluable in organic systems, particularly in crops like tomatoes, beans, and melons.

Pseudomonas fluorescens, a bacterial biocontrol agent, suppresses soilborne pathogens by producing antibiotics such as phenazines and pyoluteorin, which inhibit fungal growth. This bacterium also competes with pathogens for nutrients and colonization sites, reducing their ability to infect crops. Furthermore, *P. fluorescens* enhances soil fertility by promoting nutrient cycling and nitrogen fixation, contributing to the overall health and productivity of organic farming systems.

These biocontrol agents are applied as soil amendments, seed treatments, or root dips, depending on the crop and pathogen. For example, applying *Trichoderma spp.* to the soil during transplanting or as a seed coating ensures early root colonization and effective pathogen suppression. The integration of these agents with organic amendments, such as compost or green manure, further enhances their efficacy by providing a supportive environment for microbial activity.

Management of Lepidopteran Pests

Lepidopteran pests, including caterpillars and borers, are among the most destructive pests in organic farming systems, causing significant damage to crops such as maize, rice, and vegetables. Biological control agents like *Bacillus thuringiensis* (Bt) and *Trichogramma spp.* are widely used to manage these pests, offering targeted and environmentally friendly solutions.

Bacillus thuringiensis is a bacterial biocontrol agent that produces crystalline proteins toxic to specific insect pests. When ingested by caterpillars, these toxins disrupt the pests' digestive systems, causing mortality within a few days. Bt-based biopesticides are applied as foliar sprays, providing effective control of pests like diamondback moths (*Plutella xylostella*), cabbage loopers (*Trichoplusia ni*), and corn borers (*Ostrinia nubilalis*). The specificity of Bt ensures that non-target organisms, including beneficial insects and pollinators, remain unaffected, making it a preferred choice for organic systems.

Trichogramma spp., tiny parasitoid wasps, play a complementary role in managing lepidopteran pests. Female wasps lay their eggs inside the eggs of caterpillars, preventing the emergence of larvae. The use of *Trichogramma spp.* in crops like maize, sugarcane, and cotton has significantly reduced pest populations and minimized crop losses. These wasps are often released at regular intervals during the cropping season to ensure continuous pest suppression.

The integration of Bt and *Trichogramma spp.* with cultural practices, such as crop rotation and intercropping, enhances their effectiveness. For instance, intercropping maize with legumes reduces the abundance of lepidopteran pests by disrupting their life cycles, while providing habitats for natural enemies like *Trichogramma spp.*.

Control of Nematodes

Plant-parasitic nematodes, such as root-knot nematodes (*Meloidogyne spp.*), pose a persistent challenge in organic farming systems, causing root damage, stunted growth, and yield losses in a wide range of crops. Biological control agents like *Paecilomyces lilacinus* and *Pochonia chlamydosporia* offer effective solutions to manage nematode populations while supporting soil health.

Paecilomyces lilacinus, a fungal biocontrol agent, parasitizes nematode eggs and reduces their hatching rates, effectively disrupting their life cycle. This fungus establishes itself in the rhizosphere, where it targets nematode populations while promoting root growth. *Paecilomyces lilacinus* is particularly effective in controlling root-knot nematodes in crops like tomatoes, cucumbers, and eggplants. *Pochonia chlamydosporia* is another fungal agent that infects and destroys nematode eggs and juveniles. This fungus not only suppresses nematode populations but also enhances soil structure and nutrient availability, supporting healthy plant growth. It is commonly applied to soils in crops like bananas, potatoes, and lettuce, where nematode infestations are a significant concern.

These biocontrol agents are often integrated with organic soil management practices, such as the application of compost, crop rotation, and cover cropping, to enhance their efficacy. For example, planting marigolds as a cover crop has been shown to reduce nematode populations by releasing bioactive compounds that suppress nematode activity. When combined with fungal biocontrol agents, these practices provide a comprehensive strategy for nematode management in organic systems.

6.3.4 Challenges and Considerations in Biological Control for Organic Farming

While biological control is a cornerstone of pest and disease management in organic farming systems, its successful implementation requires strategic planning, appropriate management, and careful consideration of the unique challenges associated with organic agriculture. The efficacy of biocontrol agents depends on several factors, including environmental conditions, pest and disease pressures, and the compatibility of these agents with other organic practices. Additionally, economic and logistical barriers can pose significant challenges, particularly for small-scale farmers. These challenges and addressing them through research, training, and policy support are essential to maximize the potential of biological control in organic farming systems.

Environmental Conditions and Agent Efficacy

The effectiveness of biological control agents is heavily influenced by environmental factors such as temperature, humidity, and light conditions. Many biocontrol agents, particularly microbial antagonists and entomopathogenic fungi, have specific environmental requirements that must be met for optimal performance. For instance, the efficacy of *Beauveria bassiana* and *Metarhizium anisopliae* is closely tied to humidity levels, as these fungi require moisture to germinate and infect their targets. In dry or arid conditions, the performance of these agents may decline, necessitating adjustments in irrigation practices or microclimate management. Similarly, temperature extremes can affect the survival and activity of biocontrol agents. Predatory mites such as *Phytoseiulus persimilis* and parasitoids like *Trichogramma spp.* thrive within specific temperature ranges. In greenhouse systems, maintaining optimal temperature and humidity levels through biological ventilation or shading is critical to ensure the success of these agents. Farmers must also account for seasonal variations in environmental conditions and plan biocontrol applications accordingly.

Pest and Disease Pressures

The diversity and intensity of pest and disease pressures in organic systems can influence the choice and effectiveness of biocontrol agents. In some cases, pest populations may exceed the suppression capacity of natural enemies, requiring multiple applications or the integration of additional management practices. For instance, in regions with high pest densities, the release of *Trichogramma spp.* alone may not provide adequate control of lepidopteran pests, necessitating supplemental measures such as crop rotation or intercropping to disrupt pest life cycles. The presence of multiple pest species within a cropping system can also complicate the implementation of biological control. For example, in mixed cropping systems, farmers may need to release a combination of biocontrol agents, such as predatory mites for spider mites and parasitoid wasps for whiteflies. Ensuring that these agents coexist without interfering with each other requires careful planning and monitoring.

Compatibility with Organic Practices

The integration of biocontrol agents into organic systems must consider their compatibility with other practices, such as the use of organic amendments, crop rotation, and mechanical controls. While most biocontrol agents are highly compatible with organic farming methods, certain practices may inadvertently reduce their efficacy. For example, the application of compost or manure with high microbial activity can compete with or inhibit the establishment of introduced biocontrol agents like *Pseudomonas fluorescens* or *Trichoderma spp.*. Farmers must ensure that the timing and method of organic input applications align with the requirements of biocontrol agents. Similarly, mechanical practices such as tillage can disrupt the soil habitat of microbial and nematode biocontrol agents. Minimizing soil disturbance or adopting conservation tillage practices can help preserve the populations of these beneficial organisms. In greenhouse systems, integrating biocontrol agents with physical barriers such as insect-proof nets requires careful consideration of airflow and light conditions to maintain their effectiveness.

Economic and Logistical Challenges

The cost and availability of biocontrol agents can present significant barriers, particularly for small-scale organic farmers. Many commercially available biocontrol agents, such as *Encarsia formosa* or *Beauveria bassiana*, are relatively expensive compared to conventional chemical pesticides. This cost disparity can discourage adoption, especially in resource-limited settings. Additionally, the availability of high-quality biocontrol products is often limited in certain regions, making it difficult for farmers to access reliable supplies.

Logistical challenges, such as the need for specialized storage and transportation conditions, further complicate the use of biocontrol agents. For example, microbial formulations may require refrigeration to maintain viability, which can be a challenge in areas with limited infrastructure. Training and extension services are critical to addressing these challenges by providing farmers with the knowledge and resources needed to use biocontrol agents effectively. Government support in the form of subsidies, incentives, and research funding can play a crucial role in overcoming economic and logistical barriers. By reducing the cost of biocontrol agents and investing in local production facilities, policymakers can facilitate widespread adoption of biological control in organic systems.

Case Studies in the Application of Biological Control

1. Greenhouse Tomato Production in Europe

In European organic greenhouse systems, *Encarsia formosa* and *Eretmocerus eremicus*, both parasitoid wasps, are widely used to control whiteflies in tomato crops. These agents target different stages of the whitefly life cycle, providing comprehensive pest suppression. Farmers combine the release of these parasitoids with cultural practices such as crop rotation, biological ventilation, and the use of reflective mulches to create an integrated management system. The result has been a significant reduction in pest populations, improved crop quality, and reduced reliance on chemical inputs.

2. Organic Citrus Orchards in the United States

In organic citrus orchards, *Pseudomonas fluorescens* and *Trichoderma spp.* are applied as soil treatments to control soilborne pathogens like *Phytophthora spp.* These agents suppress root rot and enhance tree health by promoting root growth and nutrient uptake. Farmers integrate these biocontrol agents with cultural practices such as mulching, pruning, and cover cropping to maintain soil fertility and reduce disease pressure. The adoption of these practices has improved fruit quality and yield while adhering to organic certification requirements.

3. Organic Rice Systems in Asia

In organic rice production across Asia, *Trichogramma spp.* are released to manage stem borers, one of the most destructive pests in paddy fields. These parasitoid wasps are introduced at critical stages of the crop cycle to target pest eggs, preventing larval damage. Farmers combine the use of *Trichogramma spp.* with resistant rice varieties, water management practices, and the application of organic fertilizers to create a holistic management system. This approach has significantly reduced pest damage, improved yields, and enhanced the sustainability of rice farming in the region.

Chapter - 7

Biopesticides and Market Applications

Biopesticides, a subset of biological control products, are gaining significant traction in the global agricultural market as a sustainable alternative to synthetic chemical pesticides. Derived from natural organisms or their byproducts, biopesticides encompass a diverse range of products, including microbial pesticides, plant-based substances, and biochemical agents. Their growing popularity stems from their efficacy in pest and disease management, environmental safety, and alignment with consumer demand for residue-free and sustainably produced food. Microbial pesticides, which form the largest segment of biopesticides, are derived from bacteria, fungi, viruses, and other microorganisms. For example, *Bacillus thuringiensis* (Bt) is widely used to control lepidopteran pests in crops such as maize, cotton, and vegetables, while fungal agents like *Beauveria bassiana* and *Metarhizium anisopliae* are effective against a variety of insect pests. Biochemical biopesticides, including pheromones and plant growth regulators, disrupt pest behavior or physiology, providing targeted control without harming non-target organisms. Plant-based biopesticides, derived from botanical extracts like neem (*Azadirachta indica*) and pyrethrum, offer broad-spectrum activity against pests and pathogens while decomposing rapidly in the environment.

The market for biopesticides is expanding rapidly, driven by a confluence of factors, including stringent regulatory frameworks on chemical pesticides, the rising prevalence of pesticide resistance, and growing awareness of the environmental and health impacts of conventional farming practices. Regulatory agencies in regions such as North America and Europe have imposed strict limits on the use of synthetic pesticides, creating opportunities for biopesticides to fill the gap in pest management solutions. For instance, the European Union's Farm to Fork strategy emphasizes a 50% reduction in chemical pesticide use by 2030, a goal that aligns perfectly with the adoption of biopesticides. Additionally, biopesticides have been recognized for their role in Integrated Pest Management (IPM) programs, where they complement cultural, biological, and physical control methods to provide holistic and sustainable pest management. Their compatibility with IPM frameworks has further bolstered their market appeal, as farmers increasingly seek integrated solutions that reduce chemical inputs while maintaining productivity.

The environmental benefits of biopesticides are a major factor driving their market applications. Unlike synthetic pesticides, which often persist in the environment and contaminate soil, water, and air, biopesticides are biodegradable and leave minimal residues. This makes them particularly valuable in organic farming systems, where adherence to strict environmental standards is essential. Biopesticides also pose lower risks to non-target organisms, including pollinators, natural enemies, and soil microbes, ensuring that ecosystems remain balanced and functional. For example, pheromone-based biopesticides, which disrupt pest mating behaviors, are highly specific to target species and do not affect beneficial organisms, making them an ideal choice for sensitive ecosystems. These environmental advantages, coupled with their safety profile, have positioned biopesticides as a cornerstone of sustainable agriculture.

From an economic perspective, the adoption of biopesticides offers significant benefits to farmers and stakeholders across the agricultural value chain. By reducing reliance on synthetic pesticides, biopesticides help lower the costs associated with purchasing, applying, and managing chemical inputs. Furthermore, crops treated with biopesticides often meet stringent food safety standards, opening up lucrative markets for residue-free and sustainably produced food. For example, exporters targeting markets with strict residue limits, such as the European Union, benefit from the use of biopesticides to ensure compliance with Maximum Residue Limits (MRLs). Biopesticides also provide a solution to the growing issue of pesticide resistance, which has rendered many chemical pesticides ineffective against certain pests and pathogens. By introducing diverse modes of action, biopesticides help break resistance cycles, ensuring long-term pest control efficacy and reducing the economic losses associated with resistant pest populations.

The global biopesticide market is characterized by innovation and diversification, with new products and technologies emerging to address the evolving needs of farmers. Advances in biotechnology and microbial formulation have enabled the development of highly effective biopesticides with improved shelf life, stability, and application efficiency. For instance, encapsulation technologies protect microbial agents from environmental stressors such as ultraviolet radiation and desiccation, enhancing their performance in the field. Additionally, the integration of biopesticides with digital agriculture tools, such as remote sensing and precision spraying systems, allows for targeted application, minimizing waste and maximizing efficacy. Companies are also investing in research and development to expand the range of pests and diseases that biopesticides can address, ensuring their applicability across diverse cropping systems and climatic conditions.

Despite their numerous advantages, the widespread adoption of biopesticides faces challenges that must be addressed to unlock their full market potential. One of the primary barriers is the relatively high cost of production and the complexity of scaling up manufacturing processes. Unlike synthetic pesticides, which are often produced using standardized chemical synthesis techniques, biopesticides require intricate biological processes and quality control measures to ensure their efficacy and safety. Additionally, the lack of awareness and technical knowledge among farmers regarding the proper use of biopesticides limits their adoption, particularly in developing regions. Education and training programs, coupled with robust extension services, are essential to bridge this gap and promote the effective use of biopesticides. Regulatory hurdles, including lengthy approval processes and varying standards across countries, also pose challenges to market growth. Streamlining regulatory frameworks and fostering international collaboration can facilitate the registration and distribution of biopesticides, making them more accessible to farmers worldwide.

7.1 Biopesticides in Plant Pathogen Control

Biopesticides are an integral part of modern sustainable agriculture, offering environmentally friendly alternatives to synthetic chemical pesticides for the management of plant pathogens. Derived from natural sources, such as microorganisms, plant extracts, and biochemical compounds, biopesticides serve as critical tools in Integrated Pest Management (IPM) and organic farming systems. Their use not only helps control plant pathogens effectively but also aligns with the global demand for residue-free food, sustainable agricultural practices, and environmental conservation.

7.1.1 Definition and Classification of Biopesticides

Biopesticides are a pivotal innovation in modern agriculture, providing sustainable and eco-friendly alternatives to synthetic pesticides. Derived from natural materials such as microorganisms, plant extracts, and naturally occurring compounds, biopesticides are specifically formulated to manage pests, including plant pathogens, insects, weeds, and nematodes. Their application in plant pathogen control is particularly transformative, targeting fungi, bacteria, viruses, and other harmful organisms that threaten global food security. By leveraging natural mechanisms such as antibiosis, competition, and parasitism, biopesticides effectively reduce the reliance on synthetic inputs while preserving ecological integrity.

Definition of Biopesticides

Biopesticides are biologically derived agents utilized for pest and pathogen management. Unlike synthetic pesticides, which rely on chemical formulations that often disrupt ecosystems, biopesticides employ naturally occurring substances or living organisms to suppress pest populations. In the context of plant pathogen control, biopesticides are specifically developed to target disease-causing agents while minimizing harm to non-target organisms, such as beneficial insects, pollinators, and soil microbiota. Biopesticides are considered a cornerstone of Integrated Pest Management (IPM) and organic farming systems. They align with global regulatory and consumer demands for residue-free agricultural products, contributing to the shift toward environmentally sustainable food production. These agents can be applied as soil amendments, seed treatments, foliar sprays, or post-harvest treatments, making them versatile tools in agricultural practices worldwide.

Key characteristics of biopesticides include:

1. **Target Specificity**: Most biopesticides are highly specific, targeting particular pests or pathogens without harming beneficial organisms.
2. **Biodegradability**: Biopesticides decompose naturally, leaving no toxic residues in the environment or food chain.
3. **Eco-Friendliness**: Derived from natural sources, they reduce environmental contamination and conserve biodiversity.

Classification of Biopesticides

Biopesticides are broadly categorized into three types based on their origin and mode of action: microbial biopesticides, biochemical biopesticides, and plant-incorporated protectants (PIPs). Each category addresses distinct agricultural challenges, offering unique solutions for pest and disease management.

1. Microbial Biopesticides

Microbial biopesticides consist of living microorganisms—such as bacteria, fungi, viruses, and actinomycetes—that target pests or pathogens through direct or indirect interactions. These agents are among the most widely researched and utilized biopesticides due to their versatility and effectiveness across diverse cropping systems.

Examples of Microbial Biopesticides

1. **Bacterial Biopesticides**:
    - *Bacillus subtilis*: This bacterium produces secondary metabolites, including antibiotics and enzymes, which inhibit fungal pathogens like *Fusarium spp.*, *Pythium spp.*, and *Rhizoctonia spp.*.
    - *Pseudomonas fluorescens*: Known for its antagonistic properties, it suppresses soilborne pathogens through the production of phenazines and pyoluteorin, which compete with pathogens for essential nutrients like iron.
2. **Fungal Biopesticides**:
    - *Trichoderma spp.*: These fungi are highly effective against soilborne pathogens. They parasitize and degrade pathogens by producing lytic enzymes such as chitinase and glucanase.
    - *Beauveria bassiana*: While primarily an entomopathogenic fungus targeting insects, it also helps reduce pest-vector-borne diseases in plants.
3. **Viral Biopesticides**:
    - Baculoviruses: These highly specific viral biopesticides infect insect pests, indirectly reducing plant diseases spread by vectors.
4. **Actinomycetes**:
    - *Streptomyces spp.*: This group of filamentous bacteria suppresses fungal pathogens through the production of antibiotics such as streptomycin and avermectin.

Mechanisms of Action in Microbial Biopesticides

» **Antibiosis**: Microbial biopesticides produce antimicrobial compounds that inhibit the growth and reproduction of plant pathogens. For example, *Bacillus subtilis* secretes fengycin, a lipopeptide that disrupts fungal cell membranes.

» **Parasitism**: Fungal biopesticides like *Trichoderma spp.* directly parasitize pathogens, degrading their cell walls through the action of hydrolytic enzymes.

» **Competition**: Microbial agents outcompete pathogens for nutrients, colonization sites, and other resources in the rhizosphere.

» **Induced Systemic Resistance (ISR)**: Certain microbial biopesticides, such as *Pseudomonas fluorescens*, prime plants to activate their innate immune responses, enhancing resistance against future infections.

Applications of Microbial Biopesticides

- **Seed Treatments**: Bacteria like *Pseudomonas fluorescens* are coated on seeds to protect against early-season fungal and bacterial pathogens.
- **Soil Applications**: Fungal biopesticides, including *Trichoderma spp.*, are incorporated into the soil to suppress root pathogens such as *Rhizoctonia solani*.
- **Foliar Sprays**: *Bacillus subtilis* formulations are sprayed on crops to control foliar diseases like powdery mildew.

2. Biochemical Biopesticides

Biochemical biopesticides are derived from naturally occurring substances, including plant extracts, pheromones, and growth regulators. These agents disrupt pest and pathogen activity without directly killing them, making them safer for the environment and non-target organisms.

Examples of Biochemical Biopesticides

1. **Neem-Based Products**:
   - Azadirachtin, extracted from neem seeds, acts as both an antimicrobial agent and an insect growth regulator. It disrupts the life cycle of insect pests and inhibits spore germination in fungal pathogens.
2. **Chitosan:**
   - Derived from chitin, this compound triggers systemic resistance in plants and directly inhibits fungal growth. It is effective against diseases like gray mold (*Botrytis cinerea*) and anthracnose (*Colletotrichum spp.*).
3. **Salicylic Acid Derivatives:**
   - These compounds activate plant defense pathways, enhancing resistance to bacterial and fungal pathogens. They are commonly used in fruits and vegetables to prevent diseases such as blight and mildew.

Mechanisms of Action in Biochemical Biopesticides

- **Direct Antimicrobial Activity**: Compounds like neem oil disrupt the integrity of fungal cell membranes, inhibiting pathogen growth.
- **Induction of Plant Defenses**: Chitosan and salicylic acid derivatives stimulate the production of pathogenesis-related proteins and phytoalexins in plants.
- **Disruption of Pathogen Development**: Biochemicals interfere with key developmental stages of pathogens, such as spore germination or hyphal growth.

Applications of Biochemical Biopesticides

- **Foliar Treatments**: Neem oil and chitosan are sprayed on leaves to protect crops from foliar pathogens.

» **Post-Harvest Protection:** Chitosan-based coatings on fruits and vegetables prevent spoilage by suppressing post-harvest diseases.

3. Plant-Incorporated Protectants (PIPs)

Plant-incorporated protectants (PIPs) involve the genetic engineering of crops to express substances that confer resistance against pests or pathogens. These biopesticides are embedded within the plant, providing systemic and long-lasting protection.

Examples of PIPs

1. **Bt Crops:**
   - Genetically engineered crops like Bt cotton and Bt corn express toxins derived from *Bacillus thuringiensis*, which target specific insect pests while reducing secondary pathogen infections.

2. **Virus-Resistant Crops:**
   - Papaya ringspot virus-resistant papaya is an example of a crop engineered to resist viral pathogens, significantly reducing disease incidence.

Mechanisms of Action in PIPs

» **Pathogen Inhibition:** The expressed proteins disrupt key metabolic processes in pathogens, rendering them incapable of infecting the plant.

» **Reduction of Vector Populations:** By controlling insect vectors, such as aphids and whiteflies, PIPs indirectly prevent the spread of viral and bacterial diseases.

Applications of PIPs

» **Row Crops:** Bt corn and cotton are extensively used to protect against caterpillar pests.

» **Horticultural Crops:** Virus-resistant papayas and tomatoes safeguard against viral diseases.

Comparative Insights Across Biopesticide Types

Each biopesticide category provides distinct advantages:

» **Microbial Biopesticides:** Suitable for broad-spectrum control of pathogens and versatile across various environmental conditions.

» **Biochemical Biopesticides:** Environmentally safe and highly specific, making them ideal for organic and export-oriented agriculture.

» **Plant-Incorporated Protectants (PIPs):** Offer systemic and durable resistance, reducing the frequency of applications.

7.1.2 Mechanisms of Action in Plant Pathogen Control

Biopesticides have revolutionized plant disease management by leveraging natural and highly specific mechanisms to suppress plant pathogens. These mechanisms are pivotal in their ability to target harmful fungi, bacteria, viruses, and other pathogens without adversely affecting beneficial organisms or the surrounding environment. The ability of biopesticides to operate through various modes of action not only enhances their effectiveness but also reduces the likelihood of resistance development, a significant concern in the use of synthetic pesticides.

Antibiosis: A Chemical Defense Strategy

One of the most critical mechanisms employed by microbial biopesticides is antibiosis, wherein microorganisms produce secondary metabolites that inhibit or kill plant pathogens. These metabolites often include antibiotics, enzymes, or toxins, which disrupt the cellular processes of the target pathogens. For instance, *Pseudomonas fluorescens*, a widely used biopesticide, produces phenazines, pyoluteorin, and other antimicrobial compounds. These metabolites effectively suppress the growth of fungal pathogens such as *Fusarium spp.* and *Pythium spp.*, which cause root rot and damping-off diseases in crops like tomatoes, beans, and peppers. The antibiotics produced by *Pseudomonas fluorescens* interfere with critical metabolic pathways in the pathogens, preventing their reproduction and spread.

Similarly, *Streptomyces spp.*, another group of microbial biopesticides, produce streptomycin and other antibiotics that target bacterial pathogens like *Xanthomonas spp.* and *Pseudomonas syringae*. These antibiotics disrupt protein synthesis in the target bacteria, halting their growth and reducing disease incidence. The production of secondary metabolites through antibiosis is not limited to bacteria. Certain fungi, such as *Trichoderma spp.*, also produce antibiotics like gliotoxin and viridin, which inhibit the growth of competing fungal pathogens. This chemical defense mechanism makes *Trichoderma spp.* highly effective against a range of soilborne diseases.

Parasitism and Hyperparasitism: Direct Pathogen Suppression

Parasitism is another cornerstone mechanism in the efficacy of fungal biopesticides, wherein the biocontrol agent directly attacks and parasitizes the plant pathogen. Hyperparasitism, a specific form of parasitism, involves one fungus parasitizing another, resulting in the suppression or destruction of the target pathogen.

Trichoderma spp., renowned fungal biopesticides, exemplify this mechanism. These fungi attach to the hyphae of soilborne pathogens such as *Rhizoctonia solani* and *Fusarium oxysporum*, secreting hydrolytic enzymes like chitinases, glucanases, and proteases. These enzymes degrade the cell walls of the pathogenic fungi, ultimately killing them. This mechanism is particularly effective in controlling diseases like root rot, wilt, and damping-off. Hyperparasitism also occurs in fungal biopesticides such as *Ampelomyces quisqualis*, which parasitizes the powdery mildew-causing fungi. *Ampelomyces quisqualis* invades the hyphae and spores of powdery mildew pathogens, disrupting their reproductive structures and preventing further spread of the disease. This mode of action is highly specific and minimizes the risk of collateral damage to beneficial fungi in the agroecosystem.

Induced Systemic Resistance (ISR):

Induced Systemic Resistance (ISR) represents a proactive mechanism employed by biopesticides to enhance the innate immune systems of plants. Certain microbial biopesticides prime plants to activate defense-related pathways, enabling them to respond more effectively to pathogen attacks. For example, *Bacillus subtilis* is a well-documented biopesticide that induces ISR in a variety of crops. It stimulates the production of phytoalexins, pathogenesis-related (PR) proteins, and other defense compounds in plants. These molecules strengthen the plant's physical and biochemical barriers, making it more resistant to infections. This mechanism is particularly useful in managing foliar diseases like powdery mildew and bacterial blight.

Similarly, *Trichoderma harzianum* induces systemic resistance in plants by activating signaling pathways associated with jasmonic acid and ethylene. These pathways enhance the plant's ability to defend against both fungal and bacterial pathogens. The dual role of *Trichoderma harzianum* in suppressing pathogens directly and boosting plant immunity makes it a versatile tool in disease management. ISR has also been observed with biopesticides like *Pseudomonas fluorescens* and *Azospirillum spp.*, which colonize the plant rhizosphere and produce elicitor molecules that trigger immune responses in plants. The induction of ISR provides a long-lasting defense mechanism, reducing the need for repeated applications of biopesticides.

Competition for Resources and Space: Starving Pathogens

Microbial biopesticides often suppress plant pathogens through competitive exclusion, a mechanism that involves outcompeting the pathogens for nutrients, space, or colonization sites. This form of biological control is particularly effective in the rhizosphere and phyllosphere, where microorganisms compete intensively for resources.

Bacillus amyloliquefaciens, a prominent bacterial biopesticide, exemplifies this mechanism. It rapidly colonizes the root zone, forming biofilms that prevent soilborne pathogens like *Fusarium spp.* from accessing critical nutrients. By monopolizing resources such as iron, nitrogen, and carbon, *Bacillus amyloliquefaciens* creates an environment that is inhospitable to pathogens.

Fungal biopesticides such as *Trichoderma spp.* also employ competition as a mode of action. These fungi colonize root surfaces and establish symbiotic relationships with plants, effectively excluding pathogens from the root zone. The ability of *Trichoderma spp.* to thrive in diverse soil conditions further enhances their competitive advantage over pathogenic fungi. Competition for space is particularly relevant in foliar applications, where biopesticides like *Pseudomonas fluorescens* and *Bacillus subtilis* colonize leaf surfaces. By forming protective layers on plant tissues, these biopesticides inhibit the establishment and spread of foliar pathogens, such as *Alternaria spp.* and *Cladosporium spp.*.

Mycoviruses: Targeting Fungal Pathogens with Viruses

Mycoviruses, or fungal viruses, represent a unique and emerging class of biopesticides. These viral agents specifically infect fungal pathogens, weakening or killing them without affecting non-target organisms. Mycoviruses operate through several mechanisms, including disrupting fungal metabolism, interfering with reproduction, and inducing hypovirulence in the target pathogen.

One notable example is the use of hypoviruses to manage chestnut blight caused by *Cryphonectria parasitica*. Hypoviruses infect the pathogenic fungus, reducing its virulence and limiting the spread of the disease. This approach has been particularly successful in restoring chestnut forests in regions affected by blight. Mycoviruses also show potential in controlling other fungal diseases in horticultural and field crops. Research into the application of mycoviruses against pathogens like *Botrytis cinerea* and *Sclerotinia sclerotiorum* is ongoing, with promising results in reducing disease severity.

Integrated Application of Mechanisms

The effectiveness of biopesticides often lies in their ability to combine multiple mechanisms of action. For example, *Trichoderma spp.* simultaneously employ parasitism, competition, and ISR induction to suppress soilborne pathogens and enhance plant health. Similarly, *Bacillus subtilis* produces antimicrobial compounds, competes for resources, and primes plant defenses, providing comprehensive disease management. This multifaceted approach not only enhances the efficacy of biopesticides but also reduces the risk of resistance development in pathogens. Unlike synthetic pesticides, which typically act through a single biochemical pathway, biopesticides disrupt multiple aspects of pathogen biology, making it difficult for pathogens to adapt.

7.1.3 Advantages of Biopesticides in Plant Pathogen Control

Biopesticides have emerged as a transformative tool in modern agriculture, offering an array of advantages over traditional synthetic pesticides. Their integration into crop protection strategies aligns with the growing need for sustainable farming practices that address the dual challenges of environmental conservation and food security. By leveraging natural biological mechanisms, biopesticides provide a safer, more targeted, and ecologically responsible approach to managing plant pathogens.

Eco-Friendly and Biodegradable

One of the most compelling advantages of biopesticides is their eco-friendly nature. Unlike synthetic pesticides, which can persist in the environment for extended periods and contribute to soil and water contamination, biopesticides are biodegradable. They decompose rapidly under natural conditions, leaving no harmful residues in the soil, water, or harvested produce. This characteristic is especially critical in addressing the global concerns of pesticide pollution, which affects both human health and biodiversity.

The rapid breakdown of biopesticides minimizes their environmental footprint, making them particularly suitable for organic farming systems. Organic standards, which strictly regulate the use of chemical inputs, often prioritize biopesticides due to their compatibility with ecological farming practices. For example, microbial biopesticides such as *Bacillus subtilis* and *Trichoderma spp.* degrade naturally after suppressing pathogens, ensuring that soils remain uncontaminated and supportive of microbial diversity. In export-oriented agriculture, where residue levels on produce are tightly regulated, biopesticides offer a competitive advantage. Many importing countries enforce strict maximum residue limits (MRLs) for synthetic pesticides. By using biopesticides, farmers can meet these regulatory requirements while maintaining high crop quality, enabling access to premium international markets.

Specificity and Safety

Biopesticides are distinguished by their high specificity, targeting particular pathogens or groups of pathogens while sparing beneficial organisms. This specificity is a critical advantage in preserving the ecological balance of agricultural systems. Unlike broad-spectrum synthetic pesticides, which often harm pollinators, natural enemies, and soil microbiota, biopesticides act selectively, reducing collateral damage to the ecosystem.

For instance, *Pseudomonas fluorescens*, a bacterial biopesticide, suppresses fungal pathogens such as *Fusarium spp.* and *Pythium spp.* without affecting non-target organisms like earthworms and beneficial nematodes. Similarly, *Trichoderma spp.* parasitize pathogenic fungi while coexisting harmoniously with beneficial mycorrhizal fungi, which play a crucial role in nutrient uptake. The safety profile of biopesticides extends to human health, as they pose minimal risks compared to synthetic chemicals. Their natural origin and low toxicity levels make them safer for farmworkers, consumers, and communities living near agricultural areas. This attribute is particularly valuable in promoting sustainable farming practices that prioritize the well-being of all stakeholders in the agricultural value chain.

Resistance Management

The overuse of synthetic pesticides has led to the widespread development of resistance in plant pathogens, undermining the efficacy of conventional crop protection strategies. Biopesticides, with their unique modes of action, offer a powerful solution to this challenge. Unlike synthetic chemicals, which often rely on a single biochemical pathway, biopesticides employ multiple mechanisms to suppress pathogens. This multifaceted approach reduces the selection pressure for resistance development, ensuring long-term effectiveness.

For example, *Trichoderma spp.* suppress pathogens through parasitism, competition, and the production of secondary metabolites, making it difficult for pathogens to adapt. Similarly, *Bacillus subtilis* produces a suite of antimicrobial compounds while simultaneously inducing systemic resistance in plants. The diversity of these actions disrupts the evolutionary pathways that pathogens rely on to develop resistance. Biopesticides are also valuable as part of Integrated Pest Management (IPM) programs, where they can be rotated or combined with other control methods to further delay resistance development. By reducing reliance on synthetic pesticides, biopesticides help preserve the effectiveness of existing chemical products, extending their utility in agricultural systems.

Promotes Plant and Soil Health

Beyond their role in pathogen suppression, many biopesticides contribute to the overall health of plants and soils. Microbial biopesticides, in particular, are known for their ability to enhance soil fertility, stimulate plant growth, and promote beneficial microbial activity. These attributes make biopesticides a cornerstone of sustainable agriculture, where soil health is critical to long-term productivity.

Trichoderma spp. are a prime example of biopesticides that double as plant growth promoters. These fungi colonize plant roots, improving nutrient uptake and water absorption while suppressing soilborne pathogens like *Rhizoctonia solani* and *Fusarium oxysporum*. In addition to disease suppression, *Trichoderma spp.* produce phytohormones that stimulate root elongation and biomass accumulation, leading to healthier and more resilient crops.

Bacterial biopesticides such as *Pseudomonas fluorescens* and *Bacillus amyloliquefaciens* also play a critical role in nutrient cycling and soil structure improvement. These bacteria enhance the availability of essential nutrients, such as phosphorus and nitrogen, by solubilizing soil minerals and fixing atmospheric nitrogen. The resulting increase in soil fertility supports vigorous plant growth, reducing the need for synthetic fertilizers. The promotion of soil health by biopesticides aligns with regenerative agricultural practices, which seek to rebuild organic matter and restore soil biodiversity. By fostering a thriving soil microbiome, biopesticides contribute to the long-term sustainability and productivity of farming systems.

Compliance with Regulatory Standards

The global agricultural landscape is increasingly shaped by stringent regulatory requirements aimed at minimizing the environmental and health impacts of pesticide use. Biopesticides, with their natural origin and favorable safety profile, align seamlessly with these standards, enabling farmers to meet regulatory compliance while maintaining competitive market access.

In organic certification programs, such as those governed by the USDA National Organic Program (NOP) or the European Union Organic Regulation, biopesticides are often listed as approved inputs. Their use ensures that farmers can produce organic-certified crops without compromising on disease management. For example, chitosan-based biopesticides are widely used in organic fruit and vegetable production to control post-harvest diseases while adhering to organic standards. Biopesticides also help farmers comply with international residue regulations, such as the Codex Alimentarius and individual country-specific MRLs. By eliminating or reducing the need for synthetic pesticides, biopesticides minimize the risk of residue violations, which can lead to export rejections and financial losses. This compliance advantage is particularly significant for export-oriented crops like coffee, tea, and tropical fruits, where consumer preferences for residue-free products drive market demand. Furthermore, the adoption of biopesticides supports alignment with global sustainability initiatives, such as the United Nations Sustainable Development Goals (SDGs). By promoting environmentally responsible farming practices, biopesticides contribute to objectives such as climate action, biodiversity conservation, and sustainable food production.

Market Opportunities and Consumer Appeal

The growing consumer demand for eco-friendly and residue-free agricultural products has created significant market opportunities for farmers who adopt biopesticides. In both domestic and international markets, crops produced with biopesticides are often perceived as safer and more sustainable, commanding premium prices and fostering brand loyalty.

Biopesticides also support the production of organic and sustainably certified crops, which are increasingly sought after by health-conscious consumers. By enabling farmers to meet certification standards, biopesticides open doors to niche markets and value-added branding opportunities. For example, biopesticide-treated produce is prominently featured in organic retail chains and premium grocery stores, enhancing its marketability. The consumer appeal of biopesticides extends beyond food crops to ornamental plants, where the absence of chemical residues is a key selling point. Biopesticide-treated flowers and landscape plants cater to environmentally conscious consumers who prioritize sustainable gardening practices.

Economic Sustainability

While biopesticides often have higher initial costs compared to synthetic pesticides, their long-term benefits outweigh these expenses. The reduced frequency of applications, enhanced crop yields, and improved soil health associated with biopesticide use contribute to significant economic savings over time. Furthermore, the market premiums associated with biopesticide-treated crops offset the upfront investment, making them a financially viable choice for farmers. In addition, biopesticides help mitigate risks associated with regulatory non-compliance and pesticide resistance, reducing the potential for crop losses and ensuring stable income streams. Their integration into IPM programs enhances the resilience of farming systems, enabling farmers to adapt to changing environmental and market conditions.

7.1.4 Key Biopesticides in Plant Pathogen Control

Biopesticides have become an essential tool in combating plant pathogens, offering natural, effective, and environmentally sustainable alternatives to synthetic pesticides. Among the diverse types of biopesticides available, several have proven to be particularly effective in managing fungal, bacterial, and other plant pathogens. These biopesticides, derived from microorganisms, natural compounds, or inorganic substances, play critical roles in controlling diseases while promoting plant health and soil fertility.

Trichoderma-Based Biopesticides

Fungi in the genus *Trichoderma* are among the most widely studied and commercially developed biopesticides for controlling soilborne pathogens. Products based on *Trichoderma harzianum* and *Trichoderma viride* are extensively used in agriculture due to their broad-spectrum efficacy and additional benefits to plant health.

Mechanism of Action:

» *Trichoderma spp.* suppress plant pathogens through mycoparasitism, wherein they attach to the hyphae of fungal pathogens, secrete lytic enzymes (such as chitinases and glucanases), and degrade their cell walls.

» These fungi also produce secondary metabolites that act as antibiotics, inhibiting the growth of competing pathogens.

» Additionally, *Trichoderma spp.* enhance plant resilience by inducing systemic resistance (ISR), stimulating the production of pathogenesis-related proteins and other defense compounds.

Applications:

» **Soilborne Pathogen Control**: *Trichoderma spp.* are effective against pathogens like *Rhizoctonia solani*, *Fusarium oxysporum*, and *Pythium spp.*, which cause diseases such as root rot, damping-off, and wilt.

» **Crops**: Widely used in vegetables (e.g., tomatoes, peppers), field crops (e.g., rice, wheat), and fruits (e.g., bananas, strawberries) to protect against soilborne diseases.

» **Formulations**: Available as wettable powders, granules, or liquid formulations for seed treatment, soil application, or foliar sprays.

Benefits:

» Besides pathogen suppression, *Trichoderma spp.* enhance nutrient uptake, improve soil structure, and stimulate root development, contributing to higher crop yields.

Bacillus-Based Biopesticides

Bacterial biopesticides based on species of *Bacillus*, such as *Bacillus subtilis* and *Bacillus amyloliquefaciens*, are renowned for their broad-spectrum activity against fungal and bacterial pathogens. These products are widely used in crops ranging from vegetables to cereals due to their versatility and effectiveness.

Mechanism of Action:

- *Bacillus* species produce a wide array of antimicrobial compounds, including lipopeptides (e.g., iturins, surfactins) and polyketides, which disrupt the membranes of fungal and bacterial pathogens.
- They compete with pathogens for nutrients and colonization sites in the rhizosphere, effectively excluding them.
- Some species induce systemic resistance in plants, enhancing their ability to withstand infections.

Applications:

- **Fungal Diseases**: *Bacillus subtilis* is used to control diseases like rice blast (*Magnaporthe oryzae*), leaf spot, and anthracnose in vegetables and fruits.
- **Bacterial Diseases**: Effective against bacterial blight in rice and bacterial wilt in solanaceous crops.
- **Post-Harvest Protection**: *Bacillus spp.* formulations are used to prevent spoilage in fruits and vegetables.

Formulations:

- Often available as liquid suspensions, powders, or seed coatings, these biopesticides are applied as soil drenches, foliar sprays, or post-harvest treatments.

Benefits:

- In addition to pathogen control, *Bacillus spp.* improve soil health by promoting beneficial microbial activity and nutrient cycling.

Neem-Based Biopesticides

Neem-derived products, particularly those containing azadirachtin, have long been used in traditional agriculture and have gained prominence as modern biopesticides. Neem products are valued for their wide-ranging pesticidal properties, which include antifungal, antibacterial, and insecticidal activities.

Mechanism of Action:

- Azadirachtin disrupts the development and reproduction of pests and pathogens by interfering with hormonal and metabolic processes.
- Neem oil exhibits direct antifungal activity by damaging the membranes of fungal cells and inhibiting spore germination.
- Neem-based products also induce resistance in plants by stimulating defense pathways.

Applications:

- **Foliar Pathogens**: Neem oil is used to control fungal diseases such as powdery mildew and black spot in ornamental plants and vegetables.

- **Seedborne Pathogens**: Neem extracts are applied as seed treatments to protect against pathogens like *Alternaria spp.* and *Colletotrichum spp.*
- **Integrated Pest Management**: Neem products are often integrated with other biopesticides for managing complex disease-pest interactions.

Benefits:

- Neem products are biodegradable, non-toxic to humans, and safe for beneficial organisms like pollinators and natural enemies.

Chitosan-Based Products

Chitosan, a natural biopolymer derived from chitin, has gained recognition as an effective biopesticide for managing plant pathogens. It is primarily used in high-value crops where post-harvest losses are a major concern.

Mechanism of Action:

- Chitosan induces systemic acquired resistance (SAR) in plants, activating the production of phytoalexins, pathogenesis-related proteins, and other defense molecules.
- It inhibits fungal growth by disrupting cell membranes and interfering with nutrient absorption.
- Chitosan also forms protective coatings on fruits and vegetables, creating a physical barrier against microbial invasion.

Applications:

- **Post-Harvest Protection**: Widely used in fruits like citrus, apples, and bananas to prevent diseases caused by *Botrytis cinerea* and *Penicillium spp.*
- **Foliar Diseases**: Effective against fungal pathogens like powdery mildew in vegetables and ornamentals.
- **Seed Treatments**: Enhances germination rates and protects seedlings from soilborne pathogens.

Benefits:

- Chitosan is biodegradable, environmentally safe, and compatible with organic farming practices.

Copper-Based Biopesticides

Copper compounds, such as copper oxychloride, copper hydroxide, and Bordeaux mixture (a combination of copper sulfate and lime), are traditional yet effective biopesticides for managing a wide range of plant pathogens.

Mechanism of Action:

- Copper ions disrupt enzymatic activities and cellular processes in pathogens, inhibiting their growth and reproduction.

- These compounds act as contact fungicides and bactericides, requiring direct application to the pathogen or infection site.

Applications:

- **Fungal Diseases**: Copper-based biopesticides are effective against late blight in potatoes and tomatoes, downy mildew in grapes, and leaf spot in cucurbits.
- **Bacterial Diseases**: Used to control bacterial wilt in solanaceous crops and citrus canker in fruit trees.
- **Preventive Use**: Often applied as a prophylactic measure to protect plants from airborne or waterborne pathogens.

Formulations:

- Available as wettable powders, liquid concentrates, or granules, copper-based biopesticides are applied as foliar sprays or soil treatments.

Benefits:

- These products are cost-effective, long-lasting, and widely available, making them a staple in both conventional and organic farming systems.

Comparative Insights

Each biopesticide type offers unique advantages, making them suitable for specific applications in plant pathogen control:

- **Trichoderma-Based Biopesticides**: Best suited for soilborne diseases and as growth promoters.
- **Bacillus-Based Biopesticides**: Effective against both fungal and bacterial pathogens, with versatility across crops.
- **Neem-Based Biopesticides**: Ideal for foliar and seedborne diseases, especially in integrated pest management systems.
- **Chitosan-Based Products**: Essential for post-harvest protection and high-value crops.
- **Copper-Based Biopesticides**: Reliable for broad-spectrum pathogen control, particularly in horticultural crops.

7.1.5 Applications in Various Cropping Systems

Biopesticides are integral to modern agriculture, providing versatile and environmentally sustainable solutions for managing plant pathogens across a wide range of cropping systems. Their natural origins, safety for non-target organisms, and compatibility with diverse farming practices make them ideal tools for disease control in horticulture, field crops, fruit orchards, and greenhouse systems.

Applications in Horticultural Crops

Horticultural crops, including vegetables and ornamentals, are particularly susceptible to diseases due to their intensive cultivation and high-value nature. Biopesticides play a vital role in safeguarding these crops against soilborne and foliar pathogens while reducing the environmental impact of disease management.

Key Biopesticides and Their Applications:

1. **Trichoderma-Based Products**:
    - *Trichoderma spp.* are widely used to manage soilborne pathogens, such as *Fusarium spp.* and *Pythium spp.*, which cause root rot and damping-off diseases.
    - These biopesticides are typically applied as soil amendments or seed treatments to establish early root colonization and prevent pathogen infections.
2. **Bacillus-Based Biopesticides**:
    - *Bacillus subtilis* and *Bacillus amyloliquefaciens* are employed to control fungal diseases like anthracnose and bacterial diseases such as wilt.
    - Foliar sprays of *Bacillus subtilis* formulations are particularly effective in managing powdery mildew in tomatoes and cucumbers.
3. **Neem Oil:**
    - Neem-based products are applied as foliar sprays to suppress diseases like black spot and downy mildew in ornamentals and vegetables.

Crops and Diseases Managed:

» **Tomatoes**: Control of soilborne pathogens such as *Fusarium oxysporum* and foliar diseases like powdery mildew.

» **Cucumbers and Peppers**: Management of damping-off and bacterial wilt through soil and foliar applications.

» **Ornamentals**: Protection against fungal diseases like black spot in roses and powdery mildew in flowering plants.

Biopesticides not only suppress pathogens but also enhance crop quality and shelf life, making them indispensable in horticultural systems focused on high-value produce.

Applications in Field Crops

Field crops, such as cereals, pulses, and oilseeds, face significant disease pressures that can result in severe yield losses. Biopesticides are critical in managing these diseases, ensuring sustainable productivity and reducing the dependency on chemical pesticides.

Key Biopesticides and Their Applications:

1. **Bt Formulations:**
    - *Bacillus thuringiensis* (Bt) products are widely used to control caterpillar pests, such as stem borers in rice and maize. These pests often serve as vectors for plant pathogens.
    - Bt formulations disrupt the pest life cycle, indirectly preventing the transmission of fungal and bacterial diseases.

2. **Copper-Based Biopesticides:**
   - Copper hydroxide and copper oxychloride are applied as preventive measures to control fungal diseases like rust in wheat and late blight in potatoes.
   - These biopesticides act as contact fungicides, protecting plants from airborne and waterborne pathogens.
3. **Pseudomonas fluorescens:**
   - This bacterial biopesticide is applied as a seed treatment or soil drench to control sheath blight in rice and root rot in pulses.

Crops and Diseases Managed:

» **Rice**: Biopesticides like *Pseudomonas fluorescens* are employed to manage blast (*Magnaporthe oryzae*) and sheath blight (*Rhizoctonia solani*).

» **Maize**: Bt formulations are used to suppress caterpillar pests and reduce mycotoxin contamination caused by *Fusarium spp.*

» **Wheat**: Copper-based products are applied to control rusts (Puccinia spp.) and other fungal diseases.

The use of biopesticides in field crops enhances soil health, improves yield stability, and supports the adoption of sustainable farming practices.

Applications in Fruit Orchards

Fruit orchards are highly vulnerable to fungal and bacterial pathogens, which can severely impact fruit quality, yield, and marketability. Biopesticides provide an effective and residue-free solution for managing diseases in fruits such as apples, grapes, and citrus.

Key Biopesticides and Their Applications:

1. **Trichoderma-Based Biopesticides:**
   - Applied as soil treatments to suppress root diseases like citrus root rot caused by *Phytophthora spp.*
2. **Neem Oil:**
   - Neem-based products are used as foliar sprays to control fungal diseases such as powdery mildew and anthracnose in grapes and apples.
   - Neem oil also acts as a natural insecticide, reducing pest populations that can transmit diseases.
3. **Copper-Based Biopesticides:**
   - Copper compounds are widely used to manage bacterial diseases such as citrus canker and fungal diseases like scab in apples.
   - These biopesticides are particularly effective when applied preventively during the early stages of disease development.

Crops and Diseases Managed:

- **Apples**: Control of scab (*Venturia inaequalis*) and powdery mildew (*Podosphaera leucotricha*).
- **Grapes**: Management of downy mildew (*Plasmopara viticola*) and anthracnose (*Elsinoë ampelina*).
- **Citrus**: Protection against citrus canker (*Xanthomonas axonopodis*) and root rot.

Biopesticides help fruit growers meet the strict residue standards of export markets while ensuring high-quality and sustainable production.

Applications in Greenhouse Systems

Greenhouse systems, characterized by controlled environments, present unique opportunities and challenges for disease management. The high humidity and temperature levels in greenhouses often favor the rapid proliferation of pathogens. Biopesticides, with their specificity and safety, are particularly suited for use in these systems.

Key Biopesticides and Their Applications:

1. **Beauveria bassiana:**
   - An entomopathogenic fungus used to suppress insect pests that act as vectors for plant pathogens. By controlling whiteflies and aphids, *Beauveria bassiana* indirectly reduces the spread of viral diseases.
2. **Pseudomonas fluorescens:**
   - This biopesticide is applied as a soil drench to suppress fungal pathogens like *Fusarium spp.* and *Pythium spp.* in greenhouse vegetables.
3. **Neem Oil:**
   - Neem-based products are used as foliar sprays to manage powdery mildew and other fungal diseases in greenhouse crops.

Crops and Diseases Managed:

» **Tomatoes**: Control of foliar diseases like blight (*Alternaria solani*) and damping-off.

» **Cucumbers**: Management of powdery mildew and root rot.

» **Lettuce**: Suppression of fungal pathogens such as *Botrytis cinerea.*

Benefits of Biopesticides in Greenhouse Systems:

» Biopesticides help reduce the risk of chemical resistance development in pathogens and pests.

» They are safe for workers and consumers, ensuring compliance with stringent food safety standards.

Integrated Role Across Systems

The versatility of biopesticides allows them to be seamlessly integrated into diverse cropping systems. Their application is not limited to individual crops or diseases; instead, they form part of a holistic strategy that includes cultural practices, mechanical controls, and synthetic alternatives when necessary. By tailoring biopesticide use to the specific needs of each cropping system, farmers can achieve effective and sustainable disease management while maintaining ecological balance. The growing adoption of biopesticides across horticulture, field crops, fruit orchards, and greenhouse systems underscores their critical role in shaping the future of sustainable agriculture. Through continued research and innovation, biopesticides will likely expand their reach, addressing emerging challenges in crop protection and ensuring long-term food security.

7.1.6 Challenges in Biopesticide Adoption

Biopesticides, despite their numerous advantages in sustainable agriculture, face several challenges that hinder their widespread adoption. These challenges stem from factors such as variable efficacy under diverse conditions, logistical limitations, economic barriers, and regulatory complexities. Addressing these issues is essential to unlock the full potential of biopesticides in modern crop protection systems.

Variable Efficacy

The effectiveness of biopesticides can be inconsistent due to their dependency on environmental conditions. Unlike synthetic pesticides, which often maintain stable performance across a range of environmental settings, biopesticides rely on the activity of living organisms or natural compounds, making them more sensitive to factors such as:

» **Temperature**: Extreme temperatures can inhibit the growth and activity of microbial biopesticides. For instance, *Trichoderma spp.* thrive in moderate temperatures but may lose efficacy in excessively cold or hot climates.

» **Humidity**: Many fungal and bacterial biopesticides, such as *Beauveria bassiana* and *Pseudomonas fluorescens*, require specific humidity levels for optimal performance. Low humidity can reduce spore germination and microbial activity.

» **Soil pH**: Soil acidity or alkalinity can affect the survival and colonization ability of soil-applied biopesticides. For example, *Bacillus subtilis* performs best in neutral to slightly alkaline soils but may struggle in acidic conditions.

This variability discourages farmers from fully relying on biopesticides, especially in regions with unpredictable weather patterns. Overcoming this challenge requires extensive field trials to identify the optimal application strategies and environmental conditions for each biopesticide. Additionally, research into improving the resilience of biopesticides to environmental fluctuations through genetic enhancement or formulation innovations is crucial.

Short Shelf Life

The limited shelf life of many biopesticides, particularly microbial formulations, poses a significant logistical challenge. Unlike chemical pesticides, which can often be stored for years, biopesticides degrade relatively quickly due to the biological nature of their active ingredients. This degradation can occur during storage or transportation, rendering the product less effective or entirely unusable.

Key Factors Contributing to Shelf Life Limitations:

1. **Viability of Microbial Agents**: Living organisms, such as bacteria and fungi, require specific storage conditions, including temperature control and low moisture levels, to maintain viability.
2. **Exposure to Environmental Stress**: Heat, UV radiation, and oxygen can degrade the active ingredients in biopesticides, reducing their effectiveness over time.

Implications:

» Farmers in remote or resource-limited areas may struggle to access fresh, high-quality biopesticide products.

» Distributors face increased costs associated with refrigeration and rapid transportation to maintain product efficacy.

Potential Solutions:

» Developing advanced formulations, such as encapsulated or freeze-dried products, to extend shelf life.

» Establishing local production units to minimize transportation and storage times, especially in regions with limited infrastructure.

Cost and Accessibility

The relatively high production costs of biopesticides and their limited availability in certain regions are major barriers to adoption, particularly for small-scale farmers in developing countries. The production of biopesticides often involves sophisticated biotechnological processes, including fermentation, formulation, and quality control, which contribute to their higher cost compared to synthetic pesticides.

Challenges in Accessibility:

1. **High Initial Investment**: The production infrastructure for microbial and biochemical biopesticides requires significant investment, limiting their availability to a few specialized manufacturers.
2. **Supply Chain Constraints**: In regions with underdeveloped agricultural supply chains, the distribution of biopesticides is often inconsistent and unreliable.

Implications for Farmers:

» Small-scale farmers, who constitute the majority of agricultural producers in many developing countries, may find biopesticides economically inaccessible.

» The limited availability of biopesticides forces farmers to rely on readily available synthetic alternatives, perpetuating the environmental and health risks associated with chemical pesticides.

Strategies to Address Cost and Accessibility:

» Government subsidies and incentives to reduce the retail cost of biopesticides for farmers.

» Encouraging public-private partnerships to scale up biopesticide production and distribution.

» Promoting local manufacturing initiatives to reduce reliance on imported products.

Lack of Awareness and Training

The successful adoption of biopesticides requires a thorough understanding of their application, benefits, and limitations. However, many farmers lack access to the necessary training and information, resulting in suboptimal use or complete avoidance of biopesticides.

Key Knowledge Gaps:

1. **Application Techniques**: Unlike synthetic pesticides, which often require straightforward application, biopesticides may involve specific timing, dosage, and environmental conditions for optimal performance.
2. **Integration into IPM**: Farmers may be unaware of how to integrate biopesticides with other pest management practices, such as cultural and mechanical controls.
3. **Perceived Reliability**: A lack of awareness about the long-term benefits of biopesticides contributes to misconceptions regarding their efficacy and reliability.

Impact of Knowledge Gaps:

» Misapplication or overuse of biopesticides can lead to reduced efficacy and wasted resources.

» Farmers may remain skeptical of biopesticides, preferring the perceived immediacy of synthetic pesticides.

Solutions:

» Implementing targeted extension programs and farmer field schools to provide hands-on training in biopesticide use.

» Developing user-friendly guides and digital tools, such as mobile apps, to disseminate information on biopesticide application and benefits.

» Collaborating with farmer cooperatives and community organizations to promote awareness and collective learning.

Regulatory Hurdles

The regulatory framework for biopesticides often poses challenges that can delay their development, commercialization, and adoption. While regulations aim to ensure the safety and efficacy of biopesticides, the processes involved are often lengthy, complex, and resource-intensive.

Regulatory Challenges:

1. **Lengthy Approval Processes**: The registration of biopesticides typically involves extensive testing for environmental safety, human health risks, and efficacy. These tests can take years to complete, delaying market entry.

2. **High Compliance Costs**: Small and medium-sized enterprises (SMEs) involved in biopesticide development may struggle to meet the financial and technical requirements of regulatory compliance.
3. **Lack of Harmonization**: Differences in regulatory standards across countries and regions create barriers to the international trade of biopesticides.

Impact on the Biopesticide Industry:

» Prolonged approval timelines discourage innovation and investment in biopesticide development.

» Farmers face limited access to a diverse range of biopesticides, restricting their ability to implement effective disease management strategies.

Strategies to Streamline Regulation:

» Developing harmonized international standards for biopesticide registration to facilitate global trade.

» Simplifying regulatory processes for low-risk biopesticides, such as microbial and plant-based products.

» Providing financial and technical support to SMEs to help them navigate regulatory requirements.

7.2 Biopesticides Available in the Market and Their Applications

Biopesticides have become increasingly prominent in the agricultural market, offering sustainable solutions to pest and disease management. Derived from natural sources such as microorganisms, plant extracts, and inorganic substances, these products target specific pests and pathogens while minimizing environmental and health risks. A variety of biopesticides are commercially available, each tailored to address specific agricultural challenges.

7.2.1 Microbial Biopesticides

Microbial biopesticides have emerged as one of the most effective and widely used categories of biopesticides, leveraging the natural abilities of microorganisms such as bacteria, fungi, and viruses to control pests and pathogens. These products dominate the biopesticide market due to their versatility, broad-spectrum activity, and environmentally friendly profile. By utilizing mechanisms such as antibiosis, competition, parasitism, and the induction of systemic resistance in plants, microbial biopesticides play a critical role in sustainable agriculture. They are effective against a variety of plant pathogens and pests, providing targeted control while preserving non-target organisms and enhancing soil health.

Bacillus-Based Products

Among microbial biopesticides, *Bacillus* species are some of the most extensively researched and commercially developed due to their broad efficacy and adaptability. *Bacillus subtilis* and *Bacillus amyloliquefaciens* are the most prominent examples, widely used in crop protection to combat fungal and bacterial diseases.

Mechanisms of Action:

Bacillus species suppress plant pathogens through multiple mechanisms. They produce a range of secondary metabolites, including lipopeptides such as iturins, surfactins, and fengycins, which disrupt fungal cell membranes and inhibit bacterial growth. These metabolites also act as biofilm disruptors, preventing pathogen colonization on plant surfaces. Furthermore, *Bacillus* species can solubilize essential nutrients like phosphorus, enhancing plant growth and resilience.

Applications:

1. **Vegetables**: Products like Serenade, based on *Bacillus subtilis*, are used to control powdery mildew in cucumbers and tomatoes. These products are applied as foliar sprays, soil drenches, or seed treatments.
2. **Cereals**: *Bacillus amyloliquefaciens* formulations are employed to manage leaf blight in rice and sheath rot in wheat.
3. **Fruits**: In grape vineyards, Serenade has proven effective in controlling botrytis bunch rot (*Botrytis cinerea*), ensuring high-quality harvests with minimal chemical residues.

Impact:

In addition to pathogen control, *Bacillus* biopesticides promote plant growth by producing phytohormones and improving nutrient uptake. Their compatibility with organic farming practices and Integrated Pest Management (IPM) strategies makes them indispensable in modern agriculture.

Trichoderma-Based Products

Fungal biopesticides based on *Trichoderma* species, particularly *Trichoderma harzianum* and *Trichoderma viride*, are highly valued for their dual role in suppressing soilborne pathogens and enhancing plant growth. These fungi are natural inhabitants of the soil and are renowned for their ability to colonize plant roots and outcompete harmful pathogens.

Mechanisms of Action:

Trichoderma spp. employ several mechanisms to suppress plant pathogens. Mycoparasitism is their primary mode of action, wherein the fungi attach to the hyphae of soilborne pathogens like *Fusarium spp.* and *Rhizoctonia spp.*, secrete hydrolytic enzymes such as chitinases and glucanases, and degrade the pathogen's cell walls. Additionally, *Trichoderma* species compete with pathogens for nutrients and colonization sites in the rhizosphere, effectively excluding them. They also produce secondary metabolites with antimicrobial properties and trigger systemic resistance in plants.

Applications:

1. **Soilborne Diseases**: Products like RootShield and Biofungicide, based on *Trichoderma harzianum*, are applied to combat root rot, damping-off, and wilt diseases in crops such as tomatoes, cucumbers, and peppers.
2. **Horticulture**: In ornamental plants, *Trichoderma spp.* formulations help control soilborne diseases while promoting vigorous root development.

3. **High-Value Crops**: In bananas and strawberries, *Trichoderma viride* is used to manage *Fusarium wilt*, a devastating disease that significantly impacts yields.

Impact:

Beyond disease suppression, *Trichoderma spp.* enhance soil health by increasing microbial diversity and nutrient cycling. They also stimulate root growth and improve the plant's ability to absorb water and nutrients, contributing to higher yields and better stress tolerance.

Beauveria bassiana

Beauveria bassiana is a versatile entomopathogenic fungus primarily used to control insect pests, which can indirectly reduce the spread of plant pathogens. This fungus is widely used in greenhouse systems and field crops where pests like whiteflies, thrips, and aphids are prevalent.

Mechanisms of Action:

The fungal spores of *Beauveria bassiana* attach to the cuticle of insect pests, germinate, and penetrate the host's body. Once inside, the fungus proliferates, producing toxins that disrupt the pest's metabolic processes, ultimately leading to death. The infected insect cadavers become a source of fungal spores, further propagating the biopesticide's effects.

Applications:

1. **Greenhouses**: Products like Botanigard and Naturalis are applied to manage whiteflies and thrips in greenhouse tomatoes, cucumbers, and peppers.
2. **Field Crops**: In cotton and soybeans, *Beauveria bassiana* is used to target aphids and reduce viral disease transmission caused by these pests.
3. **Forestry**: The fungus is also employed in forest pest management to control beetle infestations that damage trees and transmit fungal pathogens.

Impact:

By controlling insect vectors, *Beauveria bassiana* indirectly reduces the spread of plant pathogens, making it a vital component of integrated pest and disease management strategies. Its safety for beneficial insects and pollinators further enhances its value in sustainable agriculture.

Pseudomonas-Based Products

Pseudomonas fluorescens is one of the most extensively used bacterial biopesticides for managing soilborne and foliar diseases in various crops. This bacterium is highly effective in colonizing the rhizosphere and producing metabolites that suppress plant pathogens.

Mechanisms of Action:

Pseudomonas fluorescens suppresses pathogens through the production of antibiotics (e.g., phenazines, pyoluteorin), siderophores that sequester iron, and enzymes that degrade pathogen cell walls. Additionally, it competes with pathogens for nutrients and triggers systemic resistance in plants.

Applications:

1. **Rice Cultivation**: BioSave, a *Pseudomonas fluorescens* product, is applied as a seed treatment to control sheath blight (*Rhizoctonia solani*) and bacterial leaf blight.
2. **Vegetables**: This biopesticide is used to manage damping-off diseases caused by *Pythium spp.* in cucumbers and lettuce.
3. **Turfgrass and Ornamentals**: Applied to prevent fungal diseases like brown patch and dollar spot in turfgrass.

Impact:

In addition to its antimicrobial activity, *Pseudomonas fluorescens* enhances plant growth by solubilizing phosphate and producing plant growth-promoting hormones. Its use is particularly beneficial in resource-poor regions where synthetic inputs are unaffordable or unavailable.

Viral Biopesticides

Viral biopesticides, particularly those based on baculoviruses, are used to target specific insect pests that damage crops and transmit diseases. These viruses are host-specific, making them safe for non-target organisms and beneficial insects.

Mechanisms of Action:

Baculoviruses infect insect pests by entering their gut cells, replicating, and disrupting metabolic functions. This leads to the death of the host insect, reducing pest populations and their associated impacts on crop health.

Applications:

1. **Cotton**: Helicoverpa NPV is applied to control bollworms (*Helicoverpa armigera*), reducing damage to cotton bolls.
2. **Vegetables**: Baculoviruses are used to manage caterpillar pests in tomatoes, peppers, and eggplants.
3. **Fruit Crops**: These biopesticides are employed in orchards to suppress pests like codling moths, which affect apple and pear trees.

Impact:

Viral biopesticides are integral to IPM programs, especially in organic farming systems. Their specificity and safety make them ideal for use in ecologically sensitive areas.

7.2.2 Biochemical Biopesticides

Biochemical biopesticides are derived from natural substances, including plant extracts, pheromones, and organic compounds, and are widely recognized for their safety, specificity, and environmental friendliness. Unlike synthetic pesticides, these biopesticides are biodegradable and target pests and pathogens with minimal impact on non-target organisms and ecosystems. Their ability to complement Integrated Pest Management (IPM) systems makes them essential in reducing chemical pesticide use while ensuring effective pest and disease control.

Neem-Based Products

Neem-based products are among the most well-known biochemical biopesticides, derived from the seeds of the neem tree (*Azadirachta indica*). Azadirachtin, the primary active compound in neem products, exhibits a broad spectrum of pesticidal properties, including antifungal, antibacterial, insecticidal, and nematicidal effects.

Mechanisms of Action:

Azadirachtin and other neem compounds disrupt the life cycle of pests and pathogens through various mechanisms. These include inhibiting spore germination, interfering with cellular growth and development, and disrupting hormonal regulation in pests. Neem oil also acts as a physical barrier, coating plant surfaces and preventing fungal spores from germinating or penetrating tissues.

Applications:

1. **Fruits and Vegetables**: Neem-based products like Neemix and Azamax are applied as foliar sprays to control fungal diseases such as anthracnose and powdery mildew in crops like grapes, tomatoes, and cucumbers.
2. **Ornamentals**: Neem oil is used to manage black spot and rust in roses and other ornamental plants, offering a residue-free alternative to synthetic fungicides.
3. **Pest Management**: Neem products also target insect vectors like aphids and whiteflies, which spread viral diseases, thereby indirectly reducing pathogen pressure.

Impact and Benefits:

Neem-based biopesticides are biodegradable and leave no toxic residues, making them ideal for organic farming systems and export-oriented agriculture. Their compatibility with other IPM components further enhances their utility in sustainable farming practices.

Chitosan-Based Products

Chitosan, a biopolymer derived from chitin (found in crustacean shells and fungal cell walls), is a powerful biochemical biopesticide with multiple roles in plant protection and health enhancement. Chitosan-based products like Elexa and ChitoCare are widely used in agriculture, particularly for high-value crops, due to their ability to improve crop resilience and post-harvest quality.

Mechanisms of Action:

Chitosan induces systemic acquired resistance (SAR) in plants by triggering the production of phytoalexins, pathogenesis-related (PR) proteins, and other defense compounds. This primes the plant's immune system to respond more effectively to pathogen attacks. Chitosan also disrupts fungal cell membranes, inhibiting spore germination and hyphal growth. Furthermore, it forms a protective film on plant surfaces, preventing the entry of pathogens and reducing water loss.

Applications:

1. **Berries and Citrus**: Chitosan-based biopesticides are applied as post-harvest treatments to manage diseases caused by *Botrytis cinerea* and *Penicillium spp.*, enhancing fruit shelf life and marketability.

2. **Vegetables**: In crops like tomatoes and peppers, chitosan is used to control foliar diseases and improve resistance to environmental stressors.
3. **Seed Treatments**: Chitosan coatings on seeds protect against soilborne pathogens while enhancing germination and early seedling vigor.

Impact and Benefits:

Chitosan's role in enhancing plant resistance and maintaining post-harvest quality makes it an invaluable tool for organic and sustainable farming systems. Its biodegradability and safety for humans and the environment further enhance its appeal.

Plant Growth Regulators and Essential Oils

Plant growth regulators (PGRs) and essential oils derived from aromatic plants are another significant category of biochemical biopesticides. These compounds exhibit antimicrobial properties and are particularly popular in organic farming systems for their multifunctional benefits.

Key Products:

1. **Eugenol** (clove oil): Exhibits strong antifungal activity and is used to control powdery mildew and post-harvest diseases.
2. **Citronella Oil**: Known for its antibacterial and antifungal properties, citronella oil is used to manage bacterial leaf spots and fungal diseases in vegetables and ornamentals.

Mechanisms of Action:

Essential oils and PGRs act by disrupting the cellular integrity of pathogens, inhibiting spore germination, and interfering with enzymatic processes critical to pathogen survival. They also exhibit repellent properties, deterring pests that may vector diseases.

Applications:

1. **Organic Farming Systems**: Essential oils are widely used in organic farming to control a variety of fungal and bacterial pathogens in fruits, vegetables, and herbs.
2. **Post-Harvest Treatments**: Eugenol and other essential oils are applied to fruits like apples and mangoes to prevent spoilage and decay.
3. **Integrated Pest and Disease Management**: These products are often combined with microbial and neem-based biopesticides to enhance the effectiveness of IPM programs.

Impact and Benefits:

The use of PGRs and essential oils supports residue-free agriculture, aligning with consumer preferences for safe and sustainable produce. Their natural origin and versatility make them indispensable in both conventional and organic farming systems.

Applications Across Cropping Systems

Biochemical biopesticides demonstrate versatility across various cropping systems, providing targeted solutions to pest and pathogen challenges while supporting environmental sustainability.

1. **Horticultural Crops**: Neem-based sprays and chitosan treatments are extensively used to manage foliar diseases in tomatoes, cucumbers, and peppers, enhancing yield and quality.
2. **Field Crops**: Essential oils are applied to control fungal rusts and bacterial blights in cereals and pulses.
3. **Fruit Orchards**: Post-harvest treatments with chitosan and eugenol ensure long shelf life and reduced spoilage in high-value fruits like grapes, citrus, and berries.
4. **Greenhouse Systems**: Citronella oil and neem formulations are popular in protected cultivation to manage fungal and bacterial diseases in lettuce, strawberries, and herbs.

7.2.3 Plant-Incorporated Protectants (PIPs)

Plant-incorporated protectants (PIPs) represent a revolutionary approach to pest and pathogen control, leveraging the power of genetic engineering to provide systemic and long-lasting protection against specific agricultural threats. Unlike conventional biopesticides, which are externally applied, PIPs are produced directly within the plant by expressing genes that confer resistance to pests or pathogens. This intrinsic protection ensures consistent efficacy throughout the plant's life cycle, making PIPs an essential component of modern, sustainable farming systems.

Overview of PIPs

PIPs are created by inserting genes into a plant's genome to produce specific proteins, peptides, or metabolites that target pests or pathogens. These protective substances are typically derived from natural sources, such as soil-dwelling bacteria or viruses, ensuring both environmental compatibility and high specificity. The development of PIPs has primarily focused on addressing issues such as pest-induced yield losses, pathogen resistance, and pesticide overuse. With their systemic mode of action, PIPs eliminate the need for repeated applications, thereby reducing labor, costs, and environmental impact.

Bt Crops: A Pioneer in PIPs

The development of Bt crops is one of the most significant achievements in agricultural biotechnology and remains a cornerstone of PIP technology. Bt crops are genetically engineered to express insecticidal proteins derived from *Bacillus thuringiensis* (Bt), a naturally occurring soil bacterium. These proteins are toxic to specific insect pests, providing effective control while minimizing harm to non-target organisms.

Mechanisms of Action:

Bt proteins act by disrupting the digestive systems of target insect pests. Once ingested, the protein binds to receptors in the insect's midgut, forming pores that compromise cell integrity. This leads to gut paralysis, starvation, and eventual death. Importantly, Bt toxins are highly specific to certain insect species, making them safe for beneficial insects, pollinators, and other wildlife.

Applications:

1. **Bt Cotton**:
   - Target Pest: Bollworms (*Helicoverpa armigera*), pink bollworm (*Pectinophora gossypiella*).

- Bt cotton has revolutionized pest management in cotton farming, significantly reducing losses caused by these destructive caterpillars. Additionally, by controlling these pests, Bt cotton indirectly reduces damage caused by secondary fungal infections that enter through feeding wounds.

2. **Bt Corn**:
   - Target Pest: European corn borer (*Ostrinia nubilalis*), corn rootworm (*Diabrotica spp.*).
   - Bt corn enhances yield stability by preventing stalk damage and kernel infections caused by pests. This protection reduces the risk of mycotoxin contamination from fungal pathogens like *Fusarium spp.*, which often exploit pest-induced wounds.

Impact and Benefits:

Bt crops have dramatically reduced the need for chemical insecticides, lowering production costs and environmental risks. Farmers benefit from higher yields and reduced pest management labor, while ecosystems benefit from decreased pesticide runoff and residue.

Virus-Resistant Crops: Combating Plant Viruses

Virus-resistant crops represent another transformative application of PIP technology. These crops are engineered to express genes that confer resistance to specific plant viruses, thereby preventing infections that can devastate yields and quality. This approach has been particularly successful in tropical and subtropical regions, where viral diseases pose significant challenges.

Mechanisms of Action:

Virus-resistant crops typically incorporate genes encoding viral coat proteins or other functional elements that interfere with the replication or assembly of the virus. By producing these proteins, the plant becomes effectively "immunized" against infection, even in the presence of the virus.

Applications:

1. **Papaya Ringspot Virus-Resistant Papaya:**
   - Target Pathogen: Papaya ringspot virus (PRSV).
   - This genetically modified papaya, developed in the United States and adopted in regions like Hawaii, has successfully controlled PRSV outbreaks. Before its introduction, PRSV had severely limited papaya production in affected areas.
2. **Tomato Yellow Leaf Curl Virus-Resistant Tomato**:
   - Target Pathogen: Tomato yellow leaf curl virus (TYLCV).
   - Virus-resistant tomato varieties are widely used in regions prone to TYLCV outbreaks, including the Mediterranean and Southeast Asia, ensuring yield stability and marketable quality.

Impact and Benefits:

Virus-resistant crops reduce the need for chemical sprays targeting insect vectors, such as whiteflies and aphids, which transmit viral diseases. By minimizing disease incidence, these crops also enhance yield reliability and farmer income, particularly in smallholder farming systems.

Advantages of PIPs

The adoption of PIPs offers numerous advantages that align with the goals of sustainable agriculture and food security:

1. **Systemic and Long-Lasting Protection:** PIPs provide continuous protection against pests and pathogens throughout the plant's life cycle, reducing the need for repeated pesticide applications. This systemic action ensures consistent efficacy, even in inaccessible parts of the plant, such as root systems.
2. **Reduced Environmental Impact:** By decreasing the reliance on chemical pesticides, PIPs help mitigate environmental issues such as pesticide runoff, groundwater contamination, and non-target organism toxicity. For example, Bt crops have significantly reduced the volume of synthetic insecticides used in cotton and corn production.
3. **Improved Crop Yield and Quality:** PIPs protect crops from yield losses caused by pests and pathogens while maintaining product quality. Virus-resistant papaya, for instance, produces blemish-free fruits that meet market standards, enhancing profitability.
4. **Resistance Management:** PIPs, especially those targeting pests with specific modes of action, are less likely to contribute to resistance development compared to broad-spectrum pesticides. However, strategies such as refugia planting (planting non-Bt crops alongside Bt crops) are recommended to further delay resistance.
5. **Cost-Effectiveness**: While the initial cost of PIP seeds may be higher, the reduction in pesticide use and labor inputs often results in overall cost savings for farmers. This is particularly beneficial in large-scale operations.

Challenges in PIP Adoption

Despite their advantages, PIPs face certain challenges that need to be addressed to maximize their potential:

1. **Regulatory and Public Acceptance:** The adoption of genetically modified PIPs is often hampered by regulatory hurdles and public concerns about genetic engineering. In some countries, the cultivation of GM crops is heavily restricted, limiting the availability of PIPs.
2. **Pest and Pathogen Resistance:** Continuous exposure to PIPs, such as Bt toxins, can lead to resistance development in target pests. Integrated resistance management strategies, including crop rotation and the use of non-PIP refugia, are essential to maintain their long-term effectiveness.
3. **Limited Crop Variety:** Current PIP technology is concentrated in a few major crops, such as cotton, corn, and papaya. Expanding PIP development to other economically important crops could increase their global impact.

7.2.4 Inorganic Biopesticides

Inorganic biopesticides are naturally derived mineral-based compounds that serve as effective agents against plant pathogens. These products, widely recognized for their fungicidal and bactericidal properties, are integral to both conventional and organic agricultural systems. Their affordability, broad-spectrum activity, and compatibility with environmentally friendly practices make them a preferred choice for many farmers. Inorganic biopesticides are particularly valued for their role in preventing and managing fungal and bacterial diseases in crops.

Inorganic Biopesticides

Inorganic biopesticides are composed of elemental substances or their compounds, such as copper and sulfur. These substances have been used in agriculture for centuries, long before the advent of synthetic chemical pesticides. Their enduring popularity stems from their ability to provide reliable disease control, particularly in high-value crops like fruits, vegetables, and vineyards. Inorganic biopesticides act on a wide range of pathogens, including fungi and bacteria, by disrupting critical cellular processes, making them effective tools in disease management.

Copper-Based Products

Copper-based biopesticides are among the most widely used inorganic biopesticides, valued for their versatility and effectiveness against a broad spectrum of fungal and bacterial pathogens. Products such as copper hydroxide and copper oxychloride (e.g., Kocide, Nordox) are staples in the agricultural industry.

Mechanisms of Action:

Copper ions, released from copper-based formulations, disrupt enzymatic processes in pathogens by binding to essential proteins and enzymes. This binding leads to the denaturation of proteins, disruption of cellular membranes, and interference with nutrient uptake, ultimately inhibiting pathogen growth and reproduction. Copper's broad-spectrum activity makes it effective against both fungal and bacterial diseases.

Applications:

1. **Fruit Trees**: Copper-based products are widely used in orchards to control diseases such as citrus canker (*Xanthomonas axonopodis*), apple scab (*Venturia inaequalis*), and bacterial spot in peaches.
2. **Vegetables**: These biopesticides are applied to manage bacterial wilt in tomatoes and fungal diseases like late blight (*Phytophthora infestans*) in potatoes.
3. **Vineyards**: Copper-based products are essential for managing downy mildew (*Plasmopara viticola*) and other fungal infections in grapevines.

Impact and Benefits:

Copper-based biopesticides provide a cost-effective and reliable solution for managing plant diseases, especially in organic farming systems where synthetic chemical fungicides are prohibited. Their long residual activity ensures prolonged protection, reducing the frequency of applications and lowering labor costs.

Sulfur-Based Products

Sulfur-based biopesticides, such as wettable sulfur, are another cornerstone of inorganic biopesticides. Known for their effectiveness against fungal pathogens, sulfur products are extensively used in both conventional and organic agriculture.

Mechanisms of Action:

Sulfur inhibits fungal growth by altering the metabolic processes within fungal cells. Specifically, it disrupts the synthesis of essential amino acids and enzymes, inhibiting spore germination and hyphal growth. Sulfur also acts as a physical barrier on plant surfaces, preventing pathogen entry and colonization.

Applications:

1. **Grapes**: Wettable sulfur is a primary tool for controlling powdery mildew (*Erysiphe necator*), a major fungal disease affecting grapevines.
2. **Apples and Pears**: Sulfur-based biopesticides are applied to manage scab and rust diseases in apple and pear orchards.
3. **Strawberries and Berries**: Sulfur products are used to combat powdery mildew and other fungal infections in strawberries and blueberries, ensuring high-quality fruit production.

Impact and Benefits:

Sulfur is highly effective against fungal diseases and has a long history of safe use in agriculture. It is readily available, inexpensive, and easy to apply, making it an accessible option for small-scale farmers. Furthermore, sulfur's compatibility with organic farming standards ensures its continued relevance in sustainable agriculture.

Advantages of Inorganic Biopesticides

Inorganic biopesticides offer several advantages that contribute to their widespread use in agricultural systems:

1. **Broad-Spectrum Activity**: Copper and sulfur-based products effectively target a wide range of fungal and bacterial pathogens, providing versatile solutions for disease management.
2. **Cost-Effectiveness:** These biopesticides are relatively inexpensive compared to synthetic chemical pesticides, making them accessible to farmers across diverse economic contexts.
3. **Environmental Compatibility:** Inorganic biopesticides are derived from natural materials and are generally less harmful to non-target organisms and ecosystems than synthetic pesticides. Their use aligns with the principles of organic and sustainable farming.
4. **Ease of Application:** Inorganic biopesticides are available in various formulations, including powders, granules, and liquid concentrates, making them easy to apply using conventional spraying equipment.
5. **Long Residual Activity:** Copper and sulfur provide prolonged protection against pathogens, reducing the need for frequent reapplications and lowering operational costs.

Limitations and Challenges

Despite their advantages, inorganic biopesticides face certain limitations that must be addressed to optimize their use:

1. **Phytotoxicity**: Excessive application of copper or sulfur can cause phytotoxic effects, such as leaf burn and reduced photosynthesis, particularly under high-temperature conditions.
2. **Resistance Development:** Prolonged and repeated use of copper-based biopesticides can lead to the development of resistance in certain pathogens, reducing their long-term efficacy.
3. **Environmental Accumulation:** Overuse of copper-based products can result in copper accumulation in soils, potentially affecting soil health and microbial diversity.
4. **Application Timing and Conditions:** The effectiveness of sulfur-based products depends on environmental conditions, such as humidity and temperature, requiring careful timing of applications for optimal results.

Applications Across Cropping Systems

Inorganic biopesticides are widely applied across a variety of cropping systems to manage plant diseases effectively:

1. **Horticultural Crops**: Copper-based sprays are extensively used to manage bacterial blights and fungal rots in tomatoes, peppers, and cucumbers.
2. **Field Crops**: In cereals, sulfur is applied to control rust and powdery mildew, ensuring yield stability and grain quality.
3. **Fruit Orchards**: Both copper and sulfur are integral to the management of fungal and bacterial diseases in apples, pears, peaches, and citrus.
4. **Greenhouse Systems**: Sulfur-based products are commonly used in protected cultivation to suppress fungal pathogens in lettuce, herbs, and ornamental plants.

7.2.5 Applications Across Cropping Systems

Biopesticides are versatile tools for pest and disease management across diverse cropping systems, from intensive field agriculture to specialized horticultural and protected cultivation systems. Their unique properties, including specificity, biodegradability, and compatibility with sustainable farming practices, make them suitable for addressing various challenges in agriculture

Horticultural Crops

Horticultural crops, including vegetables, fruits, and ornamental plants, are particularly vulnerable to pest infestations and diseases due to their intensive cultivation and high-value nature. Biopesticides have proven highly effective in managing these challenges, offering residue-free alternatives to chemical pesticides. Products based on microbial antagonists, plant extracts, and inorganic compounds are commonly used to control soilborne pathogens and foliar diseases.

Applications:

1. **Soilborne Disease Management:**
    - *Trichoderma spp.*, renowned for their ability to colonize plant roots and outcompete pathogens, are widely used in horticulture. They suppress diseases like damping-off, root rot, and wilt caused by pathogens such as *Pythium spp.* and *Fusarium spp.*. For example, in tomato and cucumber production, soil treatments with *Trichoderma* formulations enhance root health while reducing disease incidence.
2. **Foliar Disease Control:**
    - Neem oil, a plant-derived biopesticide, is frequently applied as a foliar spray to combat powdery mildew, anthracnose, and other fungal diseases in peppers, tomatoes, and eggplants. Neem oil's dual role as an insect repellent and fungicide makes it invaluable in managing integrated pest and disease challenges.

Benefits:

The use of biopesticides in horticulture ensures high-quality produce with minimal chemical residues, meeting consumer demand for safer, sustainably grown fruits and vegetables. Additionally, biopesticides contribute to soil health by promoting microbial diversity and nutrient cycling, ensuring long-term productivity.

Field Crops

Field crops, including cereals, pulses, and oilseeds, form the backbone of global food security. However, they are highly susceptible to pests and diseases that can significantly reduce yields. Biopesticides offer an effective and environmentally sustainable solution for managing these threats while reducing the reliance on synthetic pesticides.

Applications:

1. **Bt Formulations for Insect Control:**
    - Bt-based biopesticides are extensively used in rice, maize, and cotton production to control caterpillar pests such as stem borers and bollworms. In rice cultivation, Bt formulations protect against the rice stem borer (*Scirpophaga incertulas*), preventing yield losses and secondary infections by fungal pathogens.
2. **Management of Fungal Diseases:**
    - Copper-based biopesticides are applied to manage rust and leaf blight in wheat and other cereals. For example, copper oxychloride formulations reduce rust infections in wheat, ensuring healthier plants and higher grain quality.
3. **Seed Treatments:**
    - Bacterial biopesticides like *Pseudomonas fluorescens* are applied as seed treatments in soybean and groundnut crops to protect against root rot and damping-off diseases. These treatments enhance germination rates and early seedling vigor.

Benefits:

Biopesticides in field crops support sustainable cereal production by reducing the environmental impact of pest and disease management. Their use also helps mitigate pesticide resistance, preserving the efficacy of synthetic pesticides for critical applications.

Fruit Orchards

Fruit orchards, which produce high-value crops like apples, grapes, citrus, and berries, require effective disease management strategies to maintain yield quality and meet export standards. Biopesticides, with their specific action and eco-friendly nature, are widely used in fruit production systems to manage fungal and bacterial diseases.

Applications:

1. **Soilborne Disease Suppression:**
    - *Trichoderma spp.* formulations are applied to citrus and apple orchards to prevent root diseases caused by *Phytophthora spp.* and *Armillaria spp.*. These soil treatments enhance root health and nutrient uptake, contributing to better tree growth and fruit yields.
2. **Fungal and Bacterial Disease Management**:
    - Copper-based biopesticides are indispensable in controlling bacterial canker in citrus trees and scab in apple orchards. Foliar sprays of copper hydroxide ensure thorough coverage and prolonged protection.
3. **Post-Harvest Protection**:
    - Chitosan-based products are applied post-harvest to prevent spoilage caused by fungal pathogens such as *Botrytis cinerea*. These treatments reduce losses and improve shelf life, particularly for export-oriented fruits like grapes and berries.

Benefits:

Biopesticides improve fruit quality by reducing disease-related blemishes and maintaining the aesthetic appeal required for premium markets. Their role in residue-free disease management ensures compliance with international trade standards, boosting farmer profitability.

Greenhouse Systems

Greenhouse cultivation, characterized by controlled environments, presents unique opportunities for biopesticide application. However, the high humidity and temperature levels in greenhouses also create favorable conditions for rapid pathogen proliferation. Biopesticides, with their specificity and safety, are well-suited for greenhouse systems, where residue-free production is often a priority.

Applications:

1. **Insect Pest Control:**
    - *Beauveria bassiana*, an entomopathogenic fungus, is widely used in greenhouses to control thrips, whiteflies, and aphids. Its ability to infect and kill these pests indirectly reduces the spread of viral diseases transmitted by insect vectors.

2. **Root Pathogen Management:**
    - Bacterial biopesticides like *Pseudomonas fluorescens* are applied to manage root diseases in lettuce, herbs, and strawberries. These products suppress pathogens such as *Pythium spp.* while promoting root health.
3. **Foliar Disease Control:**
    - Neem-based sprays are used to combat powdery mildew and downy mildew in greenhouse-grown cucumbers, peppers, and tomatoes.

Benefits:

The use of biopesticides in greenhouses ensures residue-free produce, aligning with the demands of high-value and organic markets. Additionally, biopesticides minimize the risk of chemical resistance development, supporting long-term pest and disease management.

7.2.6 Market Trends and Consumer Demand

The global biopesticide market is experiencing rapid growth, driven by increasing consumer awareness of the environmental and health impacts of synthetic pesticides. With the rising demand for organic and sustainably produced food, farmers are increasingly adopting biopesticides to comply with stringent residue limits and certification standards. Products like neem oil, *Bacillus subtilis*, and chitosan are particularly favored in organic farming systems, where their natural origin and compatibility with regulatory requirements are critical.

Key Market Drivers:

1. **Consumer Preferences:**
    - Growing awareness of food safety and environmental sustainability has led to higher demand for organic produce. Biopesticides, as residue-free alternatives, cater to this market need.
2. **Regulatory Standards:**
    - Stricter regulations on pesticide residues, especially in export-oriented crops, are compelling farmers to transition to biopesticides.

Adoption Trends:

The integration of biopesticides into IPM systems is becoming standard practice in both conventional and organic farming. Innovations in formulation technology and application methods are expected to further enhance the adoption of biopesticides, making them a cornerstone of sustainable agriculture.

7.3 Regulatory Standards and Quality Control of Biocontrol Agents

The development, production, and application of biocontrol agents are subject to stringent regulatory standards and quality control measures to ensure their efficacy, safety, and environmental compatibility. Regulatory frameworks aim to minimize risks to human health, non-target organisms, and ecosystems while facilitating the adoption of biocontrol agents in sustainable agriculture. Similarly, quality control systems ensure that these agents meet consistent standards of performance, reliability, and shelf stability.

7.3.1 Regulatory Standards for Biocontrol Agents

The commercialization and widespread use of biocontrol agents are subject to rigorous regulatory oversight to ensure their safety, efficacy, and minimal environmental impact. Regulatory standards safeguard against potential risks posed by these agents to non-target organisms, including humans, pollinators, natural predators, and soil microbiota. By establishing a robust framework for evaluation and approval, these regulations enable the integration of biocontrol agents into agricultural systems while maintaining ecological balance.

National Regulatory Frameworks

Different countries have established their own regulatory authorities and standards to evaluate biocontrol agents. These frameworks focus on assessing the safety and efficacy of biocontrol products while ensuring they comply with environmental and public health standards.

United States:

In the United States, biocontrol agents are regulated by the Environmental Protection Agency (EPA) under the Federal Insecticide, Fungicide, and Rodenticide Act (FIFRA). The EPA's evaluation process for biopesticides involves a detailed examination of their toxicity, environmental persistence, and impact on non-target organisms. Biopesticides are categorized as either microbial, biochemical, or plant-incorporated protectants (PIPs), with each category requiring specific data submissions. The EPA's streamlined registration process for biopesticides, compared to synthetic pesticides, encourages innovation while maintaining safety.

Key aspects of the EPA's regulatory process include:

- **Toxicity Testing**: Acute and chronic toxicity studies are conducted to evaluate potential risks to humans, wildlife, and aquatic ecosystems.
- **Environmental Persistence**: Assessments of degradation pathways, residual activity, and potential bioaccumulation are required to minimize ecological harm.
- **Non-Target Effects**: Studies on pollinators, natural predators, and soil microbiota ensure that biocontrol agents do not disrupt beneficial ecosystems.

European Union:

In the European Union, biocontrol agents are regulated under Regulation (EC) No. 1107/2009, which governs the approval of plant protection products. This regulation emphasizes environmental safety and efficacy, requiring biocontrol agents to meet stringent criteria for authorization. The EU also promotes the use of low-risk substances, including many biopesticides, to support sustainable agriculture.

Key features of the EU framework include:

- **Approval of Active Substances**: The European Food Safety Authority (EFSA) evaluates active substances in biocontrol agents for their toxicological, ecotoxicological, and environmental profiles.

- **Mutual Recognition**: Once approved in one member state, biocontrol agents can be registered in other EU states through a simplified process, facilitating market access.
- **Low-Risk Criteria**: Substances classified as low-risk are prioritized for expedited approval, aligning with the EU's goal of reducing pesticide dependency.

India:

In India, the Central Insecticides Board and Registration Committee (CIBRC) oversees the registration of biopesticides. The regulatory process involves extensive documentation, including scientific studies on safety, efficacy, and manufacturing practices. Given India's growing emphasis on sustainable agriculture, the demand for biocontrol agents has increased, prompting stricter enforcement of regulatory standards.

Key requirements include:

- **Residue Analysis**: To ensure compliance with maximum residue limits (MRLs), biocontrol agents must undergo residue analysis for their impact on food safety.
- **Field Trials**: Multilocation field trials under diverse agro-climatic conditions are mandatory to demonstrate the efficacy and adaptability of biocontrol agents.
- **Manufacturing Standards**: Production facilities are audited to verify adherence to Good Manufacturing Practices (GMP).

International Standards

Global organizations such as the Food and Agriculture Organization (FAO) and the World Health Organization (WHO) have developed international guidelines for the use of biocontrol agents. The Codex Alimentarius, jointly developed by FAO and WHO, provides benchmarks for residue limits and safety protocols.

Key International Standards:

- **Residue Limits**: Establishing maximum residue limits (MRLs) ensures that biocontrol agents leave no harmful residues in food products.
- **Harmonized Guidelines**: The Codex encourages harmonization of regulatory requirements across countries to facilitate international trade and ensure consistent safety standards.
- **Capacity Building**: FAO and WHO support capacity-building initiatives in developing countries to improve regulatory infrastructure and promote the safe use of biocontrol agents.

Key Requirements for Regulatory Approval

The approval process for biocontrol agents involves multiple stages of evaluation to ensure their safety, efficacy, and environmental compatibility. Below are the critical components of the regulatory approval process:

1. Safety Data:

Comprehensive toxicological studies assess the potential risks of biocontrol agents to humans, animals, and the environment. These studies include:

- **Acute Toxicity**: Evaluations of immediate adverse effects following exposure.
- **Chronic Toxicity**: Long-term studies to identify potential carcinogenic, mutagenic, or reproductive effects.
- **Allergenicity**: Tests to determine whether biocontrol agents pose risks of allergic reactions.

2. Efficacy Trials:

Field and laboratory trials are conducted to demonstrate the effectiveness of biocontrol agents against target pests and pathogens under various conditions. These trials evaluate:

- **Control Efficacy**: The percentage reduction in pest populations or disease incidence.
- **Crop Safety**: Assessments to ensure that biocontrol agents do not cause phytotoxicity or yield reductions.
- **Environmental Adaptability**: Trials in diverse agro-climatic zones to determine performance consistency.

3. Non-Target Testing:

Biocontrol agents must be evaluated for their impact on beneficial organisms, such as pollinators, predators, and soil microbiota. For example:

- **Impact on Pollinators**: Studies on honeybees and other pollinators assess the safety of biocontrol agents in pollination-dependent crops.
- **Soil Health**: Assessments of microbial diversity and nutrient cycling ensure that biocontrol agents do not disrupt soil ecosystems.

4. Manufacturing Compliance:

Production facilities are required to adhere to Good Manufacturing Practices (GMP) to ensure consistent quality and safety of biocontrol agents. This includes:

- **Quality Assurance**: Regular testing of active ingredients and formulations to meet specified standards.
- **Traceability**: Documentation of raw materials, production processes, and distribution chains for regulatory audits.
- **Packaging and Labeling**: Compliance with labeling requirements to provide clear information on product composition, application guidelines, and safety precautions.

7.3.2 Quality Control in Biocontrol Agent Production

Quality control in biocontrol agent production is an essential process aimed at ensuring that products meet stringent standards of efficacy, viability, stability, and safety. Biocontrol agents, often comprising living microorganisms or biologically active compounds, require meticulous monitoring and testing throughout the production pipeline to maintain their effectiveness and reliability. The biological nature of these products introduces unique challenges in quality assurance, necessitating advanced methodologies and protocols to address variability in microbial activity, environmental adaptability, and shelf-life concerns.

Microbial Viability and Purity

The viability and purity of microbial biocontrol agents are fundamental to their success. High microbial viability ensures that sufficient active organisms are available to suppress pests or pathogens effectively, while purity eliminates contamination risks that could compromise efficacy or pose harm to crops and ecosystems.

Key Considerations for Microbial Viability:

1. **Monitoring Colony-Forming Units (CFUs):**
   - Regular assessment of CFUs is essential to confirm that microbial populations meet or exceed the concentrations specified on product labels. For instance, *Trichoderma spp.* formulations are tested to ensure they contain adequate propagules for effective colonization of plant roots.

2. **Optimized Cultivation Conditions:**
   - Cultivation parameters, such as temperature, nutrient availability, and oxygen levels, are carefully controlled to maximize microbial growth during production. Any deviations from these parameters can lead to reduced viability or contamination.

Ensuring Purity:

Purity testing involves rigorous screening to eliminate contamination by non-target microorganisms, including harmful pathogens or opportunistic species. Molecular techniques, such as polymerase chain reaction (PCR) and next-generation sequencing (NGS), are employed to verify the genetic identity of the target organism and detect contaminants. For example, *Bacillus subtilis* biopesticides are screened to ensure they do not harbor harmful bacterial strains.

Stability and Shelf Life

The stability and shelf life of biocontrol agents determine their commercial viability and practicality for end-users. Given the biological nature of these products, ensuring consistent efficacy over extended storage periods is a significant challenge.

Stability Testing Under Diverse Conditions:

1. **Environmental Stressors:**
   - Biocontrol agents are subjected to varying temperature, humidity, and light conditions to assess their resilience. For instance, spore-forming agents like *Bacillus spp.* tend to exhibit better stability under fluctuating environmental conditions compared to non-spore-forming microorganisms.

2. **Formulation Integrity:**
   - The physical and chemical stability of formulations is evaluated to ensure that ingredients do not degrade, separate, or interact adversely over time. Emulsifiable concentrates and wettable powders are tested for their dispersion properties and homogeneity.

Shelf-Life Determination:

Shelf-life studies involve monitoring the viability and efficacy of biocontrol agents over predetermined intervals. For example, chitosan-based biopesticides are evaluated for their ability to induce systemic resistance in plants even after extended storage periods, ensuring reliability during application.

Efficacy Testing

Efficacy testing simulates real-world conditions to evaluate the performance of biocontrol agents against target pests and pathogens. These tests are crucial for verifying product claims and ensuring consistent effectiveness across diverse agricultural scenarios.

Laboratory and Field Trials:

1. **Controlled Laboratory Studies:**
   - Initial testing involves controlled environments where variables such as pest population density, environmental conditions, and application rates are systematically manipulated. For instance, *Pseudomonas fluorescens* formulations are tested for their ability to suppress *Fusarium* in soil samples under laboratory conditions.
2. **Field Trials:**
   - Field evaluations are conducted across multiple agro-climatic zones to ensure that the biocontrol agent performs effectively under varying environmental conditions. Trials for *Trichoderma* biopesticides, for example, assess their adaptability to different soil types and climates.

Performance Metrics:

Key performance indicators include:

» **Pest or Pathogen Suppression**: Reduction in disease incidence or pest population compared to untreated controls.

» **Crop Yield and Quality**: Improvement in plant health, yield, and marketability as a result of biocontrol agent application.

Formulation Testing

The formulation of biocontrol agents is critical for optimizing their delivery, stability, and ease of application. Formulation testing ensures that products are compatible with existing agricultural practices and provide consistent performance.

Physical and Chemical Properties:

1. **Particle Size and Dispersion:**
   - Formulations are tested for particle size uniformity and dispersion in water or other carriers. Fine particles improve the coverage and efficacy of sprays, especially in foliar applications.
2. **pH and Viscosity:**
   - The pH of liquid formulations is adjusted to enhance microbial stability, while viscosity

testing ensures compatibility with various application equipment, such as sprayers and irrigation systems.

Compatibility with Farming Systems:

Biocontrol agents must integrate seamlessly into diverse farming systems, including organic, conventional, and protected cultivation. Compatibility testing evaluates whether the formulation interacts adversely with other inputs, such as fertilizers or pesticides. For instance, *Beauveria bassiana* formulations are tested for compatibility with insect growth regulators in integrated pest management (IPM) programs.

Safety Testing

Safety testing is a cornerstone of quality control, ensuring that biocontrol agents pose no harm to non-target organisms, including beneficial insects, earthworms, and aquatic species. This step is vital for regulatory compliance and public acceptance of biocontrol products.

Toxicity Assessments:

1. **Non-Target Organisms:**
   - Laboratory studies evaluate the impact of biocontrol agents on non-target species. For example, *Trichoderma spp.* and *Bacillus subtilis* formulations are tested for their effects on honeybees, which are critical pollinators.
2. **Soil and Aquatic Ecosystems:**
   - Biocontrol agents are assessed for their influence on soil microbial diversity and aquatic organisms, ensuring that their use does not disrupt ecological balance.

Human Safety:

Tests for allergenicity, dermal irritation, and inhalation toxicity ensure that biocontrol agents are safe for handlers and consumers. For instance, chitosan-based biopesticides are evaluated for potential allergic reactions before market release.

Technological Innovations in Quality Control

Advances in biotechnology and data analytics are transforming quality control processes in biocontrol agent production. Key innovations include:

1. **Genomic Tools:**
   - Techniques such as DNA barcoding and whole-genome sequencing enable precise identification and characterization of microbial strains, ensuring consistency and purity.
2. **Automated Monitoring:**
   - Automated systems for CFU counting, environmental control, and product testing enhance efficiency and reduce human error during quality control procedures.
3. **Data Integration:**
   - Digital platforms integrate data from production, storage, and distribution, enabling real-time tracking and traceability of biocontrol agents.

7.3.3 Labeling and Certification Standards

The labeling and certification of biocontrol agents are critical components of their commercialization and usage, ensuring that these products are applied correctly, safely, and effectively. Labeling provides essential information to end-users, including farmers, agronomists, and agricultural professionals, enabling them to understand the composition, usage protocols, and safety requirements of biocontrol products. Certification standards, on the other hand, validate the compliance of these agents with specific regulations, such as those for organic farming or environmental safety, thereby facilitating market access and consumer confidence. Together, these aspects ensure that biocontrol agents meet both regulatory and practical requirements.

Importance of Labeling for Biocontrol Agents

Labeling serves as a bridge between manufacturers and end-users, providing detailed information about the product's nature, purpose, and proper usage. Since biocontrol agents often involve living organisms or biologically active compounds, their efficacy depends heavily on precise application methods and adherence to specific storage and handling conditions. Inadequate labeling can lead to suboptimal results, reduced efficacy, and even potential harm to crops or the environment.

Key Elements of Labeling:

1. **Active Ingredients:**
    - Labels must clearly specify the active microorganism or compound, including its scientific name, strain, and concentration. For instance, a *Trichoderma harzianum*-based product would detail the strain (e.g., T-22) and its concentration in terms of colony-forming units (CFUs) per gram or milliliter.
    - The inclusion of this information allows users to evaluate the product's compatibility with their specific crop and pest management needs.
2. **Target Pests and Pathogens:**
    - Each label must identify the pests or diseases that the product is designed to control. For example, a neem-based product might specify its efficacy against powdery mildew, aphids, and whiteflies.
    - This clarity helps users select appropriate products for specific agricultural challenges, reducing the risk of misuse.
3. **Application Guidelines:**
    - Instructions for dosage, timing, and methods of application are crucial to maximizing efficacy. Labels should include detailed recommendations, such as the concentration for foliar sprays, seed treatments, or soil applications.
    - For example, a *Bacillus subtilis* product might specify the dilution rate for spraying crops and the frequency of applications needed for optimal disease suppression.
4. **Storage and Handling Instructions:**
    - Biocontrol agents, particularly those involving living microorganisms, are sensitive to

environmental conditions such as temperature, humidity, and light. Labels must provide guidance on proper storage, such as refrigeration requirements for microbial formulations or temperature ranges for shelf stability.

- Handling precautions, including protective gear recommendations and disposal methods for unused product, are also essential to ensure safety for users and the environment.

5. **Safety and Environmental Impact Information:**
   - Labels should include information on the product's impact on non-target organisms, such as pollinators or aquatic species. For instance, a copper-based biopesticide label might caution against overapplication to avoid soil accumulation.

Certification Standards for Biocontrol Agents

Certification validates that biocontrol agents comply with specific regulatory and market requirements, enhancing their credibility and acceptability. Certification standards often address criteria such as natural origin, environmental safety, and compliance with organic farming protocols.

Key Certification Programs:

1. **Organic Farming Certification:**
   - Biocontrol agents must meet stringent criteria to qualify for use in organic agriculture. These criteria typically include:
     - ◊ **Natural Origin**: The product must be derived from natural sources, such as microorganisms, plant extracts, or minerals.
     - ◊ **Residue Limits**: Certification programs ensure that the product does not leave harmful residues on crops or in the environment.
     - ◊ **Environmental Safety**: The product must demonstrate minimal risk to non-target organisms and ecosystems.
   - Products approved for organic farming often bear certification marks, such as those from the USDA Organic Program or the EU Organic Farming Certification.
2. **Global Gap and Fair Trade Standards:**
   - For export-oriented agricultural systems, biocontrol agents must comply with international standards like Global GAP (Good Agricultural Practices) or Fair Trade certifications. These standards emphasize sustainability, worker safety, and traceability.
3. **National Certification Programs**:
   - Many countries have their own certification schemes to regulate biocontrol agents. For example, India's National Program for Organic Production (NPOP) certifies products for use in organic agriculture, ensuring compliance with national residue and safety standards.

Labeling and Certification: Challenges and Innovations

Despite their importance, labeling and certification processes face challenges that can hinder the effective adoption of biocontrol agents. Addressing these issues through innovations and harmonization efforts is critical for the growth of the biocontrol industry.

Challenges:

1. **Complexity of Regulations:**
   - Variability in labeling and certification requirements across regions can complicate market entry for biocontrol agents. For example, a product approved in the United States may require additional documentation and testing for acceptance in the European Union.
2. **Knowledge Gaps:**
   - Farmers and distributors may lack awareness of how to interpret labels or verify certifications, leading to misuse or underutilization of biocontrol agents.
3. **Cost of Certification:**
   - Certification processes, especially for organic farming, can be costly and time-consuming, particularly for small-scale manufacturers.

Innovations:

1. **Digital Labels and QR Codes:**
   - QR codes on product packaging can link users to digital resources, including application videos, compatibility charts, and safety guidelines.
2. **Harmonization of Standards:**
   - International efforts to harmonize certification and labeling requirements, such as those led by the FAO, can streamline global trade and facilitate the wider adoption of biocontrol agents.
3. **Blockchain for Traceability:**
   - Blockchain technology can ensure transparency in the production, certification, and distribution of biocontrol agents, building trust among stakeholders.

7.3.4 Challenges in Regulatory and Quality Control

Despite their growing importance in sustainable agriculture, biocontrol agents face significant challenges in regulatory approval and quality control processes. These challenges stem from the complexity of their biological nature, the variability in global regulatory frameworks, and gaps in production and application knowledge. Addressing these challenges is essential to ensure the widespread adoption and consistent performance of biocontrol agents.

Lengthy Approval Processes

The regulatory approval process for biocontrol agents is often lengthy and complex, posing a significant barrier to their commercialization. Unlike synthetic pesticides, biocontrol agents involve living organisms or biologically active compounds, necessitating extensive data submissions to demonstrate their safety and efficacy.

Key Issues:

1. **Comprehensive Data Requirements**:
   - Regulatory authorities typically require a wide range of data, including toxicology, environmental impact assessments, efficacy trials, and non-target testing. For example, microbial agents like *Trichoderma spp.* must undergo multiple studies to assess their effects on soil microbiota, pollinators, and aquatic ecosystems. These requirements can significantly prolong the approval timeline.
2. **Multi-Stage Evaluations:**
   - The approval process often involves multiple stages, including laboratory testing, field trials, and peer reviews by regulatory committees. Each stage adds to the time and cost of bringing a new biocontrol agent to market.
3. **Bureaucratic Inefficiencies**:
   - In some regions, inefficient regulatory systems with overlapping jurisdictions further delay approvals, discouraging innovation and investment in biocontrol technologies.

Impact:

The extended timelines for regulatory approval hinder the commercialization of innovative biocontrol agents, delaying their availability to farmers and slowing the adoption of sustainable pest and disease management practices.

Variability in Standards

The lack of harmonized regulatory standards across countries poses significant challenges to the global trade and adoption of biocontrol agents. While efforts are underway to create uniform guidelines, disparities in regulatory frameworks remain a major obstacle.

Key Issues:

1. **Differences in Data Requirements**:
   - Regulatory bodies in different countries often have unique requirements for biocontrol agents. For instance, while the United States focuses on non-target toxicity and environmental safety, the European Union emphasizes ecological risk assessments and low-risk classification. This variability complicates the submission process for companies seeking multi-national registrations.
2. **Inconsistent Definitions:**
   - The definition of biocontrol agents varies across regulatory frameworks, leading to inconsistencies in classification and approval criteria. For example, some countries classify neem-based products as biopesticides, while others consider them plant-derived substances requiring different regulatory pathways.
3. **Barriers to Market Access:**
   - The lack of standardized regulations creates trade barriers, making it difficult for

biocontrol products approved in one region to gain acceptance in another. This limits the availability of effective agents in key agricultural markets.

Impact:

Variability in standards hampers the international scalability of biocontrol agents, restricting their potential to address global agricultural challenges. Harmonization of regulatory requirements is critical to overcoming these barriers.

uality Control in Decentralized Production

In regions with decentralized production systems, particularly in developing countries, small-scale manufacturers often lack the resources and expertise to implement rigorous quality control measures. This results in inconsistencies in product performance and reduces farmer confidence in biocontrol agents.

Key Issues:

1. **Resource Constraints:**
   - Small-scale producers may not have access to advanced technologies or facilities required for quality control, such as automated CFU counting systems or molecular diagnostic tools for purity testing.
2. **Inconsistent Formulations:**
   - Variability in microbial concentrations, stability, and shelf life is common in decentralized production systems. Products may not meet the labeled specifications, leading to reduced efficacy in the field.
3. **Contamination Risks**:
   - Inadequate sterilization and handling protocols can introduce contaminants, including non-target microorganisms, into biocontrol formulations, compromising their safety and effectiveness.

Impact:

Poor quality control diminishes the reliability of biocontrol agents, eroding trust among farmers and limiting their adoption. This is particularly problematic in resource-poor regions where biocontrol agents could play a critical role in reducing pesticide dependency.

Education and Training

A lack of awareness and training among farmers, distributors, and agricultural advisors poses significant challenges to the effective use of biocontrol agents. Unlike synthetic pesticides, which often have broad-spectrum activity, biocontrol agents require precise application techniques and adherence to specific conditions for optimal performance.

Key Issues:

1. **Knowledge Gaps:**
   - Farmers may not understand the biological nature of biocontrol agents or the importance

of proper storage, handling, and application. For instance, microbial formulations often require specific storage temperatures to maintain viability, which may be overlooked in the supply chain.

2. **Inappropriate Application Practices:**
   - Incorrect application rates, timing, or methods can lead to suboptimal results. For example, applying *Beauveria bassiana* during hot, dry conditions may reduce its efficacy, as the fungal spores require high humidity to germinate and infect pests.
3. **Distributor Training:**
   - Distributors often lack the technical knowledge to provide accurate guidance to farmers, resulting in misapplication or misuse of biocontrol agents.

Impact:

Inadequate education and training reduce the efficacy of biocontrol agents, leading to farmer dissatisfaction and reluctance to adopt these products. Expanding training programs and extension services is essential to bridge these knowledge gaps.

Addressing Challenges and Future Directions

Overcoming the challenges in regulatory approval and quality control requires coordinated efforts from policymakers, researchers, and industry stakeholders. Key strategies include:

1. **Streamlining Approval Processes:**
   - Simplified regulatory pathways, such as fast-tracking approvals for low-risk biocontrol agents, can accelerate commercialization. Collaborative efforts to harmonize international standards, such as those led by the FAO and OECD, are also critical.
2. **Capacity Building for Quality Control:**
   - Investments in infrastructure, training, and technology transfer can enhance quality control in decentralized production systems. Public-private partnerships can play a vital role in providing technical support to small-scale manufacturers.
3. **Education and Awareness Campaigns:**
   - Expanding extension services and farmer training programs can improve the understanding and adoption of biocontrol agents. Digital platforms, mobile apps, and demonstration farms can serve as effective tools for knowledge dissemination.

7.3.5 Future Directions in Regulation and Quality Control

The future of regulation and quality control for biocontrol agents lies in leveraging advancements in biotechnology, data analytics, and collaborative frameworks to address existing challenges and streamline processes. The growing adoption of biocontrol agents, driven by the demand for sustainable agriculture, necessitates innovative solutions that enhance their efficacy, reliability, and accessibility.

Biotechnology for Enhanced Characterization and Efficacy

Advancements in biotechnology, particularly in genomics, proteomics, and metabolomics, offer transformative opportunities for improving the characterization and efficacy of biocontrol agents. These tools enable precise identification, functional analysis, and optimization of biocontrol organisms and compounds.

Key Innovations:

1. **Genomic Tools and Molecular Markers:**
   - Whole-genome sequencing and molecular markers allow for the detailed characterization of microbial strains, ensuring accurate identification and differentiation from closely related species. For example, *Trichoderma harzianum* strains can be precisely characterized for their biocontrol traits, such as mycoparasitism or secondary metabolite production.
   - Genomic data also support the discovery of novel genes responsible for antifungal or antibacterial activity, enabling the development of next-generation biocontrol agents.
2. **Synthetic Biology and Metabolic Engineering:**
   - Synthetic biology techniques can enhance the performance of biocontrol agents by engineering microbial strains to overproduce bioactive compounds or improve stress tolerance. For instance, engineering *Bacillus subtilis* to produce higher levels of lipopeptides can increase its efficacy against fungal pathogens.
   - Metabolic engineering of plant-associated microbes, such as rhizobacteria, can improve their compatibility with specific crops, enhancing their colonization and biocontrol potential.
3. **Advanced Screening Platforms:**
   - High-throughput screening technologies enable rapid evaluation of microbial libraries for biocontrol activity. These platforms combine robotics and data analytics to identify promising strains or compounds with minimal time and resources.

Data-Driven Regulatory Frameworks

The integration of data analytics into regulatory frameworks can streamline approval processes and enhance transparency, facilitating the global adoption of biocontrol agents.

Key Strategies:

1. **Digital Platforms for Traceability and Compliance:**
   - Blockchain technology can provide end-to-end traceability of biocontrol agents, ensuring compliance with regulatory standards throughout the production, distribution, and application chain. This transparency builds trust among stakeholders, including regulators, manufacturers, and farmers.
   - Digital platforms can also enable real-time monitoring of production parameters, such as microbial viability and formulation quality, improving regulatory oversight.

2. **Artificial Intelligence (AI) in Risk Assessment:**
   - AI algorithms can analyze complex datasets from toxicity, efficacy, and environmental impact studies to predict the safety and performance of biocontrol agents. This can reduce the time required for risk assessments while maintaining rigorous evaluation standards.
3. **Harmonization of International Standards**:
   - Global collaboration on data-sharing platforms can harmonize regulatory requirements across regions. For instance, a centralized database of approved biocontrol agents and their safety profiles can facilitate mutual recognition of regulatory approvals, reducing redundant evaluations.

Innovations in Quality Control

Emerging technologies are revolutionizing quality control processes, ensuring consistent performance and reliability of biocontrol agents.

Technological Advancements:

1. **Automated Quality Testing:**
   - Robotics and automated systems can perform routine quality control tasks, such as colony-forming unit (CFU) counting, purity testing, and stability analysis. These systems improve efficiency and reduce human error in quality assurance.
2. **Molecular Diagnostics:**
   - Real-time PCR and other molecular diagnostic tools enable rapid detection of contaminants and verification of microbial identity in biocontrol formulations. These tools ensure that products meet purity and concentration standards throughout the production process.
3. **Nano-Encapsulation for Stability**:
   - Nano-encapsulation technologies can improve the stability and shelf life of biocontrol agents by protecting active ingredients from environmental stressors. Encapsulation also enables controlled release, enhancing field performance under variable conditions.
4. **Big Data Integration:**
   - Integrating data from production, storage, and application phases allows manufacturers to identify trends and optimize processes. Predictive analytics can be used to anticipate and mitigate quality issues before they affect product performance.

Capacity Building for Stakeholders

The successful adoption of biocontrol agents depends on equipping stakeholders, including farmers, manufacturers, and regulators, with the knowledge and resources needed to implement these technologies effectively.

Key Areas for Investment:

1. **Farmer Training and Education:**
   - Expanding extension services and training programs is critical to improving farmers' understanding of biocontrol agents, their application techniques, and storage requirements. Digital tools, such as mobile apps and e-learning platforms, can provide accessible and tailored guidance.
2. **Support for Small-Scale Manufacturers:**
   - Capacity-building initiatives for small-scale manufacturers can improve their ability to implement rigorous quality control measures. Public-private partnerships can provide technical assistance, access to advanced testing facilities, and funding for infrastructure upgrades.
3. **Regulatory Training Programs:**
   - Training programs for regulatory authorities can improve the consistency and efficiency of biocontrol agent evaluations. Workshops on new technologies, such as molecular diagnostics and AI risk assessment tools, can enhance regulatory capacity.

Public-Private Partnerships for Innovation

Collaboration between public institutions, private companies, and research organizations is essential for overcoming the regulatory and quality control challenges associated with biocontrol agents.

Potential Contributions:

» **Research and Development:** Joint research initiatives can accelerate the discovery of novel biocontrol agents and innovative production methods.

» **Infrastructure Development:** Investments in shared facilities, such as quality testing laboratories and genomic sequencing centers, can support small-scale manufacturers and streamline regulatory evaluations.

» **Policy Advocacy:** Collaborative efforts can influence policy reforms that reduce regulatory bottlenecks and promote harmonization of international standards.

Future Prospects for Biocontrol Agents

The future of biocontrol agents lies in integrating advanced technologies, streamlined regulatory processes, and robust quality control systems. Key areas of focus include:

1. **Precision Agriculture Integration:**
   - Biocontrol agents will increasingly be integrated with precision agriculture tools, such as drones and IoT devices, for targeted application and real-time monitoring.
2. **Climate-Resilient Biocontrol Solutions:**
   - Research into climate-resilient strains and formulations will address the challenges posed by changing environmental conditions.

References

Baker, K. F., & Cook, R. J. (1974). *Biological Control of Plant Pathogens*. W.H. Freeman and Company.

Cook, R. J., & Baker, K. F. (1983). *The Nature and Practice of Biological Control of Plant Pathogens*. American Phytopathological Society.

Campbell, R. (1989). *Biological Control of Microbial Plant Pathogens*. Cambridge University Press.

Mukerji, K. G., & Garg, K. L. (1988). *Biocontrol of Plant Diseases*. CRC Press.

Gnanamanickam, S. S. (2002). *Biological Control of Crop Diseases*. CRC Press.

Boland, G. J., & Kuykendall, L. D. (1998). *Plant-Microbe Interactions and Biological Control*. CRC Press.

Whipps, J. M., & Lumsden, R. D. (2001). *Biological Control of Plant Diseases: Progress and Challenges for the Future*. Springer.

Fravel, D. R. (2005). *Commercialization and Implementation of Biocontrol*. Annual Review of Phytopathology, 43, 337-359.

Harman, G. E., & Kubicek, C. P. (1998). *Trichoderma and Gliocladium, Volume 2: Enzymes, Biological Control and Commercial Applications*. Taylor & Francis.

Pal, K. K., & McSpadden Gardener, B. (2006). *Biological Control of Plant Pathogens*. The Plant Health Instructor.

Compant, S., Duffy, B., Nowak, J., Clément, C., & Barka, E. A. (2005). *Use of Plant Growth-Promoting Bacteria for Biocontrol of Plant Diseases: Principles, Mechanisms of Action, and Future Prospects*. Applied and Environmental Microbiology, 71(9), 4951-4959.

Larkin, R. P., & Fravel, D. R. (1998). *Efficacy of Various Fungal and Bacterial Biocontrol Organisms for Control of Fusarium Wilt of Tomato*. Plant Disease, 82(9), 1022-1028.

Weller, D. M. (2007). *Pseudomonas Biocontrol Agents of Soilborne Pathogens: Looking Back Over 30 Years*. Phytopathology, 97(2), 250-256.

Hoitink, H. A. J., & Boehm, M. J. (1999). *Biocontrol Within the Context of Soil Microbial Communities: A Substrate-Dependent Phenomenon*. Annual Review of Phytopathology, 37, 427-446.

Handelsman, J., & Stabb, E. V. (1996). *Biocontrol of Soilborne Plant Pathogens*. The Plant Cell, 8(10), 1855-1869.

Alabouvette, C., Olivain, C., & Steinberg, C. (2006). *Biological Control of Plant Diseases: The European Situation*. European Journal of Plant Pathology, 114(3), 329-341.

Haas, D., & Défago, G. (2005). *Biological Control of Soil-Borne Pathogens by Fluorescent Pseudomonads*. Nature Reviews Microbiology, 3(4), 307-319.

Spadaro, D., & Gullino, M. L. (2005). *Improving the Efficacy of Biocontrol Agents Against Soilborne Pathogens*. Crop Protection, 24(7), 601-613.

Thomashow, L. S., & Weller, D. M. (1996). *Current Concepts in the Use of Introduced Bacteria for Biological Disease Control: Mechanisms and Antifungal Metabolites*. In Plant-Microbe Interactions (Vol. 1, pp. 187-235). Springer.

Kloepper, J. W., & Ryu, C. M. (2006). *Bacterial Endophytes as Elicitors of Induced Systemic Resistance*. In Microbial Root Endophytes (pp. 33-52). Springer.

Choudhary, D. K., & Johri, B. N. (2009). *Interactions of Bacillus spp. and Plants – With Special Reference to Induced Systemic Resistance*. Microbiological Research, 164(5), 493-513.

Saravanakumar, D., & Samiyappan, R. (2007). *ACC Deaminase from Pseudomonas fluorescens Mediated Induction of Systemic Resistance Against Fusarium Wilt of Banana*. Phytopathology, 97(7), 867-875.

Van Loon, L. C., Bakker, P. A. H. M., & Pieterse, C. M. J. (1998). *Systemic Resistance Induced by Rhizosphere Bacteria*. Annual Review of Phytopathology, 36, 453-483.

Jetiyanon, K., & Kloepper, J. W. (2002). *Mixtures of Plant Growth-Promoting Rhizobacteria for Induction of Systemic Resistance Against Multiple Plant Diseases*. Biological Control, 24(3), 285-291.

Harman, G. E. (2000). *Myths and Dogmas of Biocontrol: Changes in Perceptions Derived from Research on Trichoderma harzianum T-22*. Plant Disease, 84(4), 377-393.

Howell, C. R. (2003). *Mechanisms Employed by Trichoderma Species in the Biological Control of Plant Diseases: The History and Evolution of Current Concepts*. Plant Disease, 87(1), 4-10.

Benítez, T., Rincón, A. M., Limón, M. C., & Codón, A. C. (2004). *Biocontrol Mechanisms of Trichoderma Strains*. International Microbiology, 7(4), 249-260.

Baker, R., & Paulitz, T. C. (1996). *Theoretical Basis for Microbial Interactions Leading to Biological Control of Soilborne Plant Pathogens*. In Plant-Microbe Interactions (Vol. 1, pp. 50-79). Springer.

Khan, Z., & Doty, S. (2009). *Endophyte-Assisted Growth Promotion in Plants: The Potential Mechanisms and Applications*. Microbial Ecology, 58(4), 567-573.

Sarma, B. K., Yadav, S. K., Singh, S., & Singh, H. B. (2015). *Microbial Consortium-Mediated Plant Defense Against Phytopathogens: Readdressing for Enhancing Efficacy*. Soil Biology and Biochemistry, 87, 25-33.

Timmusk, S., & Nevo, E. (2011). *Plant Growth Promoting Rhizobacteria: Prospects for Sustainable Agriculture*. Antonie Van Leeuwenhoek, 100(3), 507-519.

Shoresh, M., Harman, G. E., & Mastouri, F. (2010). *Induced Systemic Resistance and Plant Responses to Fungal Biocontrol Agents*. Annual Review of Phytopathology, 48, 21-43.

Mendes, R., Garbeva, P., & Raaijmakers, J. M. (2013). *The Rhizosphere Microbiome: Significance of Plant Beneficial, Plant Pathogenic, and Human Pathogenic Microorganisms*. FEMS Microbiology Reviews, 37(5), 634-663.

Pertot, I., Vinale, F., & Pascale, A. (2017). *The Microbial Community's Role in Sustainable Disease Control: Current Knowledge and Future Directions*. Plant Disease, 101(2), 252-259.

Bardin, M., Ajouz, S., Comby, M., & Nicot, P. C. (2015). *Critical Factors in Biocontrol Success with Microbial Pathogens of Insects and Pathogens*. Biological Control, 91, 1-9.

Singh, H. B., & Prabha, R. (2019). *Microbial Interventions for Sustainable Agriculture: A Pathway Toward Climate-Smart Farming*. Springer.

Köhl, J., Postma, J., Nicot, P., Ruocco, M., & Blum, B. (2011). *Stepwise Screening of Microbial Antagonists for Biological Control of Plant Pathogens: From Lab to Field Applications*. Biological Control, 57(1), 1-12.

Pieterse, C. M., van der Does, D., Zamioudis, C., Leon-Reyes, A., & Van Wees, S. C. (2012). *Hormonal Modulation of Plant Immunity*. Annual Review of Cell and Developmental Biology, 28, 489-521.

Haas, D., & Défago, G. (2005). *Biological Control of Soilborne Pathogens by Fluorescent Pseudomonads*. Nature Reviews Microbiology, 3, 307-319.

Siddiqui, Z. A. (2006). *PGPR: Biocontrol and Biofertilization*. Springer.

Velivelli, S. L., De Vos, P., Kromann, P., Declerck, S., & Prestwich, B. D. (2014). *Biological Control Agents: From Field to Market, Problems, and Challenges*. Trends in Biotechnology, 32(10), 493-496.

Chandler, D., Bailey, A. S., Tatchell, G. M., Davidson, G., Greaves, J., & Grant, W. P. (2011). *The Development, Regulation, and Use of Biopesticides for Integrated Pest Management*. Philosophical Transactions of the Royal Society B: Biological Sciences, 366(1573), 1987-1998.

Harman, G. E. (2006). *Overview of Mechanisms and Uses of Trichoderma spp.*. Phytopathology, 96(2), 190-194.

Vinale, F., Sivasithamparam, K., Ghisalberti, E. L., Marra, R., & Woo, S. L. (2008). *Trichoderma-Plant-Pathogen Interactions*. Soil Biology and Biochemistry, 40(1), 1-10.

Islam, M. T., & Hossain, M. M. (2015). *Biological Control of Plant Diseases by Microbial Antagonists: A Historical Review and Future Challenges*. Biological Control, 91, 111-124.

Patel, H. (2019). *Microbial Approaches for Crop Improvement*. Academic Press.

Whipps, J. M. (2001). *Microbial Interactions and Biocontrol in the Rhizosphere*. Journal of Experimental Botany, 52(suppl_1), 487-511.

Bulgarelli, D., Schlaeppi, K., Spaepen, S., Ver Loren van Themaat, E., & Schulze-Lefert, P. (2013). *Structure and Functions of the Bacterial Microbiota of Plants*. Annual Review of Plant Biology, 64, 807-838.

Lugtenberg, B., & Kamilova, F. (2009). *Plant-Growth-Promoting Rhizobacteria*. Annual Review of Microbiology, 63, 541-556.

Pal, K. K., & Gardener, B. M. (2006). *Biological Control of Plant Pathogens*. The Plant Health Instructor.

Heydari, A., & Pessarakli, M. (2010). *A Review on Biological Control of Fungal Plant Pathogens Using Microbial Antagonists*. Journal of Biological Sciences, 10(4), 273-290.

Kumar, S., & Gupta, R. (2017). *Fungi and Their Role in Sustainable Development: Current Perspectives*. Springer.

Raaijmakers, J. M., De Bruijn, I., & Nybroe, O. (2009). *Natural Functions of Lipopeptides From Bacillus and Pseudomonas: More Than Surfactants and Antibiotics*. FEMS Microbiology Reviews, 34(6), 1037-1062.

Tariq, M., & Yasmin, S. (2016). *Rhizobacteria Mediated Stress Management in Plants: Molecular and Biotechnological Advances*. Springer.

Glick, B. R. (2012). *Plant Growth-Promoting Bacteria: Mechanisms and Applications*. Scientifica.

Arora, N. K., & Verma, M. (2017). *Microorganisms in Sustainable Agriculture and Biotechnology*. Springer.

Babalola, O. O. (2010). *Beneficial Bacteria of Agricultural Importance*. Biotechnology Letters, 32(11), 1559-1570.

Singh, B., & Gaur, S. (2018). *Agricultural Microbiology: A Laboratory Manual*. CRC Press.

Domenech, J., Reddy, M. S., & Kloepper, J. W. (2006). *Plant-Associated Microorganisms for Biological Control*. Springer.

Ahemad, M., & Kibret, M. (2014). *Mechanisms and Applications of Plant Growth Promoting Rhizobacteria: Current Perspective*. Journal of King Saud University-Science, 26(1), 1-20.

Campbell R. (1989). *Biological Control of Microbial Plant Pathogens*. Cambridge Univ. Press,

Cambridge.

Cook RJ and Baker KF. (1983). *Nature and Practice of Biological Control of Plant Pathogens*. APS, St. Paul, Minnesota.

Fokkemma MJ. (1986). *Microbiology of the Phyllosphere*. Cambridge Univ. Press, Cambridge.

Gnanamanickam SS (Eds). (2002). *Biological Control of Crop Diseases*. CRC Press, Florida.

Heikki MT and Hokkanen James M. (Eds.). (1996). *Biological Control – Benefits and Risks*.

Cambridge Univ. Press, Cambridge.

Mukerji KG, Tewari JP, Arora DK and Saxena G. (1992). *Recent Developments in Biocontrol of Plant Diseases*. Aditya Books, New Delhi.

For Product Safety Concerns and Information please contact our
EU representative GPSR@taylorandfrancis.com Taylor & Francis
Verlag GmbH, Kaufingerstraße 24, 80331 München, Germany